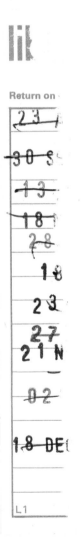

SEABED MECHANICS

Scientific Committee for IUTAM/IUGG Symposium

Prof. Bruce Denness (Chairman)
Sir Alan Muir Wood
Prof. Akio Nakase
Prof. Paul Novak
Dr. Adrian Richards

Members of Study Group for Programme Arrangements

Prof. Bruce Denness
Prof. Len Maunder
Sir Alan Muir Wood
Prof. Akio Nakase
Dr. Adrian Richards

Local Organisation

Prof. Bruce Denness
Mrs. Susan M. Davidson

Selective financial assistance provided by the International Union of Theoretical and Applied Mechanics, the International Union of Geology and Geodesy and the British Council

Bureau of the International Union of Theoretical and Applied Mechanics

Prof. Daniel C. Drucker (President)
College of Engineering
University of Illinois
Urbana, Illinois 61801
USA

Prof. Jan Hult (Secretary-General)
Chalmers University of Technology,
S-412 96 Gothenburg
SWEDEN

Prof. Ernst Becker (Treasurer)
Institut für Mechanik
Technische Hochschule
Hochschulstrasse 1
D-6100 Darmstadt
BRD

SEABED MECHANICS

Edited Proceedings of a Symposium, sponsored jointly by the International Union of Theoretical and Applied Mechanics (IUTAM) and the International Union of Geodesy and Geophysics (IUGG), and held at the University of Newcastle upon Tyne, 5—9 September, 1983.

Edited by
Prof. Bruce Denness

Published by
Graham & Trotman

First published in 1984 by
Graham & Trotman Limited
Sterling House
66 Wilton Road
London
SW1V 1DE

British Library Cataloguing in Publication Data

Seabed Mechanics: edited proceedings of a
    symposium, sponsored jointly by the International
    Union of Theoretical and Applied Mechanics
    (IUTAM) and the International Union of Geodesy
    and Geophysics (IUGG), and held at the
    University of Newcastle upon Tyne, 5–9 September 1983.
    1. Marine geotechnique
    I. Denness, Bruce   II. International Union of
    Theoretical and Applied Mechanics
    III. International Union of Geology and Geophysics
    624.1'5136   GC380.15

    ISBN 0-86010-504-0

Typeset in Great Britain by The Castlefield Press, Moulton, Northants.
Printed and bound in Great Britain by Robert Hartnol Ltd., Bodmin.

Contents

(Papers presented by M. Perlow and J. T. Dette, *Geotechnical Investigations using Vibratory Coring Systems and Shallow Penetration in in situ Testing* (Section 3), P. Holmes and P. C. Barber, *Nearshore Hydrodynamics Related to Sediment Transport,* and W. R. Parker, *Behaviour of Fine Sediments in Estuaries and Nearshore Waters* (Section 6) are not reproduced here.)

Foreword

Symposia do not just happen, and those which venture into new territory happen even less frequently. Nevertheless, faced by the challenge of organising such an event, many of us volunteer eagerly for the task. I am not one of this select band.

Confronted by what at first sight was but a casual invitation to consider whether I might be interested in commenting on the possibility I attended an informal (as I then thought) meeting at Newcastle University in 1980 to discuss the subject of Seabed Mechanics. Professor Len Maunder, Head of Mechanical Engineering here at the University of Newcastle upon Tyne, and the University's IUTAM representative, had convened the meeting to explore the idea of Newcastle's responding to an IUTAM initiative to sponsor such a venture.

I should have been suspicious from the outset. There were only three of us there, Len, myself and Professor Paul Novak, then Head of Civil Engineering and the only one apart from me who knows where the seabed is! During the softening up stage while I merely supped my coffee Paul imperceptibly manoeuvred himself into a situation from which it was obviously impossible for him to absorb the extra burden that would be imposed by organising a symposium. I woke up to hear Len telling me that it was about time I did something to earn my keep anyway.

While inwardly acknowledging that he was right, I strenuously resisted Len's bull-dozing only to be told — in reply to my protestations that I really knew nothing about Seabed Mechanics, being but a simple environmental lad who carries the misleading title "Ocean Engineer" by accident — that my only possible way out lay through the office of Sir James Lighthill, Provost of University College, London and the British representative on the Bureau of IUTAM. I was on the train like a shot armed to the teeth with excuses.

I rehearsed my speech all the way down. I had studiously opted for the environmental side of seabed studies. I could prove it and produced papers on the correlation of geotechnical properties of the seabed with the populations of bugs that live there. Elsewhere in my defence package were items on landslips, climatology, cluster analysis applied to business development — anything I could find to illustrate my dissociation from Seabed Mechanics proper.

Sir James, being the type who leads commandos to capture strategic objectives hundreds of miles behind enemy lines, is faster than a speeding bullet, and was not fooled in the least. His immediate response was totally disarming in both its charm and cunning: he suggested that my breadth of interest would provide a springboard to approach successfully the wide range of experts needed to do justice to such a diverse subject and that he saw my position on its fringe as implying absence of partiality and prejudice which, if present, sometimes endanger a subject by inducing unwanted bias in a particular direction. I couldn't win!

Well, was Sir James right? It is now up to the reader to judge. But beforehand, let me explain the construction of this proceedings which differs from conventional volumes as much as did the organisation and the conduct of the meeting itself.

This volume follows the same order as the programme at the symposium. It was designed to lead the reader of the proceedings logically upwards from the sediment 10 metres below the seabed to the waters 10 metres above. In more detail, it sets about first defining the characteristics of subseabed materials, particularly those of interest to the engineer — strength and consolidation characteristics — before going on to consider their implications for submarine slope stability and site investigation requirements. Special problems relating to pore fluid pressures are then encompassed before adding the complicating phenomenon of cyclic loading induced by wave action. Sediment transport follows as we move above the fixed subseabed: the influence of waves, currents, sediment type and remote meteorological forcing agents are all considered in order. Finally, at appropriate intervals through the volume, are items reflecting our understanding of the impact of manmade structures on both subseabed strata and seabed erosion.

By a stroke of near genius, matched only by my inherent modesty, I hit upon the idea of passing on much of the editorial burden to the chairmen of each of the sessions. These gentlemen generously agreed to contribute a paper in their own right to summarise, after the meeting, the events within their sessions in the light of the state-of-the-art as they saw it. The reader will immediately recognise the status of these people and thus realise that this device also represents, in effect, a second level of refereeing. In turn, it should be said that it was considered neither necessary nor polite to have the chairmens' comments further reviewed. The first level of refereeing rested with the Scientific Committee with myself, as its chairman, having the final say in the event of occasional disagreement among referees.

Before concluding, I should like to recount some of the memorable incidents from the meeting for those readers who were not present. This I do partially out of self-gratification to relish further what may well be my only sally into the centre of the convention arena and partially to satisfy a wicked urge to expose my friends, old and new to the same ridicule that I so enjoy. More especially I do it as an open gesture of gratitude to all those whose good humour made the meeting itself an outstanding social success at a venue which might, when coupled with my organisational mismanagement, have so easily resulted in disarray.

We began to assemble on Sunday, September 4th at Henderson Hall, a student residence at Newcastle University. I had arranged through the good offices of my secretary Susan Maisie Davidson to have a bevvy of delightful young ladies, with whom she frequently consorts, draped around the reception desk. Among the first arrivals was one Armand J. Silva (who features prominently in the scientific record of the meeting which follows). Not for him the shy, modest, retiring approach. Within seconds he broke the ice with an invitation to join him in a jogging session which confused the otherwise willing receptionists somewhat.

Late arrivals kept pouring in throughout the week necessitating a continuous presence at the reception desk. So it was then that Susan was again amused by Armand's return to the desk next morning clad only in his underpants and pretending to jog around the foyer — he had accidentally been locked out of his bedroom by an over-zealous cleaner who had not realised that he had only gone to the bathroom and thus locked the door after making the bed. That Armand took it with such good grace when I recounted this story at the Symposium Dinner — I still have the broken arm to prove it — is a lasting tribute to his fine sense of the ridiculous and good humour.

There were many other incidents which amused those of us with particularly offbeat mentalities. Among them are the recollection of: Pat Kemp, another contributor to this volume, being congratulated warmly by the then irate Susan Davidson for noticing that the first page of the preprints for the third day was pinned to the back of the hand-out; Gideon Almagor, yet another contributor, being thanked for his attempting to write the entire proceedings single-handedly (his original contribution extended to 62 pages); Susan Davidson explaining her drinking habits to Gideon (again) in terms of the legendary British expression "legless" only to be told by him that he had been thus afflicted literally since losing his in the 6-day Middle East war; Adrian Richards, one of our Scientific Committee and a Session Chairman, and his wife, Efrosine, adjusting the bedroom furniture of two neighbouring single student rooms to achieve an effect never anticipated by the original architect; Mike Adye, our guest speaker, attempting an Irish joke; etc. I hope these people will still speak to me.

Beyond all of these memories is to be recorded our gratitude to Councillor Arthur Stabler, the Lord Mayor of Newcastle, for his generosity in inviting us all to a civic reception at the Mansion House. Not only that but he went on to regale us with traditional Geordie humour and to arrange for alcohol to be poured all over us in the time-honoured way.

To return to this volume you, dear reader, must now make of it what you will. For my part I am satisfied with the assembly of so much individual effort from all its contributors into the pattern intended. Therefore, I must publicly eat my original words — in the event, I actually enjoyed it!

Bruce Denness
Professor of Ocean Engineering
University of Newcastle upon Tyne

IUTAM '83 Symposium — Seabed Mechanics
Opening address

In his Opening Address to the Symposium, the Dean of Engineering at Newcastle University (Professor J. B. Caldwell) welcomed delegates, and especially those from overseas, to Newcastle and to its University. He considered this to be a particularly appropriate venue for this first jointly-sponsored (by IUTAM and IUGG) Symposium, not only because of the long involvement of North-East England in marine work of the traditional kind, but especially also because in recent years the stimulus of marine resource recovery has caused a significant shift in emphasis, particularly in R & D work, at the University.

Professor Caldwell outlined briefly the very substantial initiative taken by the Science and Engineering Research Council in creating in the past five years a co-ordinated national programme of marine technological research in British Universities. Current funding was around £5M p.a., of which over £1M is related directly to seabed work, in which the major components were for work on geophysical and geotechnical aspects, sediment transport, and on piling and foundations.

At Newcastle University the importance of ocean developments was recognized by the creation of the country's first Chair of Ocean Engineering, to which Professor Bruce Denness was appointed in 1975. He has subsequently built up an active research group, whose contributions have included:—

- development of penetrometers and correlation techniques for marine sediments;
- studies of the mechanics and effects of breaking waves in the near-shore environment;
- prediction of climates;
- pockmark studies, automated bore-hole analysis, and other related topics.

But marine technology research is essentially multi-disciplinary; and the work at Newcastle (for which a major grant award of £1.5M was recently announced by SERC) includes projects in Civil, Electronic, Marine and Agricultural Engineering, Naval Architecture and Ship-building, Physics, and Plant Biology. Many aspects of this work proceed from the belief that the use of the seabed — for supporting or anchoring structures and systems, for resource recovery, or for laying and burying artefacts — will form a growing part of marine work in future. For this, a better knowledge and understanding of the seabed is clearly essential.

The Symposium could therefore hardly be more timely or relevant; and the organizers and Authors were to be complimented for having put together a comprehensive and international programme of papers and events which, Professor Caldwell hoped, would ensure a stimulating, instructive and enjoyable stay in Newcastle for all concerned.

SERC Marine Technology Programme

It is a considerable honour to have been invited to make a brief presentation dealing with the interest of the UK's Science and Engineering Research Council (SERC) in ocean-related research and development activities. First, however, I must congratulate IUTAM and the organisers of this conference for their imagination in using the seabed as a medium for bringing together at the one time those who more often meet in separate groups to consider what occurs immediately above or below this level. In many areas of science and technology there is a need to transcend disciplinary boundaries to achieve an all round understanding. This approach has been an inherent feature of a special initiative taken by SERC some years ago to develop a marine technology programme.

It was in 1970, at a time when the potential for offshore oil and gas production from UK waters was beginning to be recognised, that SERC decided to convene an expert panel to examine the need for future support in marine technology. The main outcome was a recommendation that the Council should increase its support substantially above the modest level than pertaining, and offer positive encouragement to universities to allow them to make appropriate contribution, particularly of a multi-disciplinary nature, to the growing national research effort in this area. The panel's report generated relatively little response at the time, but interest had been awakened: a further investigation some years later was greeted with considerable enthusiasm, and led SERC to decide on the development of a marine technology programme.

The stated aim of the programme was to develop in British Universities and polytechnics a coordinated programme of research and training in marine technology which would attract industrial support and which would enable academy to assume a proper national role in the developing use of the oceans. The marine technology programme and its associated Directorate have been in existence since 1977, during which time the Council's annual expenditure in this area has grown from a few hundred thousand pounds to about £5 million. A thriving academic community with concentration in seven centres is now working together on projects of relevance to the needs of industry, and attracting increasing industrial support.

In developing the programme, SERC has seen its role as being progressively to:—

- support high quality research proposals against a developing framework of funding;
- attract to the programme academics of stature, drawn from a wide range of disciplines;
- involve industrialists in the selection and guidance of research aspirations, and provide them with the opportunity to act both as proxy and real customers;
- integrate SERC funded work with the more focussed programmes of other government agencies;
- arrange cohesion of projects within the overall programme by subject matter;
- help to develop in universities a technical capability that can be demonstrated to potential customers;
- encourage in grant holders more self reliance and less dependence on SERC funding.

Many of those who attended this symposium have at one time or another received support through the SERC marine technology programme, and I am glad to see that one or two of the presentations deal with results arising from SERC research grants. The fact that such work may have been funded as part of a cohesive pro-gramme addressed to seabed problems, or fluid loading, or a more general activity in coastal engineering, is important, as it indicates SERC's view on the most effective way to organise its programme of support. For the purposes of this conference, however, that is by the way: it is the results and the interactions between disciplines that are important rather than the funding arrangement. A blinkered approach contained entirely within one discipline rarely leads to a successful out-come in ocean-related activities, where the potential value of a particular technical contribution has usually to be developed in a wider context if it is to be properly realised. I applaud the intent of the organisers of this symposium to bring together leading experts with a variety of interests pertinent to the seabed (± 10 metres), and hope that everyone has benefited from the result-ing interactions.

A. M. Adye
Director,
SERC Marine Technology Programme

Part I

Sub-seabed Characteristics

Section 1 — Strength and Consolidation

Introduction
Modelling and the Consolidation of Marine Soils

Adrian F. Richards

Fugro B. V., Leidschendam, The Netherlands

An essay is presented on some problems of laboratory and *in situ* modelling, on the relationship of excess pore pressure to the effective overburden pressure and on the vertical distribution of the overconsolidation ratio in seabed soils. The question is asked whether hypotheses resulting from laboratory investigation of marine soils are likely to have validity *in situ*. Use of a new paradigm that excess pore pressure will always be found in cohesive marine soils infers that these soils cannot be called 'normally consolidated'. Calculation of effective overburden pressures must take into account the presence of excess pore pressure to have validity. Invertebrate mucous, causing a proposed reduction in marine soil permeability, may help explain the high water content of some deep sea soils. The reported significant underconsolidation of some deep sea soils below the top soils may be related to the probable presence of excess pore pressures, which may increase with depth.

Introduction

The symposium sessions on strength and consolidation of marine soils demonstrated that while there is much known about these properties there is also much that is poorly known or unknown. Rather than a review, an essay is offered as a means of commenting on two general themes directly or indirectly arising out of the strength and consolidation papers in this section. The first theme deals with certain aspects of laboratory versus *in situ* models and modelling, and the second theme with certain aspects of consolidation. In this essay, the words 'marine soils' always refer to soils of the seabed covered with water.

Laboratory and *in situ* modelling

A great amount of laboratory testing, particularly that using sophistocated testing equipment and procedures, has been performed over the years in order to derive models that try and explain the behaviour of soils in the field. Model verification in the field on land is fairly straightforward, and it has often been done. Model verification in the field at sea, on the other hand, is neither straightforward, simple nor cheap, and it has rarely been done. It is very difficult to obtain soil behaviour models from *in situ* tests that are similar to the models obtained from triaxial tests in the laboratory; consequently, it is convenient, as well as traditional, to model soil behaviour resulting from laboratory tests and to assume that the resulting model bears a significant resemblance to *in situ* behaviour. Field verification, particularly in the ocean, of laboratory models does not always occur as quickly as one might like, often because of constraints that include time and money. A number of points are raised below that are sometimes overlooked or minimized by investigators working in the laboratory with marine soil samples.

(1) It is perhaps worth emphasizing in a compendium dealing with seabed mechanics that natural marine soils are, usually, a complex mixture of (1) electrolyte, (2) solids, and, often, (3) gas. In water depths of less than the shelf-break (about 100–150 m), the concentration of the dissolved constituents in the pore fluid may vary between zero (fresh water) to well over 3.5% salinity in tropical lagoons which have restricted access to the open sea. The solids range from clay minerals to carbonate or quartz and feldspar sands, and may also

include calcareous or siliceous oozes of biogenic origin. Gas may be dissolved, in a free state, or a solid (clathrate) under conditions of deeper burial. Soil has structure and this property and the *in situ* state of stress is always disturbed or changed when a volume of marine soil *in situ* is sampled and transported to the laboratory.

(2) In many instances, laboratory experimenters may have a limited comprehension of just how their samples were collected from the seafloor and how badly disturbed they may be. Disturbance is particularly likely if the samples were collected by drilling and wireline sampling methods. A summary of sampling methods and some of the many problems occuring in sampling has been made by Richards and Parker (1968), Subcommittee on Soil Sampling (1981) and Richards and Zuidberg (1983). The magnitude of sample changes or disturbances, when compared to *in situ* conditions, usually is not well known. This suggests that verification in the field should be assigned a higher priority.

(3) It is particularly emphasized that soil is a non-homogeneous material, and its properties, behaviour and response to loading of one soil may be quite different compared to another. The index properties do not describe a soil in a unique way, unlike similar descriptions of metal or plastic. This may make comparisons difficult. Should it also be stated how well laboratory loads approximate environmental loads in magnitude, vector and time? In addition, testing procedures may unexpectedly influence test results. For example, when performing certain laboratory tests, such as Atterberg limits or when back-pressuring consolidation samples, any fluid added to the sample being tested must be electrolytically identical to the pore fluid in the sample if the solids in the sample are affected by electrolyte concentration. Otherwise, the material properties and behaviour of interest may be significantly changed.

(4) The distinction between natural and unnatural or anomalous soil behaviour may be poorly known or unknown in many environments. The author is reminded that several of his initial hypotheses explaining soil behaviour that were based on assumed relationships had to be discarded in one locality when the statistical dependence of geotechnical properties was determined by a cross-correlation matrix and appropriate confidence levels (Hulbert and Richards, 1980). It would be highly desirable to indicate the specifications of 'average' or normal soil behaviour, preferably by the use of statistics. In other words, what criteria should be used to differentiate between similar or different types of soil material?

In summary, a plea is made for the presentation of sufficient information to evaluate the methodology of laboratory testing and the significance of any resulting models, particularly in relation to similar well-studied soils. The minimum required information includes: how the samples were collected, transported and stored; soil classification; and, test descriptions or published standards. A description of how remoulded soil was made would be helpful to anyone wishing to replicate an experiment. In other words, quality control needs to be established or improved. It would be useful to try and answer the following questions, in so far as possible:

(1) What is the quality of samples that are investigated in the laboratory for strength and consolidation? Have criteria been developed to evaluate sample quality with regard to sampling equipment, method of sampling, *in situ* stress relief, degassing, transportation, storage, extrusion method, laboratory sample preparation, etc.?

(2) What criteria have been developed to evaluate whether laboratory analytical procedures, hypotheses and assumptions are likely to be valid for the same soil *in situ*?

(3) What *in situ* verification is proposed to test any laboratory hypothesis?

Consolidation

Three aspects of consolidation in cohesive seafloor soils will be briefly discussed: (1) the problem of excess pore pressure as it relates to the effective overburden pressure *in situ*; (2) a new hypothesis to explain why deep sea soils have unusually high water contents; and (3) the use of the overconsolidation ratio as a key to defining the state of consolidation. For reasons that follow, it will be useful to proceed with a few basic ASCE and ASTM definitions (ASTM, 1983) that are well known:

(1) Normally consolidated soil deposit — a soil deposit that has never been subjected to an effective pressure greater than the existing overburden pressure.

(2) Overconsolidated soil deposit — a soil deposit that has been subjected to an effective pressure greater than the present overburden pressure.

(3) Underconsolidated soil deposit — a deposit that is not fully consolidated under the existing overburden pressure.

(4) Effective stress (effective pressure) (intergranular pressure) — the average normal force per unit area transmitted from grain to grain of a soil mass. It is the stress that is effective in mobilizing internal friction.

Ambient Excess Pore Pressure and the Effective Overburden Pressure

A paradigm is proposed, which has evolved out of over fifteen years of research on the *in situ* measurement of ambient excess pore pressures and from the discussions with many persons in different parts of the world.

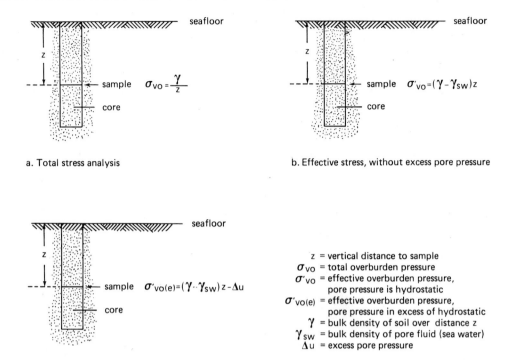

a. Total stress analysis

b. Effective stress, without excess pore pressure

c. Effective stress, with excess pore pressure

z = vertical distance to sample
σ_{vo} = total overburden pressure
σ'_{vo} = effective overburden pressure, pore pressure is hydrostatic
$\sigma'_{vo(e)}$ = effective overburden pressure, pore pressure in excess of hydrostatic
γ = bulk density of soil over distance z
γ_{sw} = bulk density of pore fluid (sea water)
Δu = excess pore pressure

Fig. 1. Three example methods showing the computation of overburden pressure or stress on a seabed sample of water-saturated soil to be collected for consolidation testing.

Pore pressure in excess of hydrostatic pressure has always been measured *in situ* in cohesive seabed soils, having a grain-size finer than most sands, at depths below a metre or two of the soil—water interface.

This generalization admittedly is based on only a relatively small number of *in situ* piezometer measurements of ambient pore pressure, primarily in soils found on the continental shelf. Nevertheless, should the paradigm be found to be valid when more data have been amassed, the implication is that there may be little, if any, cohesive seabed soil that is (by definition) 'normally consolidated' regardless of water depth.

An aspect of the problem can be graphically illustrated (Fig. 1). It is assumed for purposes of simplification that a truly homogenous soil is to be cored without disturbance for a consolidation test in an oedometer or a triaxial cell. A sample is obtained from the same depth, z, in the three examples shown. In example a of Fig. 1, the vertical overburden stress, σ_{vo}, is expressed in terms of total stress. In example b, the effective overburden stress is calculated on the assumption that excess pore pressure does not exist (only hydrostatic pressure on the sample). In example c, the effective overburden stress is calculated on the basis that an excess pore pressure (above hydrostatic) is present. These considerations lead to the conclusion that

it is probable that in most, if not all, calculations to date in which the effective overburden pressure has been determined, the ambient *in situ* excess pore pressure has not been taken into account (example b in Fig. 1). However, based on the information given above, example b is probably invalid. The vertical distribution of excess pore pressure in typical seabed soils is too poorly known to be summarized at present. It is, however, reasonable to expect the excess pore pressures to vary with burial depth, partly as a result of the natural heterogenity of most soils, rather than to be a constant.

Hamilton (1964) clearly described what is now called 'apparent overconsolidation'. In a surface sample from off the coast of Baja California, he presented data that this author has used to calculate an overconsolidation ratio of 156. The overconsolidation was reported to be caused by strong interparticle forces and/or cementation (Hamilton, 1964). It was well recognized early (Richards and Hamilton, 1967) that most surficial seabed cohesive soils are in a state of overconsolidation and that this state does not represent the removal of overburden by erosion or by any desiccation in shallow water.

Invertebrate Mucus Hypothesis to Explain High Water Contents in Deep Sea Soils

The author devised an additional hypothesis in the

early 1970s to explain the fact that many deep sea soils tend to have higher water contents than can be explained by the existing overburden. The hypothesis was that the high water contents were due, at least in part, to the probable reduction of soil permeability caused by mucus from the epifauna and infauna. These invertebrates obtain their metabolic energy by ingesting organic matter contained in cohesive soil. When the remains of a 'meal' are expelled from the gut or cloaca, it seems reasonable to expect that the fecal matter will contain some mucus, which may appreciably help to decrease soil permeability by acting like a 'glue' between soil particles or groups of particles. This, in turn, will retard the expulsion of pore water from the soil as the overburden increases due to sedimentation. The mucous affect may be greatest in the top soils, which are the most overconsolided. Several years of study by post-doctoral geochemists in the author's laboratory at Lehigh University suggest that the influence most likely was in the gel size-fraction, much less than $2\,\mu$m, and might be due to colloids. This would be consistant with a mucous hypothesis. Unfortunately, relatively little biogeochemical knowledge of benthic marine invertebrate mucus exists than can be used by geotechnologists.

The importance of infauna bioturbation on the benthic bottom layer (top 10 m of soil) may be considerable, by the way, for both shallow and deep water environments. Principal effects include mechanical disturbance and biogeochemical changes (Richards and Parks, 1976), which are conditions that are difficult to model in the laboratory.

The Overconsolidation Ratio and Consolidation

Several other interesting phenomena also exist with regard to the relationship of the overconsolidation ratio or OCR to depth. This is irrespective of whatever single or multiple hypotheses may be valid for any given seabed cohesive soil in shallow or deep water to explain surficial overconsolidation. The OCR is defined as p_c/σ_{vo}, where p_c is the preconsolidation pressure, usually derived from a Casagrande construction, and σ_{vo} may be defined in one of the three ways (Fig. 1), but almost always in the past and present as indicated in Fig. 1b. In many soils, both in shallow and deep water, a common relationship is for the OCR to approach or equal unity at some depth below the seafloor, usually about 5–15 m, and then to remain at unity as the depth further increases (Fig. 2). This relationship appears to be a paradox when compared to the paradigm, presented earlier, that normally consolidated soils are uncommon or non-existent because all cohesive seabed soils are assumed to have excess pore pressures.

In some, if not most, very deep-water soils located

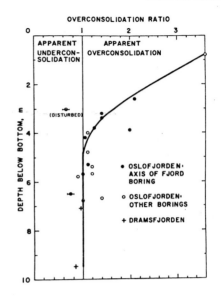

Fig. 2. Depth below the seabed surface related to the overconsolidation ratio (OCR) for Oslofjorden and Dramsfjorden, Norway, soil samples. Note that core disturbance decreases the OCR (Richards, 1976).

in the ocean basins, a distinctly different OCR-burial depth relationship occurs compared to that discussed above. The OCR decreases more-or-less uniformly from high values near the surface to values much less unity at greater depths. It appears that an OCR of 1 has no special significance for these soils, or as Silva *et al.* (1984) has stated, the depth of normal consolidation is transitional from overconsolidation to underconsolidation. This phenomenon is clearly illustrated in the Atlantic Ocean soils described by Silva *et al.* (1984). Although Silva *et al.* (1984) reported OCRs slightly above 1 at depths greater than about 5 m for north central Pacific soils, earlier data (Silva *et al.*, 1976) from the same general vicinity of the north-central Pacific also show significant underconsolidation (Fig. 3). It is suggested that the magnitude and the occurrence of underconsolidation from so many areas excludes the likelihood that core disturbance (Fig. 1) is a causative factor. The expected increase of excess pore pressures with increasing depth to some limiting depth, may account for some or all of the underconsolidation, except perhaps in certain cases.

The existence of underconsolidation at depth in ocean basin soils is not new (Bryant *et al.*, 1974; Silva *et al.*, 1976). Bryant *et al.* (1981) have discussed a number of Deep Sea Drilling Project (DSDP) examples from the western Pacific and the Gulf of Mexico that show underconsolidation. There is also another aspect to this problem to be noted.

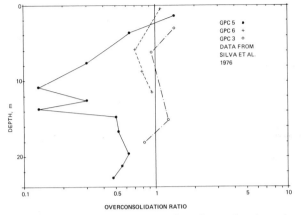

Fig. 3. Depth below the seabed surface related to the logarithm of the overconsolidation ratio for samples from three cores collected in the north central Pacific Ocean. Graphs made from tabular data published by Silva *et al.*, 1976.

Two of the three cores for the Pacific (see Fig. 3) and those for the Bermuda Rise Plateau (Silva *et al.*, 1984) show a decrease of the OCR from near the seabed surface to some minimum value at depths ranging from a few metres to about 12 m and then an increase of the OCR, but not as high as unity, with increasing depth. Very detailed OCR data for a DSDP (Leg 86, Hole 576A) in the north central Pacific by some members of the Geotechnical Consortium (San Diego State University, Texas A & M University and the University of Rhode Island) were presented recently (Noorany, 1984). These data showed that variations of OCR with depth had a similar profile trend with a minimum OCR, ranging from about 0.3 to slightly less than one, occurring at depths of about 28 to 55 m. The minimum OCR approximately corresponded to a specific soil layer. At a depth of about 55 m, the OCR increased to values of about two. The shape of the profile of these new detailed data is similar to core GPC 5 (Fig. 3), although the OCR minima occur at very different depths.

In summary, it is believed that the transition from overconsolidation to underconsolidation probably is real, although there is a possibility that it is an artifact of the consolidation test or computation of the effective overburden and preconsolidation pressures. There appears to be a limiting depth for the underconsolidation below which the OCR again approaches unity. It is suggested that the validity and extent of the overconsolidation—underconsolidation—'normal?' consolidation phenomena be investigated, that hypotheses be formulated to explain the phenomena and that the hypotheses be tested or verified with reference to *in situ* testing if at all possible. It is anticipated that the phenomenon may be important in the understanding of the consolidation process in deep sea soils, and perhaps also in soils sedimented in lesser water depths.

Conclusions

(1) Soil is disturbed by sampling and marine soils probably are more disturbed by sampling than terrestrial soils. Drilling and wireline sampling, in particular, may disturb soil more than is generally realized. As a consequence, there is a greater need for models derived from laboratory tests on marine soils to be verified by *in situ* testing on the seabed.

(2) Quality control may be deficient in laboratory testing of marine soils, and this may directly or indirectly degrade models resulting from laboratory tests.

(3) It would be beneficial to use more often a statistical approach in differientating similarities and differences of marine soils.

(4) It would also be beneficial to ask, and attempt to answer, a number of specified questions, including whether laboratory procedures, hypotheses and assumptions are likely to be valid for the same soil *in situ.*

(5) A paradigm is proposed that excess pore pressures will always be found *in situ* in cohesive marine soils, at least to some limiting depth.

(6) Use of the paradigm infers that marine soils cannot be normally consolidated using the existing definition.

(7) Effective overburden pressure calculations must take into account excess pore pressures *in situ* if they are to have validity.

(8) A hypothesis that invertebrate mucous may decrease the permeability of some marine soils is proposed. Mucous may be an important influence, along with interparticle bonding and cementation, causing high water contents in marine soils, especially in the top soils that are overconsolidated (OCR > 1).

(9) It is proposed that high, excess pore pressures may be the cause of significant underconsolidation (OCR < 1) in soils under the top soils, particularly in soils from great water depths (ocean basins).

Acknowledgements

Fruitful discussions on some of the essay topics were held with Messrs. Richard Lyons, Harry Kolk and Nicholas Withers, and Drs. Ronald Chaney and Gilliane Sills. Prof. Bruce Denness probably is to be held responsible for whatever is good or bad in this essay because he asked for a session one review contribution. Piet van Goudoever and Herman Zuidberg reviewed a draft of the essay and offered constructive criticisms.

References

ASTM, 1983. Standard definitions of terms and symbols relating to soil and rock mechanics. *Soil and Rock; Building Stones (standard D 653-82)*, Annual Book of ASTM Standards, Vol. 04.08, American Society for Testing and Materials, Philadelphia, PA, pp. 170–198.

Bryant, W. R., Bennett, R. H. and Katherman, C. E., 1981. Shear strength, consolidation, proosity, and permeability of oceanic sediments. In: C. Emiliani (Ed.) *The Ocean Lithosphere*, The Sea, Vol. 7, John Wiley, New York, pp. 1555–1615.

Bryant, W. R., Deflache, A. P. and Trabant, P. K., 1974. Consolidation of marine clays and carbonates. In: A. L. Inderbitzen (Ed.) *Deep-Sea Sediments Physical and Mechanical Properties*. Plenum Press, New York, pp. 209–244.

Hamilton, E. L., 1964. Consolidation characteristics and related properties of sediments from experimental mohole (Guadalupe site). *J. Geophys. Res.*, 69(20) 4257–4269.

Hulbert, M. and Richards, A. F., 1980. Investigation of geotechnical and geochemical relationships by parameter cross-correlation methods, Oslofjorden and Dramsfjorden, Norway. *Mar. Geotechnol.*, 4(2) 163–180.

Nakase, A. and Kamei, T., 1984. *In situ* void ratio, strength and overburden pressure anomalies in seabed clays. This volume, pp. 9–16.

Noorany, I., 1984. Shear strength of deep sea pelagic clays. Proc. ASTM Symp. *Laboratory and In Situ Determination of the Strength of Marine Soils,* San Diego, 26–27 January.

Richards, A.F., 1976. Marine geotechnics of the Oslofjorden region. In: N. Janbu, F. Jørstad and B. Kjaerasli (Eds.) *Contributions to Soil Mechanics* Laurits Bjerrum Memorial Volume. Norwegian Geotechnical Institute, Oslo, pp. 41–64.

Richards, A. F. and Hamilton, E. L., 1967. Investigations of deep-sea sediment cores, III. Consolidation. In: A. F. Richards (Ed.) *Marine Geotechnique* University of Illinois Press, Urbana, IL, pp. 93–117.

Richards, A. F. and Parker, H. W. 1968. Surface coring for shear strength measurements. In: *Civil Engineering in the Oceans,* American Society of Civil Engineers, N.Y., pp. 445–489.

Richards, A. F. and Parks, J. M., 1976. Marine geotechnology: average sediment properties, selected literature and review of consolidation, stability, and bioturbation-geotechnical interactions in the benthic boundary layer. In: I. N. McCave, (Ed.) *The Benthic Boundary Layer.* Plenun Press, N.Y., pp. 159–181.

Richards, A. F. and Zuidberg, H. M., 1984. Sampling and *in situ* geotechnical investigations offshore. Proc. Symp. *Marine Geotechnology and Nearshore/ Offshore Structures,* Shanghai (in press).

Sills, G.C. and Thomas, R.C., 1984. Settlement and consolidation in the laboratory of steadily deposited sediment. This volume, pp. 41–50.

Silva, A. J., Hollister, C. D., Lain, E. P. and Beverly, B. E., 1976. Geotechnical properties of deep sea sediments: Bermuda Rise. *Mar. Geotechnol.*, 1(3) 195–232.

Silva, A. J. and Jordan, S. A., 1984. Consolidation properties and stress history of some deep sea sediment. This volume.

Sub-committee on Soil Sampling, International Society for Soil Mechanics and Foundation Engineering (Eds.), 1981. *International Manual for the Sampling of Soft Cohesive Soils.* Tokai University Press, Tokyo, pp. 1–129.

Yasuhara, K., Hirao, K., Fujiwara, H. and Ue, S., 1984. Undrained shear behaviour of quasi-overconsolidated seabed clay induced by cyclic loading. This volume pp. 17–24.

1

In situ Void Ratio, Strength and Overburden Pressure Anomalies in Seabed Clays

Akio Nakase and Takeshi Kamei

Department of Civil Engineering, Tokyo Institute of Technology, Japan

Seabed clays are said, in general, to be in an apparently overconsolidated state due to ageing effects. Mostly the *in situ* relationship between effective overburden pressure p, void ratio e and undrained shear strength c_u, has been considered not much different from that in normally consolidated young clays, i.e. c_u increases and e decreases with depth.

Four case records of *in situ* state of seabed clays in Japan are presented. In each case, c_u increases approximately linearly with depth; however, the distribution of e and water content w with depth are somewhat different from those anticipated from the p vs. e relationship obtained by oedometer tests. The *in situ* soil state at Kikai Bay is an extreme case, where w increases with depth although c_u also increases at depth. An embankment was constructed on the seabed at Kikai Bay and the consolidation phenomena observed for over ten years. Observations show that the change in soil state during consolidation by the embankment was not much different from that anticipated from oedometer test result, although the soil state before consolidation was very much different.

Introduction

Seabed clays in continental shelves are said, in general, to be in an apparently overconsolidated state due to ageing effects (Noorany and Gizienski, 1970). It is also considered that the eustatic sea level change, i.e. change in the groundwater pressure may result in an apparently overconsolidated state (Kenney, 1964; Parry and Wroth, 1981). In studies of such an apparently overconsolidated state, the main interest has been in mechanical states of soils, e.g. relationship between the precompression stress p_c and effective overburden pressure p, and distribution of undrained shear strength c_u with depth.

According to the concept of compression curve obtained by consolidation tests, the void ratio e in the ground is to decrease with depth, except for dried crusts, since the effective overburden pressure increases with depth. In the report of the results of soil investigation in the Mohole Project, however, Hamilton (1964) emphasized the fact that the *in situ* void ratio showed very little decrease with depth in spite of an appreciable increase in c_u with depth. Nacci and Huston (1969) state that the *in situ* void ratio of clays in ocean floors

does not necessarily decrease with depth. Meade (1963) has presented surprising data obtained from seabed soil layers in California, in which the *in situ* void ratio increased appreciably down to the depth of 500 m.

In the present contribution, four case records of *in situ* state of seabed clays in Japan are presented. In each case, c_u increases approximately linearly with depth; however, distribution of void ratio e, or water content w, with depth in three cases are somewhat different from those anticipated from p vs. e relationship obtained by oedometer tests.

Case record of Osaka Bay

The investigation site is located in an area where a huge man-made island is going to be constructed for a new international airport. The top 20 m of soft clay is a recent alluvial deposit and is underlain by stiffer diluvial soils and water depth at the site is 18 m. Though plasticity indices I_p fluctuate along the depth to some extent as shown in Fig. 1(a), the top alluvial deposit seems to be of the same type of clay. Liquidity indices I_L in the top

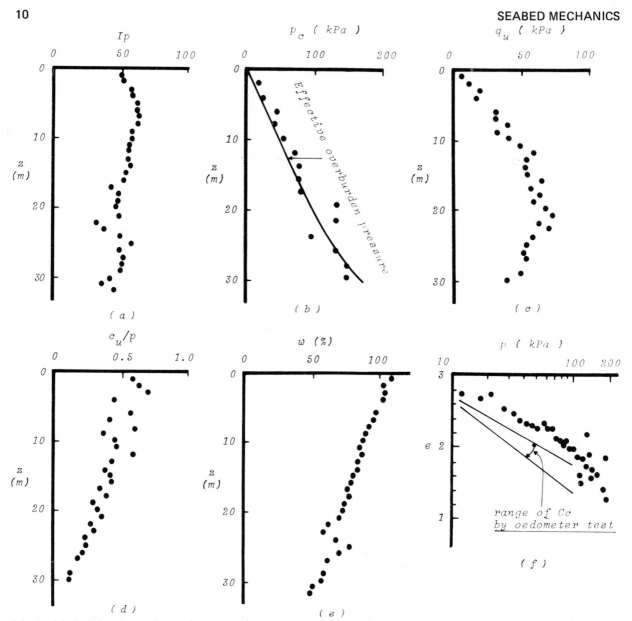

Fig. 1. Osaka Bay case study: (a) plasticity index vs. depth; (b) precompression vs. depth; (c) unconfined compression strength vs. depth; (d) ratio of undrained shear strength to effective overburden pressure vs. depth; (e) water content vs. depth; (f) *in situ* void ratio vs. effective overburden pressure.

of layer are in the range from 0.9 to 1.0 for the depth from 5 m through 20 m; above that the values are about 1.15.

Precompression stress p_c obtained by oedometer tests is compared with the effective overburden pressure p in Fig. 1(b). It is seen in the figure that the clay down to the depth of 20 m is in a slightly overconsolidated state with an overconsolidation ratio (OCR) in the range from 1.0 to 1.2. Unconfined compression strength q_u increases approximately linearly with depth down to the depth of 20 m as shown in Fig. 1(c). Ratio of the undrained shear strength c_u to p, which is obtained as

$q_u/2p$, decreases with depth as shown in Fig. 1(d).

Water content w decreases with depth as shown in Fig. 1(e). The *in situ* relationship between e and p is shown in Fig. 1(f), where the straight lines show a range of e vs. p relationship anticipated from oedometer test results, i.e. range of the value of compression index C_c.

Looking at the *in situ* state of seabed clay at this particular site, it may be said that the top 20 m of the recent alluvial clay is in either normally consolidated state or slightly overconsolidated state.

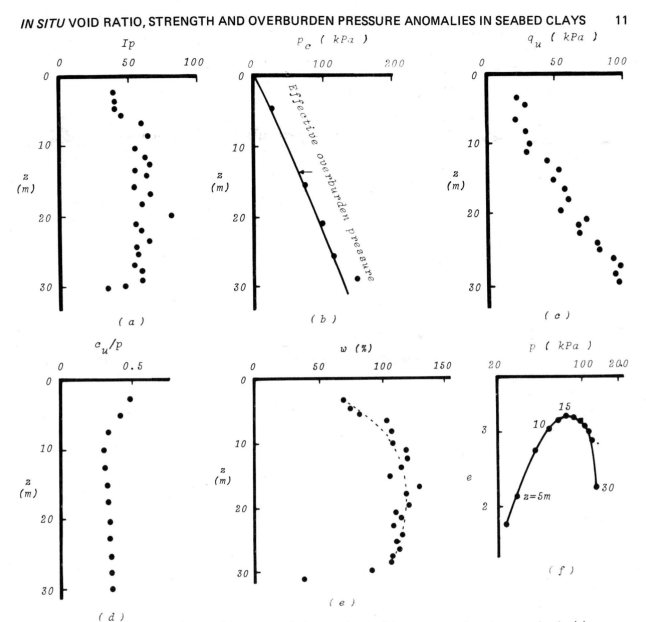

Fig. 2.   Tokyo Bay 'A' case study: (a) plasticity index vs. depth; (b) precompression stress vs. depth; (c) unconfined compression strength vs. depth; (d) ratio of undrained shear strength to effective overburden pressure vs. depth; (e) water content vs. depth; (f) void ratio vs. effective overburden pressure.

Case records

Tokyo Bay 'A'

Figure 2 shows various soil parameters of a recent alluvial seabed clay at Tokyo Bay. Water depth at the investigation site is 5.5 m. From Fig. 2(a), the clay at depths from 7 m to 28 m seems to be homogeneous and of similar type. Values of liquidity index I_L show some scatter along the depth; however, most of the values exceed unity. The precompression stress p_c obtained by oedometer tests is found to be practically equal to

the effective overburden pressure p, i.e. OCR = 1.0, as shown in Fig. 2(b).

Unconfined compression strength q_u increases approximately linearly with depth as shown in Fig. 2(c). The *in situ* value of c_u/p, i.e. $q_u/2p$, is seen to be practically constant below the depth of 7 m shown in Fig. 2(d). All of these results seem to imply that this particular clay is in a normally consolidated state.

Distribution of water content w with depth is shown in Fig. 2(e). Water content w increases with depth down to the depth of 20 m. This type of water content distribution with depth may be encountered at a very early

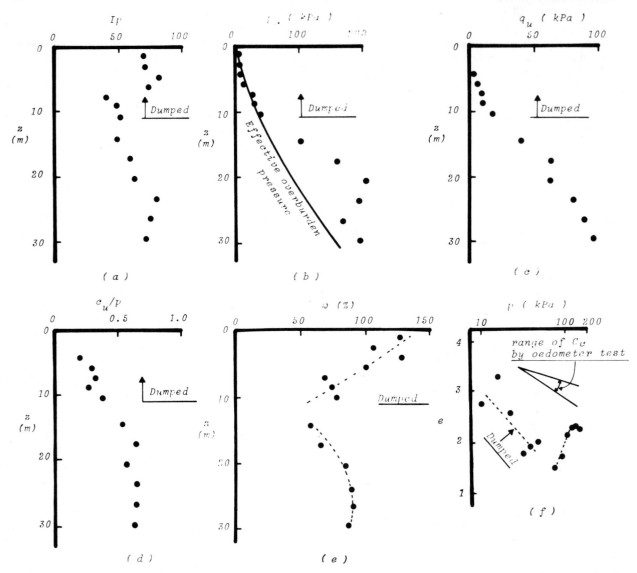

Fig. 3. Tokyo Bay 'B' case study: (a) plasticity vs. depth; (b) precompression stress vs. depth; (c) unconfined compression strength vs. depth; (d) ratio of undrained shear strength to effective overburden pressure vs. depth; (e) water content vs. depth; (f) void ratio vs. effective overburden pressure.

stage of consolidation under an extremely heavy load. At this particular site, however, no such a loading has been done. *In situ* e vs. p relationship looks anomalous as shown in Fig. 2(f), where the value of e is calculated from w as represented by the smooth curve shown in Fig. 2(e). As shown in Fig. 2(a), the clay type of the surface layer, of some 7 m thickness, is considered to be a little different from the clay below, however, an anomalous relationship between e and p being clearly seen in the clay for depths from 7 m to 20 m.

Tokyo Bay 'B'

Subsoil at the site consists of two kinds of soils of

different age, i.e. very young clay in the surface layer and relatively aged clay below. Remoulded clay derived from maintenance dredging at nearby fairways has been dumped on the site. The original seabed soil is a recent alluvial clay, covered by a sand layer of 2 m thick. The dumping started in 1969 and ended in 1980, and the present ground surface is at the low water level. At the investigation site, the newly dumped clay is 11 m in thickness, and is underlain by the recent alluvial clay down to a depth of 30 m. It has been suggested that both of the young and the aged clays are essentially the same soil, and surface layer of the seabed has been remoulded by pump-dredging and dumped over the nearby seabed.

Plasticity index I_p of the surface clay scatters as shown in Fig. 3(a). Liquidity index I_L in the surface layer is in the range from 0.9 to 1.2, whereas the value in the underlying alluvial clay is in the range from 0.6 to 0.8. By comparing the precompression stress p_c with the effective overburden pressure p, as shown in Fig. 3(b), it seems that the newly dumped clay is in a slightly underconsolidated state, with an OCR in the range of 0.9 to 1.1, and the underlying aged clay is in a slightly overconsolidated state, with the OCR of 1.2 to 2.2. A similar sort of difference in the consolidation state is also seen in the distribution of q_u and the *in situ* value of c_u/p with depth as shown in Figs. 3(c) and (d).

The distribution of w with depth is shown in Fig. 3(e). Water content w in the surface layer shows much scatter, although it clearly decreases with depth. Water content vs. depth relationship in the underlying clay, on the other hand, seems anomalous as in the previous case record of Tokyo Bay 'A'.

The *in situ* relationship between e and p is shown in Fig. 3(f), where e is calculated from w in Fig. 3(e). In Fig. 3(e) and (f) both w vs. depth (z) and e vs. p relationships are approximated by the dashed lines. The two straight lines in Fig. 3(f) correspond to the range of compression indices C_c obtained by oedometer tests for the surface layer of clay. It is interesting to see that *in situ* e vs. p relationship in the surface young clay is very similar to that anticipated from oedometer test results. A marked difference in the *in situ* e vs. p relationship may be due mainly to the difference in the age of clay.

Kikai Bay

The seabed soil at Kikai Bay is soft, recent alluvial clay of some 20 m thickness, which is underlain by much stiffer diluvial soil. Water depth here is 3 m. A reclamation levée was constructed on the seabed. In the construction, 12-m vertical sand drains were installed so as to increase the bearing capacity of the clay. The construction started in 1956 and was completed within a period of two years.

Results of soil investigation before the construction showed that q_u increased linearly with depth, so it was assumed that the clay was in a normally consolidated state. Results of observation of consolidation settlement and also the increase in the undrained shear strength supported the initial assumption that the clay was in a normally consolidated state without ageing effects.

Distribution of water content with depth in the seabed clay before the construction, however, was much more anomalous than those in the case records of Tokyo Bay 'A' and 'B' sites.

The technical standard of boring and sampling at the time of construction is not considered satisfactorily high. In the present report, therefore, the more reliable data from a soil investigation in 1965 has been used for assessing the *in situ* state of the clay before consolidation by the levée load. The soil investigation in 1965 was carried out at the site, at which the influence of the levée load was considered negligible. Original ground level of the construction site was EL -2.5 m, whereas the ground surface at the soil investigation site in 1965 was EL -1.8 m. A little change in the soil state, due to surface desiccation, was anticipated over a 5-year period. Soil sampling, therefore, was done below the depth of 4 m in order to avoid any possible influence of desiccation. As for the state of the clay after consolidation, results of soil investigation made in 1967 are used, which is 3920 days after start of consolidation.

Figure 4(a) shows the distribution of w with depth in the clay before and after consolidation by the levée load. As shown in the figure, w increases with depth in both cases. Figure 4(b) compares distribution of q_u with depth before and after consolidation. *In situ* value of $\Delta c_u/\Delta p$, based on measurements during the construction period, was observed to be 0.4 at the clay surface and gradually decreased to 0.3 at the bottom.

The *in situ* relationship between e and p, based on Fig. 4(a), is shown in Fig. 4(c), where results of three borings are plotted. The straight line in the figure corresponds to the compression index C_c obtained by oedometer tests. As seen in the figure, the *in situ* e vs. p relationship is opposite to that anticipated from the oedometer tests. A similar sort of anomalous relationship is also seen in the e vs. q_u relationship in Fig. 4(d).

In order to investigate the change in soil state by consolidation, it is necessary to know the elevation of corresponding soil element before and after consolidation. For this purpose, the clay strata before and after consolidation were divided into five parts of equal thickness, and the corresponding soil elements $(1-4)$ were assumed as shown in Fig. 4(a) and (b). This crude method is based on an assumption that the clay stratum has compressed uniformly for an 11-year period, because the consolidation by vertical sand drains has been practically completed within 3 years.

The *in situ* q_u vs. e relationship of each soil element is shown in Fig. 4(e), where the states before and after consolidation are shown. The two compression curves in the figure are those obtained by oedometer tests of samples before and after consolidation. It will be seen in the figure that, in spite of anomalous q_u vs. e relationship in the intact condition, the mode of change in the relationship of q_u vs. e throughout the course of consolidation is not much different from that anticipated from the oedometer test result. Figure 4(f) shows the change in the e vs. p relationship, where the value of p after consolidation is back-calculated from the measured values of increment in q_u and the *in situ* value of $\Delta c_u/\Delta p$ during the construction period.

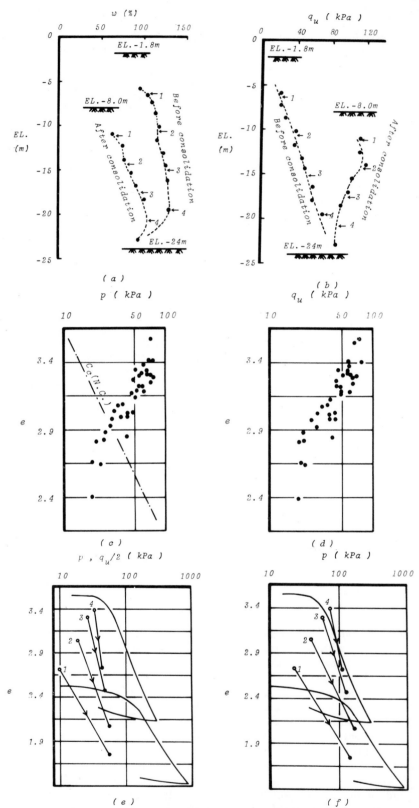

Fig. 4. Kikai Bay case study: (a) water content vs. depth; (b) unconfined compression strength vs. depth; (c) void ratio vs. effective overburden pressure; (d) void ratio vs. unconfined compression strength; (e) unconfined compression strength vs. void ratio for each soil element (1—4), (○) before and (●) after consolidation; (f) void ratio vs. effective overburden pressure for each soil element (1—4).

Concluding remarks

Four case records of *in situ* soil state of recent alluvial seabed clays in Japan have been presented. The seabed clay at Osaka Bay seems to be practically in a normally-consolidated state. In the other three cases, however, the *in situ* relationship between water content, or void ratio, undrained shear strength and effective overburden pressure is somewhat anomalous.

The case record of Tokyo Bay 'B' clearly indicates that this sort of anomalous relationship is not seen in very young clays. This fact implies the importance of ageing effects, such as delayed compression and cementation, on the *in situ* state of seabed clays. The influence of repeated shearing by waves (Henkel, 1970), and also the influence of the eustatic sea level change will have to be taken into consideration in the case of seabed soils. At the moment, however, none of these factors can fully explain the existence of such an anomalous relationship between void ratio, undrained shear strength and effective overburden pressure.

Fortuitously, the case record for Kikai Bay showed that, in spite of anomalous void ratio, strength and overburden pressure relationships in the intact state, the mode of change in the soil state through consolidation by an additional loading is not much different from that anticipated from the consolidation test results.

Acknowledgements

The authors wish to thank the Soils Division of Port and Harbour Research Institute and the 2nd District Bureau of Port Construction, Ministry of Transport, Japanese Government, for kindly offering the data of soil investigation described here.

References

Hamilton, E. L., 1964. Consolidation characteristics and related properties of sediments from experimental Mohole (Guadalupe Site). *J. Geophys. Res.* **69** (20) 4257–4269.

Henkel, D. J., 1970. The role of waves in causing submarine landslides. *Geotechnique,* **20**(1) 75–80.

Kenney, T. C., 1964. Sea-level movements and the geologic histories of the post-glacial marine soils at Boston, Nicolet, Ottawa and Oslo. *Geotechnique* **14** (3) 203–230.

Mead, R. H., 1963. Factors influencing the pore volume of fine-grained sediments under low-to-moderate overburden loads *Sedimentology,* **2,** 235–242.

Nacci, V. A. and Huston, M. T., 1969. Structure of deep sea clays. *Proc. Symp. Civil Engineering in the Ocean II.* ASCE, pp. 599–619.

Noorany, I. and Gizienski, S. F., 1970. Engineering properties of submarine soils – A state-of-the-art review. *Proc. ASCE,* **96** (SM5) 1735–1762.

Parry, R. H. G. and Wroth, C. P., 1981. Shear stress–strain properties of soft clay. In: E. W. Brand and R. P. Brenner (eds.), *Soft Clay Engineering.* Elsevier Scientific Publishing Co., Amsterdam, pp. 309–364.

2

Undrained Shear Behaviour of Quasi-overconsolidated Seabed Clay Induced by Cyclic Loading

Kazuya Yasuhara†, Kazutoshi Hirao†, Haruo Fujiwara and Syunji U-He*

†Department of Civil Engineering, Kanda, Japan
*Department of Civil and Architectural Engineering, Tokuyama Technical College, Takajyo, Japan

Seabed soil lying at shallow depth sometimes exhibits abnormal geotechnical and geological behaviour, particularly with regard to the void ratio—effective overburden pressure—undrained strength relation. This behaviour may possibly be connected with the cyclic loading induced by wind and wave forces acting on the seabed. The authors have considered the seabed in this circumstance as quasi-overconsolidated soil, the behaviour of which is assumed to be equivalent to that of overconsolidated soil due to release of overburden pressure. That is, for predicting the undrained strength of quasi-overconsolidated clay, the authors employ the simple relation which holds good between the overconsolidated clay due to release of effective stress and normally consolidated clay.

The prediction of undrained behaviour falls into two categories: one is concerned with the quasi-overconsolidated state generated by cyclically induced pore-water pressure; the other is relevant to the state with drainage due to dissipation of cyclically induced pore-water pressure. The former may be in accordance with the decrease in undrained strength, while the latter may be concerned with the increase in undrained strength. The predictable relations are modified from the relation proposed independently by Nakase *et al.* (1971) and Mayne (1980), and are given as the ratio between the undrained strength of quasi-overconsolidated soil and that of normally consolidated soil. An explanation of the applicability of the relations proposed here is attempted on the basis of the results obtained by cyclic triaxial compression tests on a highly plastic marine clay. Comparison of calculated and observed values for change in undrained strength has indicated that the theoretical relations are readily predictable even for highly plastic seabed clay.

Introduction

Seabed subsoil is often exposed to overconsolidation originating from the release of effective stress due to excavation, erosion and rising of the groundwater level. On the other hand, clay subjected to delayed compression behaves as though it lay under the overconsolidation induced by so-called diagenesis and cementation. In addition, it is known that cyclic loading also causes saturated clay to behave like the overconsolidated soil, since it brings about a decrease or an increase in the effective stress. In particular, cyclic loading of clay involves not only the earthquake response, but also wave action on seabed clay under off shore structures. In this sense this type of soil may be described as apparently overconsolidated soil. It is therefore a very interesting and important geotechnical problem to decide how both types of overconsolidated soil are interrelated with each other.

In the present paper the authors have attempted to describe the undrained shear behaviour of quasi-overconsolidated clay induced by cyclic loading, on the basis of research on overconsolidated clay which is due to release of the effective stress.

Stress history and $e - \log p'$ relation

As described in the previous section, there are several ways in which soil may become overconsolidated (Fig. 1). Among them, quasi-overconsolidation may be caused by several geological factors, as well as release of overburden pressure. As shown in Fig. 1(b), for instance, some clays have been subjected to the ageing or cementation effect after sedimentation. Bjerrum (1967) called this type of soil 'normally consolidated aged clay'. He found that the degree of overconsolidation of 'normally consolidated aged clay' depended upon the time of deposition and the plasticity index.

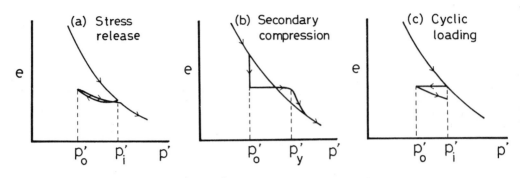

Fig. 1. $e - \log p'$ Representation of stress histories.

Other soils have accepted the change in effective stress due to cyclic or transient loading such as earthquake and wave actions as shown in Fig. 1(c). From the e–$\log p'$ representation shown in Fig. 1, a better understanding can be obtained of the difference between overconsolidated and quasi-overconsolidated soils. It is of special importance to ascertain the difference and the similarity between these overconsolidations due to different geological histories, particularly in the construction of the constitutive relation for clays. In the present contribution apparent overconsolidated soil is assumed, as a first approximation, to behave like overconsolidated soils due to the release of effective stress.

Quasi-overconsolidated soil will be classified into two types — apparent overconsolidated soils which are caused by delayed compression and/or cyclic loading. It is possible for seabed clays to be laid down under both circumstances. The present paper, however, deals with quasi-overconsolidated clay subject to wave-induced cyclic loading. Undrained strength of quasi-overconsolidated soft clay due to secondary compression has been dealt with elsewhere (Yasuhara and Ue, 1983).

Undrained shear behaviour under cyclic loading

The demands imposed by construction of civil engineering structures on stiff soils have led to increasing research into undrained behaviour of overconsolidated clay. Nakase and co-workers (1967, 1971), for instance, have made an important contribution to the estimation of decrease in undrained shear strength due to rebound, and then have proposed the following relation between the undrained strength, $(c_u)_{NC}$, of normally consolidated clay and the undrained strength $(c_u)_{OC}$, of overconsolidated clay:

$$\frac{(c_u)_{OC}}{(c_u)_{NC}} = (1-r)n^{-\lambda} + rn^{-\mu} \tag{1}$$

where n is the overconsolidation ratio, and γ, λ, μ are experimental constants. A similar relation to equation (1) has been found by Mitachi and Kitago (1976), Mayne (1980) and Ohta et al. (1981) as follows:

$$\frac{\left(\dfrac{c_u}{p}\right)_{OC}}{\left(\dfrac{c_u}{P}\right)_{NC}} = n^{\Lambda_0} \tag{2}$$

where Λ_0 is an experimental constant. Several methods are proposed for determination of Λ_0, such as

$$\Lambda_0 = 1 - C_s/C_c \text{ (Mayne, 1980; Ohta et al., 1981)} \quad 3(a)$$

$$\Lambda_0 = \frac{1}{\ln 2}\left(\frac{M}{2m}\right) \text{ (Mitachi and Kitago, 1982)} \tag{3b}$$

$$\Lambda_0 = 0.805(1 - C_s/C_c) + 0.0305 \text{ (Mayne, 1980)} \tag{3c}$$

where C_s, C_c are swelling and compression indices, respectively, M is a critical state parameter and m is the strength increment ratio in normally consolidated soil. Equation (2) is rewritten as

$$\frac{(c_u)_{OC}}{(c_u)_{NC}} = n^{\Lambda_0 - 1} \tag{4}$$

In the present paper, we employ equation (4) in order to estimate the change in undrained strength due to cyclic loading.

Consider the case shown in Fig. 2, where the quasi-overconsolidation occurs in saturated clay as a result of the excess pore-water pressure generated by transient loading and/or arrested secondary compression (Shen et al., 1976). It is required that the undrained shear strength of quasi-overconsolidated soil is estimated from the given undrained strength, c_{ui}, at normal consolidation. Here both the overconsolidation ratio (OCR), n, and the quasi-overconsolidation ratio, n_q, are defined by

$$n = \frac{p_c'}{p_0'} \tag{5a}$$

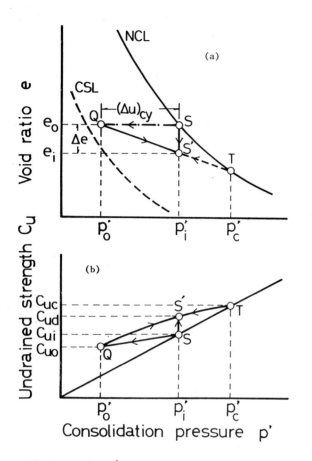

Fig. 2. $e-\log p' - c_u$ Relation of clay subjected to cyclic stress–strain history.

$$n_q = \frac{p_i'}{p_0'} \tag{5b}$$

Both are interrelated by

$$n = (n_q)^{\frac{1}{1-\lambda}} \tag{6}$$

Referring to Fig. 2, the undrained shear strength ratio between quasi-overconsolidation and normal consolidation is

$$\frac{c_{u0}}{c_{ui}} = n^{\Lambda_0 - (1-\lambda)} \tag{7}$$

According to equation (7), we have

$$c_{u0} = c_{ui} \tag{8}$$

in the case where $\Lambda_0 = 1 - \lambda$. This means that the strengths corresponding to quasi-overconsolidation and normal consolidation do not differ when we use equation (3b) to determine the value of Λ_0. Since OCR in equation (7) is given by equation (6), we have

$$n = \left\{ \frac{1}{1 - \dfrac{(\Delta u)_{cy}}{p_i'}} \right\}^{\frac{1}{1-\lambda}} \tag{9}$$

Substitution of equation (9) into equation (7) leads to

$$\frac{c_{u0}}{c_{ui}} = \left\{ \frac{1}{1 - \dfrac{(\Delta u)_{cy}}{p_i'}} \right\}^{\frac{\Lambda_0 - 1}{1-\lambda}} \tag{10}$$

On the other hand, when the accumulated pore-water pressure, $(\Delta u)_{cy}$, dissipates after removal of the applied cyclic loads, the undrained shear strength will increase from c_{u0} to c_{ud}, in accordance with the process QS' shown in Fig. 2. It is natural to consider that the void ratio change, Δe, due to drainage during QS should exhibit a secondary time dependence under constant vertical pressure, p_i'. From this consideration, we obtain another relation for estimating the change in undrained strength after draining, which is expressed by

$$\frac{c_{ud}}{c_{ui}} = \left\{ \frac{1}{1 - \dfrac{(\Delta u)_{cy}}{p_i'}} \right\}^{\lambda \frac{\Lambda_0}{1-\lambda}} \tag{11}$$

The important point in estimating the undrained strength ratio by using equations (10) and (11) is to determine the exact value of Λ_0 included in both equations.

Test programme

A series of undrained cyclic triaxial compression tests was performed on a reconsolidated seabed clay. Details of the cyclic triaxial equipment have previously been described (Yasuhara et al., 1982).

The tests used are listed in Table 1. In the type A tests pulsating loading (one-way loading) was imposed on each sample, while reverse loading (two-way loading) was employed in the type B tests. The frequency was 1.0 Hz in every test, except A-13 and A-14, which were performed to investigate the effect of frequency on undrained behaviour.

It should be pointed out that reverse loading results in the rotation of the principal stress and this may give rise to the different behaviour of clay compared with pulsating loading (one-way loading).

The tests of types C and D involve drainage subsequent to cyclic loading, before undrained shear as shown in Fig. 3. In addition, undrained shear behaviour after cyclic loading is affected by the pre-consolidation, cyclic loading and draining processes. Therefore, these

TABLE 1 Programme of cyclic triaxial compression test

Test no.	σ_c (kPa)[a]	σ_r (kPa)[b]	N (cycles)	f (Hz)	w_i (%)
A-1	200	40	3 600	1.0	95.6
A-2	200	80	3 600	1.0	93.3
A-3	200	100	3 600	1.0	90.6
A-4	200	120	3 600	1.0	93.8
A-5	200	140	3 600	1.0	90.5
A-6	200	160	3 600	1.0	90.8
A-7	200	100	172 800	1.0	92.5
A-10	200	120	172 800	1.0	100.0
A-11	200	160	91 800	1.0	92.0
A-12	200	100	60	1.0	90.5
A-13	200	100	3 600	0.1	94.1
A-14	200	100	3 600	3.0	95.0
B-1	200	40	3 600	1.0	88.9
B-2	200	80	3 600	1.0	93.2
B-3	200	100	3 600	1.0	92.1
B-4	200	120	3 600	1.0	93.8
B-5	200	160	3 600	1.0	90.9
B-6	200	120	172 800	1.0	92.1
B-7	200	160	172 800	1.0	90.9
C-1	200	100	3 600	1.0	90.9
C-2	200	120	172 800	1.0	91.7
C-3	200	140	3 600	1.0	92.7
D-1	200	100	3 600	1.0	84.6
D-2	200	120	172 800	1.0	94.5
D-3	200	140	3 600	1.0	90.5

[a] Consolidation pressure.
[b] Cyclic load intensity.

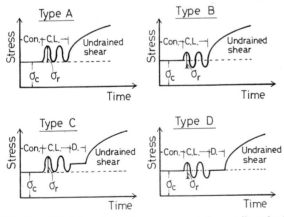

Fig. 3. Sketch of loading sequence in cyclic triaxial test: Con., consolidation; C.L., cyclic loading; D., draining.

factors are combined in the test schemes, listed in Table 1.

Types A and B, without draining after cyclic loading, are concerned with the loss of undrained strength, while series C and D, with drainage in the test sequence, are accompanied by the gain in undrained strength. In the sequence of tests as shown in Fig. 3, strain-

TABLE 2 Index properties of Ariake seabed clay

Specific gravity, G_s	2.65
Liquid limit, w_L (%)	123
Plasticity index	69
Unified soil classification	CH
Compression index, C_c	0.700
Swelling index, C_s	0.163
Coefficient of secondary compression, C_α	0.027
Critical state parameter, M	1.68

controlled triaxial compression tests were carried out on the samples after application of cyclic loads with a strain rate of 0.1 per cent/min. In the present work cyclic strength is defined as half the peak deviatoric stress in static triaxial tests after cyclic stress–strain history. Index properties of the soft marine clay used are summarized in Table 2. This clay is classified as a typical soft and sensitive marine clay in Japan.

Thus, in summary, the testing sequence is divided into the following two categories:

(1) consolidation → undrained cyclic loading
→ undrained shear
(2) consolidation → undrained cyclic loading
→ draining → undrained shear

Results and analysis

Prior to describing the influence of cyclic stress–strain history on undrained shear behaviour, conventional oedometer and triaxial compression preliminary tests were conducted on a reconsolidated clay and mechanical constants required for analysis were determined from these test results.

Undrained Shear Behaviour under Monotonic Static Loading

In all triaxial tests the reconsolidated marine clay samples were consolidated one-dimensionally from a slurry at over the liquid limit to a very effective stress of 58 kPa and the samples, 3.5 cm in diameter and 8.75 cm in height, were mounted between polished and greased platens.

Monotonic triaxial compression tests were performed by subjecting a sample to shear distortion without overall volume change. Immediately after isotropic consolidation to 200 kPa over 24 h, the normally consolidated sample was subjected to the changes of the principal stress difference under undrained conditions. Overconsolidated samples were sheared after swelling from 200 kPa with 1–20 of the overconsolidation ratio (OCR). Isotropic consolidation and swelling duration was 24 h in every test.

The stress path for overconsolidated samples is shown in Fig. 4. Figure 5 depicts the stress–strain curves

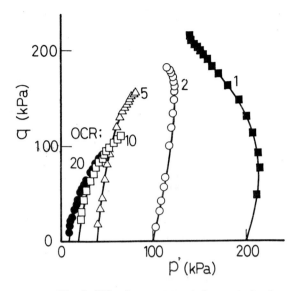

Fig. 4. Effective stress path for undrained overconsolidated samples.

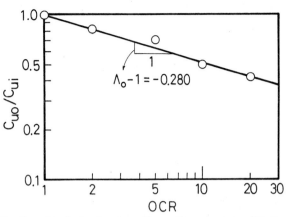

Fig. 6. Undrained strength ratio—overconsolidation ratio relationship of overconsolidated samples.

Fig. 6, we obtain a value of 0.720 for Λ_0. The compression index, C_c, and swelling index, C_s, were determined from oedometer and isotropic triaxial compression tests. Since the swelling index differed between oedometer and isotropic consolidation tests, the average value was adopted.

Stress Path of Cyclically-loaded Clay

Figures 7—9 show the stress paths obtained from triaxial compression tests of each cyclically-loaded specimen. The influence of the intensity of cyclic loads on the $p'-q$ space is shown in Fig. 7. The number of load cycles was maintained at 3600 in every test. The full thick lines indicate the stress paths of normally consolidated and overconsolidated clay samples which have not experienced the cyclic stress—strain history. The stress paths of the cyclically-loaded samples agree favourably with the path of the overconsolidated samples, as pointed out by Wood (1976). Besides this similarity, most of the curves lie between the curves of normally consolidated clay and of overconsolidated clay (OCR = 2), and every plot at failure lies on the critical state line or in its vicinity.

In another test, in which an incremental load intensity of 100 kPa had been cycled over a long period (172 800 cycles), we investigated the influences of not only the intensity of cyclic loads, but also the rotation of the principal stress on the stress path (Fig. 8). Although it can be seen from Fig. 8 that some plots exceed the critical state line, the reason for this phenomenon has not been clarified so far.

Figure 9 shows the influence of the number of load cycles on the $p'-q$ space. As the number of load cycles increases, the effective stress path travels towards the left side in the space, and this trend is more marked in the samples under one-way loading than in those under two-way loading.

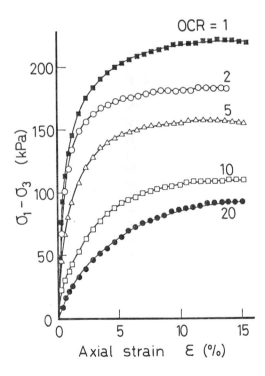

Fig. 5. Stress—strain curves of undrained overconsolidated samples.

of samples corresponding to the stress path in Fig. 4. By plotting half of the peak principal stress difference designated by c_{ui} and c_{u0}, for normally consolidated and overconsolidated samples, respectively, against OCR in

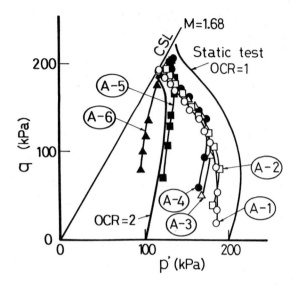

Fig. 7. Influence of cyclic load intensity on effective stress path

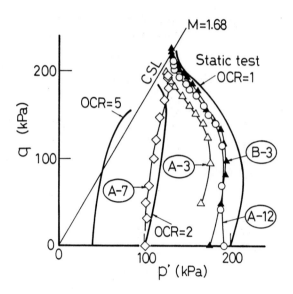

Fig. 9. Influence of number of cycles on effective stress path.

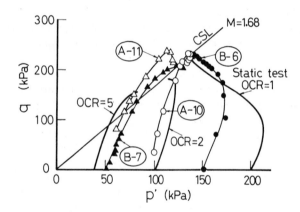

Fig. 8. Influence of loading type and load intensity on effective stress path.

The agreement between the behaviour of over-consolidated clay and that of quasi-overconsolidated clay was assumed in the theoretical treatment in a previous section. As far as the test results described above are concerned, we conclude that this assumption should be correct, because there is no difference in stress path pattern between the overconsolidated and the quasi-overconsolidated samples.

In summary, the end points of the stress path in clay samples after cyclic loading without drainage shows every plot is gathered near the critical point. Also characteristic is the fact that failure points in clay samples under two-way cyclic loading are located a little above the Hvorslev surface.

Stress—Strain and Undrained Strength due to Cyclic Loading

In cyclic loading of saturated clay two factors play a part in the change in undrained strength. One is the increase or decrease in the effective stress and another is the thixotropic or remoulding effect. The latter factor may cause either softening or hardening in clay particle structures. Those two factors seem to run counter to each other. However, note that in the present paper the theory described earlier does not consider the thixotropic effect of clay soil.

Figure 10(a) provides an example of the difference in stress—strain curves between samples without cyclic stress—strain history, under undrained cyclic loading and after dissipation of cyclic-induced pore-water pressure. No more considerable difference in undrained strength is observed than was expected in highly plastic clay. On the contrary, the difference in undrained strength between clay samples without and with cyclic stress—strain history was observed by Matsui *et al.* (1980) in clay of low plasticity (Fig. 10b) on which cyclic loading tests were run using the triaxial equipment.

As mentioned before, the normally consolidated saturated clay loses the effective stress due to cyclic loading, while it returns to the original effective stress (p_i', for instance, in Fig. 2) by draining and then gains the undrained strength in the same manner as clays subjected to secondary compression, as discussed by Matsui *et al.* (1980).

Changes in these undrained strengths were predicted by applying equations (10) and (11) to clay samples subjected to cyclic loading by comparison with normally

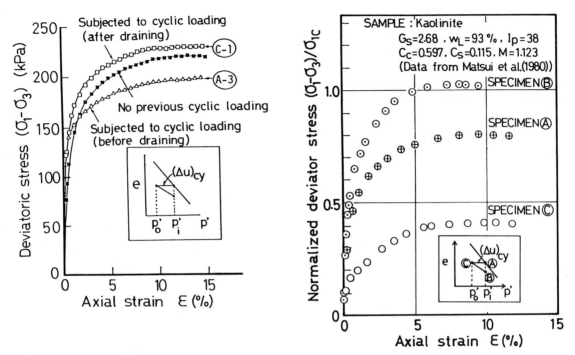

Fig. 10(a). Undrained stress–strain curves of marine clay samples with and without previous cyclic loading history; (b) undrained stress–strain curves of kaolinite samples with and without previous cyclic loading history.

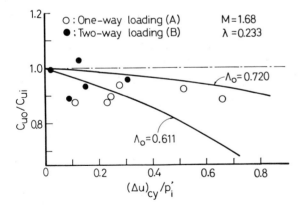

Fig. 11. Predicted and measured undrained strength ratio without drainage.

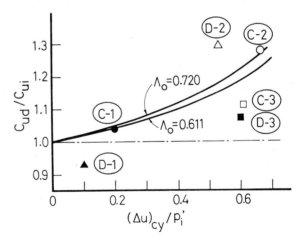

Fig. 12. Predicted and measured undrained strength ratio with drainage.

consolidated clay samples. Figure 11 shows the predicted and measured values of strengths due to decrease of the effective stress in clay under cyclic loading. Variations of the undrained strength ratio, c_{uo}/c_{ui}, are plotted to the cyclic pore-water pressure, normalized by the initial confining pressure, $(\Delta u)_{cy}/p_i'$. The measured values are shown by points, and curves calculated by equation (10) are shown by the solid line. Two calculated curves are drawn for $\Lambda_0 = 0.611$ from equation (3b) and $\Lambda_0 = 0.720$ from triaxial compression tests on normally consolidated and overconsolidated clay samples, respectively. Both curves seem to fit the averaged measured

strength ratio, in which there is a slight scatter of test data. At all events, both observation and prediction of undrained strength warn that the Ariake clay loses 10–20% of undrained strength owing to cyclic loading in cases where a sample has not reached fatigue failure during cyclic loading.

As has been previously noted, clays permitting dissipation of cyclic-induced pore-water pressure acquire undrained strength. This case is shown in Fig. 12 in which predicted and measured values of strength are compared. The calculated strength ratios agree better with those measured in triaxial tests. However, it is suggested by Figs. 11 and 12 that more extensive investigation on clays with a different plasticity index is needed to clarify the mechanism of clay behaviour under cyclic loading, besides the effect of thixotropy.

Conclusion

The effect of cyclic stress–strain history on the shear behaviour of a typical seabed soft clay was investigated in cyclic triaxial compression tests. Test results were considered in terms of both the effective stress and the total stress. Shear behaviour of quasi-overconsolidated clay induced by cyclic loading was compared with behaviour of overconsolidated clay due to release of the effective stress. A method for predicting the change in undrained strength due to cyclic loading was proposed and discussed. The following conclusions were drawn from the present study.

(1) The undrained shear behaviour of quasi-over-consolidated seabed clay induced by cyclic loading favourably agrees with that of overconsolidated clay. In particular, the quasi-overconsolidated clay displays a similar pattern of effective stress path to that of the overconsolidated clay.

(2) Test data of the cyclically loaded clay plotted by the effective stress fall close to the critical state line.

(3) End points in the normalized $p'-q$ space of clay samples subjected to two-way loading are located a little above the critical state line.

(4) The number of load cycles and the frequency to influence the effective stress parameters were not observed, particularly in clay samples with two-way cyclic loading.

(5) Owing to cyclic loading without drainage, the clay loses the undrained strength, while it gains the strength due to recovery of the effective stress by draining after cyclic loading.

(6) The methods proposed for the decrease and the increase accompanied by cyclic loading agree well with the test results observed with a highly plastic and soft seabed clay.

References

Andersen, K. H. 1976. Behaviour of clay subjected to undrained cyclic loading. *Proc. BOSS'76*, pp. 392–403.

Bjerrum, L. 1967. Engineering geology of Norwegian normally-consolidated marine clays as related to settlements of buildings. *Geotechnique*, **17**, 63–117.

Matsui, T. *et al.* 1980. Cyclic stress-strain history and shear characteristics of clay. *Proc. ASCE*, **106**, (GT10), 1101–1120.

Mayne, P. W. 1980. Cam-clay prediction of undrained strength. *Proc. ASCE*, **106**, (GT11), 1219–1242.

Mitachi, T. and S. Kitago 1976. Change in undrained shear strength characteristics of saturated remoulded clay due to swelling. *Soils Fndns*, **16**, (1), 45–58.

Nakase, A. 1967. Decrease in undrained strength of saturated marine clays due to rebound. *Proc. 3rd Asian Reg. Conf. SMFE*, Vol. 1, pp. 227–230.

Nakase, A. *et al.* 1971. Prediction of decrease in c_u due to rebound. *Proc. 4th Asian Reg. Conf. SMFE*, Vol. 1, pp. 47–50.

Ohta, H. *et al.* 1981. Some considerations on the undrained shear strength of saturated clay. *16th Japan National Conf. SMFE*, pp. 321–324.

Schofield, A. W. and C. P. Wroth 1969. *Critical State Soil Mechanics*. McGraw-Hill, London.

Shen, C. K. *et al.* 1973. Secondary compression and its effect on the undrained strength of a marine clay. *Proc. ASCE*, **109** (GTI), 95–111.

Wood, D. M. 1976. Comments on cyclic loading of clay. *Proc. BOSS'76*, pp. 1–7.

Yasuhara, K. *et al.* 1982. Cyclic strength and deformation of normally-consolidated clay. *Soils Fndns*, **22**, (3), 78–91.

Yasuhara, K. and S. Ue 1984. Increase in undrained strength due to secondary compression. *Soils Fndns*. (in press).

3

Consolidation Properties and Stress History of Some Deep Sea Sediments

Armand J. Silva and Stephen A. Jordan

Marine Geomechanics Laboratory, University of Rhode Island, Narragansett, USA

This paper focuses on the consolidation properties and analysis of stress conditions of deep sea sediments in two regions: the North Western Atlantic in the vicinities of the Bermuda Rise and Blake Bahama Outer Ridge, and the North Central Pacific between the Murray and Mendocino Fracture Zones. One-dimensional consolidation tests have been conducted on over 100 samples obtained from a variety of different types of core samplers, including box, standard piston, giant piston (i.e. large-diameter), large-diameter gravity and Kasten, taken to depths of over 30 m. The sediments from the Atlantic sites are primarily illitic clays with varying amounts of calcium carbonate, smectite and silt content. In the Pacific sites the sediments range from illite-rich to smectite-rich pelagic and authigenic clays.

In general, the upper 2–4 m of the sediment in all deep sea regions examined exhibit overconsolidation ratios (OCR) substantially greater than unity with a logarithmic relationship between OCR and depth. The high preconsolidation stresses, which are fairly constant for each region, in this zone of 'apparent' overconsolidation are the result of high interparticle bonding and possibly some cementation effects. Therefore caution must be used in interpretation of compressibility data from the upper few metres of deep sea clays. Implications of this apparent overconsolidation on the state of stress in the underlying strata are examined.

The illites in the upper 4 m of the North Central Pacific region have compression indices of slightly less than 1.0, whereas the smectites below 7 meters have very high compression indices of between 2.5 and 3.0. The rate of sedimentation in this region is very low (<1 mm/1000 years) and the sediment column is normally consolidated.

The rates of sedimentation for the abyssal hill regions of the North Western Atlantic are generally quite high (over 20 cm/1000 years) and the compression indices range from 0.5 to 2.2. The distal abyssal plain sediments have compression indices of less than unity. Most of the North Western Atlantic sites examined have OCRs considerably less than 1.0 at depth, indicating an underconsolidated condition below the zone of apparent overconsolidation. Possible explanations for this phenomenon are examined. A consolidation test from a core taken in a 10 m deep channel showed an OCR value which confirmed that this feature was an erosional furrow.

Introduction

In contrast to the wealth of information on compressibility of terrestrial soils, relatively few data exist on deep sea sediments, especially in terms of areal coverage and depth into the strata. The paucity of information on consolidation properties can be attributed in part to the difficulties and expense involved in obtaining suitable 'undisturbed' samples in deep water and the limited engineered utilization of these seafloor regimens. However, improvements in coring technology, renewed interest in deep sea applications and the use of geotechnical analyses in helping to understand geological/sedimentary processes have provided some impetus for recent studies.

The search for acceptable sites for disposal of nuclear wastes has been extended to consideration of certain geologically stable deep sea sediment regimens which possess favorable physical property characteristics (Anderson *et al.*, 1980). The permeability characteristics and other geotechnical parameters are important inputs for assessment of the sediments as an effective containment of solidified waste buried below the seafloor. As part of this analysis, it is important to determine the state of consolidation or stress history of the sediment column. Under the auspices of the Subsea-bed Disposal Program, preliminary investigations were undertaken in the North Western Atlantic and detailed geotechnical studies have been conducted on sediments from the North Central Pacific.

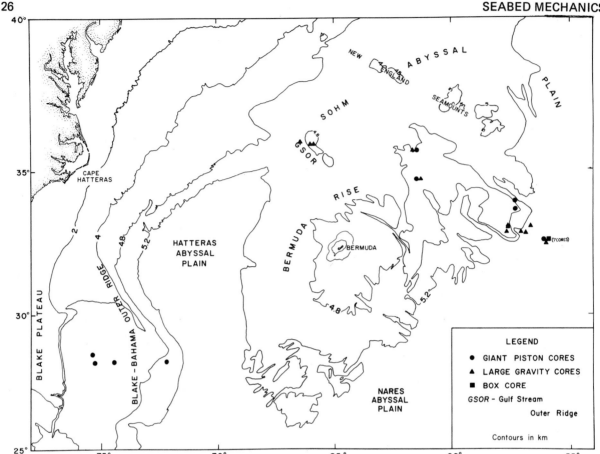

Fig. 1.   Location of 26 cores from four North Western Atlantic study areas: North Bermuda Rise, Gulf Stream
Outer Ridge, Blake–Bahama Outer Ridge and Eastern Shom Abyssal Plain (east of NBR).

In addition to the obvious need for compressibility properties in designing bottom-supported structures or instrument packages, other interests have included sub-bottom acoustic characteristics, design of sampling and testing equipment, deep sea mining, and geological processes such as mass wasting, erosion and heat flow. Recent and planned exploration for hydrocarbons in deep water (over 1500 m) is certain to increase the need for information on engineering behavior of sediments, including compressibility and stress history.

The pervasiveness of apparent overconsolidation in the upper 4 m of surficial materials (Richards and Hamilton, 1967; Bryant *et al.,* 1974; Richards, 1976; Silva *et al.,* 1976; Silva, 1979) makes it imperative to obtain samples to greater depths for studies on stress history. However, the standard piston corer used in most oceanographic studies is of small diameter (6 cm), with penetration depths typically 15 m or less, and sample disturbance can be significant.

This paper summarizes results of 180 consolidation tests on samples from 52 cores taken with a variety of samplers in deep sea regimens of the North Western

Atlantic and North Central Pacific. Most of the samplers were of large cross-sectional area (over 10 cm diameter) and attention was given to improving field techniques and reducing structural disturbance of the sediments. Good-quality samples have been recovered to depths in excess of 25 m in several locations. The sediments were primarily fine-grained clays and silty clays with the predominant clay mineral being illite; however, the presence of smectite and calcium carbonate in some samples had significant influence on the properties.

Geologic setting and sediment types

North Western Atlantic

The sediment samples were from a large region to the east and southeast of Cape Hatteras, USA, and can be categorized in terms of four general physiographic regimens (Fig. 1): the Northern Bermuda Rise, which is further subdivided into the central 'plateau' areas and 'slope' or scarp features; the Gulf Stream Outer

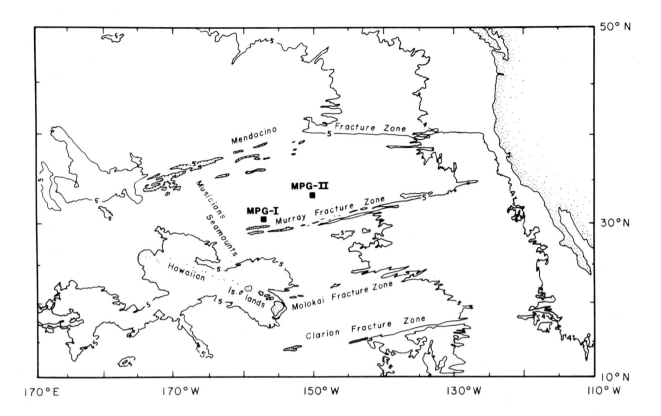

Fig. 2. Location of two Subseabed Disposal Program study sites (MPG-I and MPG-II) in North Central Pacific.

Ridge, which is contiguous to the Northern Bermuda Rise on the west; the Eastern Sohm Abyssal Plain adjacent to the Northern Bermuda Rise on the east; and the Blake Bahama Outer Ridge, which lies under the Western Boundary Under Current southeast of Cape Hatteras.

The Northern Bermuda Rise has a thick sequence of rapidly deposited hemipelagic clays of thickness exceeding 1000 m (Ewing *et al.*, 1973; Laine *et al.*, 1973). The plateau regions, at water depths of less than about 4800 m, have a cap of acoustically well-stratified sediments approximately 200–300 m thick, with an average accumulation rate of 30–40 cm/1000 years. These sediments are primarily illite-rich clays containing a variable calcium carbonate component that fluctuates between 5 and 40%. This laminated sequence is underlaid by an acoustically transparent sediment which outcrops on an eastward scarp or slope and in other localized areas of erosion (Laine *et al.*, 1980). The acoustically transparent material seen in the outcrops (below a thin cap of hemipelagic material) is an unfossiliferous brown clay with significant amounts of montmorillite (Silva *et al.*, 1976).

Although the Gulf Stream Outer Ridge is contiguous to the western side of the Bermuda Rise, it appears that this feature was once part of the Hatteras Outer Ridge system to the southwest and was severed from it by erosional events approximately 2.5–3.0 million years ago (Tucholke and Laine, in press). The western flank exhibits truncated stratified sediments suggesting considerable erosion. Only a few relatively shallow cores (< 4 cm) have been taken in this region, but it appears that the sediments are very similar to those of the laminated materials on the plateau areas of the Northern Bermuda Rise.

The Eastern Sohm Abyssal Plain region to the east of the Bermuda Rise is near the distal reaches of the Shom Abyssal Plain in water depth of about 5500 m. The upper few meters consist primarily of alternating layers of very soft silty clay, clayey silt and thin layers of silt. Coarser sands and silts, more typical of turbidites, have been found at depths of about 10 m (Silva *et al.*, 1981a).

The Blake Bahama Outer Ridge is a topographic rise in water depth of about 4900 m (Fig. 1) with two distinct ridges: the Bahama Ridge at 29 °N 73 °W and the Blake Ridge at 29 °N 75 °W. Except where acoustic returns produce hyperbolic acoustic profiles because of erosional features, the sediment is acoustically laminated

(Hollister *et al.,* 1974). The formation mechanism and source of sediment is still somewhat controversial (Silva and Hollister, 1979) but the majority of information seems to indicate that the ridge sediments have been transported from the north by the southerly-flowing Western Boundary Under Current (Hollister and Heezen, 1972). The dominant clay mineral is illite, with clay fraction generally between 60 and 70%; but calcium carbonate of some layers exceeds 50% and there are zones with significant montmorillonite (18%).

North Central Pacific

The North Central Pacific area considered here is an abyssal hill region north of the Hawaiian Chain between the Mendocino and Murray Fracture Zones (Fig. 2) having a water depth of about 5800 m and very low sedimentation rates (0.3–4.5 mm/1000 years). Two areas have been studied in some detail: Mid-Plate Gyre-I (MPG-I) is between latitudes 30 and 31° 30'N and longitudes 157 and 159°W; and MPG-II is about 700 km northeast, between 33 and 33° 31'N and 150° 30'N and 151°W. Small-scale relief is of the order of 100–200 m and sediment thickness is generally 35–45 m in MPG-I and 15–20 m in MPG-II. The sediments are inorganic pelagic clays which are gray to brown (illite-rich) or chocolate-brown (smectite-rich) in color. These 'red clays' are primarily of terrestrial origin deposited through atmospheric dust and suspended particles circulated by ocean currents, but significant fractions of authigenic and volcanic materials are present (Heath *et al.,* 1980; Silva *et al.,* 1980).

Equipment and procedures

The two primary coring devices used to obtain samples for these studies were the giant piston corer (GPC), with an inside diameter of 11.4 cm (Silva *et al.,* 1976), and the large-diameter gravity corer (LGC), with an inside diameter of 10.2 cm. However, the standard piston corer (6.4 cm) and box corer (0.5 x 0.5 x 0.5 m) were used in a few locations. The sampling equipment, procedures and core processing techniques were developed to minimize disturbance to the sediments. In some cases cores were extruded on board the ship immediately after recovery, with subsamples taken for laboratory testing. The samples were sealed and transported in refrigerated sea-water, and larger core sections were stored in a refrigerated room.

Most of the laboratory equipment has been modified or especially fabricated for use with soft sediments and sea-water. Water contents and associated index properties are corrected for 35 p.p.t. salt content (Noorany, in press). The appropriate American Society for Testing and Materials (ASTM) 1982 specifications were used for Atterberg limits, specific gravity and standard incremental consolidometer tests. The consolidometers have been specially outfitted with Teflon-lined stainless steel units for use with marine sediments. Set-up procedures for grain size analysis were according to ASTM specifications; however, measurements were performed by the pipette method. Backpressured consolidometers are described by Silva *et al.* (1981b).

Casagrande's (1936, 1938) graphical procedures to determine preconsolidation stress and coefficient of consolidation (logarithm of time) were utilized. In general, a load increment ratio equal to 0.5 was used for consolidation testing.

North Western Atlantic results

Index Properties

The geotechnical properties of the Northern Bermuda Rise (NBR) and Blake Bahama Outer Ridge (BBOR) sediments have been discussed in some detail in other publications (Silva *et al.,* 1976, 1980; Silva and Hollister, 1979) and therefore will only be briefly reviewed here along with more recent data. The data on the Gulf Stream Outer Ridge (GSOR) and Eastern Sohm Abyssal Plain (ESAP) have not been previously published. Summaries of some properties are listed in Table 1.

Cores on the 'plateau' region of the NBR show extreme vertical variability in water content, with excursions from about 70% to over 200% sometimes occurring within a few centimeters: a fairly typical profile of these acoustically laminated sediments is shown in Fig. 3 for a GPC sample. A very persistent zone of higher water content (generally over 180%) which begins within 1.0–3.0 m sub-bottom and is 0.5–2.0 m thick seems to pervade throughout the NBR. The Atterberg limits follow the same trends as the water content, with liquidity index being substantially greater than unity (over 2.0 in the zone of high water content) in the upper 15 m and approaching unity at about 25 m. As indicated in Fig. 3 and from other cores recovered in these laminated sediments, the undrained shear strength does not show a simple monotonic increase with depth, although overall this trend does exist. In some cases — for example, in the upper high water content zone previously mentioned the shear strength decreases significantly, but there are other similar deeper zones where the strength increases. On the basis of the limited data it is not possible to draw any general conclusions on strength. However, the upper 25 m of these laminated sediments have fairly low strength compared with the North Central Pacific region, to be discussed later.

The sediments of the outcrop (or 'slope') region of

TABLE 1 Summary of geotechnical properties

Gen. Loc.	Region	Zone or sed. type	z (m)	Ave. w_o (%) (fr. profile)[f]	e_o[g]	w_L (%)	I_p (%)	Silt (%)	Clay (%)	I_L	C_c	$\dfrac{C_c}{1'+e_o}$	$c_v \times 10^{-4}$ (cm²/s)	OCR	No. cores
NBR[a]	Plateau	Acoust. Lam. Hemi-pelagic Clay	0–4	117 / 77–204	3.10 / 2.2–5.61	74 / 56–100	42 / 25–62	38 / 13–54	59 / 46–86	1.97 / 1.27–2.46	1.30 / 0.58–2.87	0.30 / 0.24–0.40		3.7 / 0.6–37.5	7
			4–10	118 / 70–200	3.00 / 2.20–3.04	61 / 31–129	47 / 25–93	37 / 29–57	59 / 43–71	1.58 / 1.39–1.78	0.88 / 0.66–1.08	0.22 / 0.22–0.23		0.6 / 0.1–1.6	3
			>10	95 / 73–197	2.76 / 2.36–3.38	77 / —	41 / —	38 / 17–57	59 / 43–83	1.08 / —	1.06 / 0.78–1.75	0.27 / 0.21–0.34		0.4 / 0.4–0.5	2
	Slope	Acoust. Transp. Clay	0–4	97 / —	2.62 / —	77 / —	40 / 31–51	43 / 3–55	53 / 42–64	1.55 / —	1.02 / 0.51–1.71	0.32 / 0.29–0.36		3.2 / 1.0–12.3	3
			4–10	97 / 87–105	3.50 / 2.68–4.32	68 / —	36 / —	36 / 16–42	61 / 48–83	2.63 / —	1.31 / 0.94–2.04	0.33 / —		0.9 / 0.7–1.1	2
			>10	96 / 88–105	2.57 / —	115 / 110–122	76 / 70–80	50 / 27–72	49 / 28–73	0.74 / —	1.16 / 1.07–1.26	0.32 / —		0.9 / —	1
GSOR[b]	Eroded Rise	Eroded, hemi-pelagic	0–4	103 / 75–121	2.56 / 0.93–2.80	82 / 62–103	43 / 33–62	40 / 36–49	55 / 43–64	1.57 / 1.21–1.92	0.99 / 0.48–1.36	0.27 / 0.18–0.35		4.2 / 0.7–9.2	3
BBOR[c]	Rise lam.	Acoust. lam. Hemi-pelagic Clay	0–4	128 / 98–159	3.52 / 3.2–3.84	110 / 95–136	73 / 63–92	30 / 16–42	67 / 56–80	1.53 / 1.16–1.63	1.80 / 1.31–2.22	0.38 / 0.31–0.45		4.1 / 1.2–7.0	4
			4–10	127 / 102–151	3.76 / 3.76	93 / 90–96	57 / 54–60	30 / 15–55	65 / 30–81	1.86 / 1.77–1.95	1.72 / 1.42–2.02	0.42 / —		0.6 / 0.4–0.8	4
			>10	101 / 54–148	2.06 / 2.06	86 / 52–103	45 / 12–61	38 / 17–65	57 / 9–78	1.54 / 0.89–3.08	0.93 / 0.45–2.18	0.27 / —		0.7 / 0.2–1.5	4
ESAP[d]	Distal abys. plain	Clays, Silts, Sands	0–4	93 / 70–190	2.49 / 2.06–3.43	66 / 37–94	33 / 14–50	33 / 22–77	66 / 23–75	1.90 / 1.19–3.17	1.10 / 0.68–1.06	0.23 / 0.19–0.29	40 / 23–55	2.0 / 0.4–4.0	9
			4–10	74 / 71–87	1.96 / 1.18–2.68	59 / 37–77	34 / 14–39	38 / 26–66	61 / 34–74	1.42 / 1.15–1.84	0.74 / 0.50–1.00	0.27 / 0.19–0.42	19 / 3–19	0.4 / 0.3–0.4	3
			>10	48 / 22–87	1.41 / 0.66–2.22	55 / 37–62	30 / 19–36	33 / 26–35	24 / 9–65	1.07 / 0.77–1.44	0.92 / 0.57–1.28	0.13 / 0.07–0.19	300 / —	0.3 / —	3
NCP[e]	Abys. hills	I[h]	0–4	110 / 98–137	3.05 / 2.56–3.33	90 / 81–101	50 / 42–71	33 / —	66 / —	1.6 / 2.0–1.0	0.88 / 0.46–1.50	0.22 / 0.15–0.30	3.2[i] / —	10.6 / 1.2–48.0	26
		T	4–10	147 / 100–220	4.15 / 2.82–6.20	160 / 110–240	100 / 60–140	26 / —	74 / —	0.90 / 0.65–1.03	1.75 / 0.88–2.71	0.34 / 0.24–0.44	2.1 / —	1.4 / —	3
		S	>10	203 / 139–254	5.90 / 4.47–7.12	240 / 200–294	140 / 105–152	31 / —	68 / —	0.76 / 0.51–1.00	2.80 / 2.46–3.10	0.41 / 0.38–0.45	2.7 / —	1.2 / —	1

[a] NBR = North Bermuda Rise.
[b] GSOR = Gulf Stream Outer Ridge.
[c] BBOR = Blake-Bahama Outer Ridge.
[d] ESAP = Eastern Sohm Abyssal Plain.
[e] NCP = North Central Pacific.
[f] From profile plots.
[g] From consolidation samples.
[h] I = illite; T = transitional; S = smectite.
[i] Exclusive of upper 2 m.

Note on data format: $\dfrac{\text{ave.}}{\text{min.—max.}}$

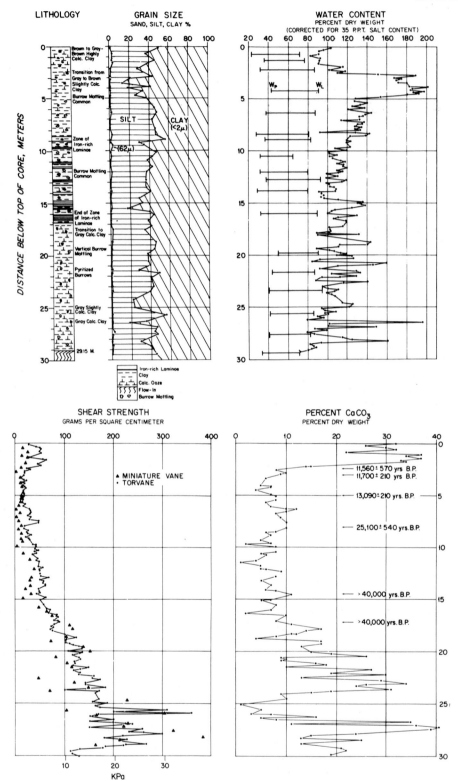

Fig. 3.   Profiles of some physical properties of acoustically laminated sediments in plateau region of North Bermuda Rise. This was a Giant Piston Core (11.4 cm inside diameter) taken at 33° 41.2'N, 57° 36.9'W in water depth of 4583 m.

the NBR are considerably different from those of the adjacent plateau. The water contents are much lower (Table 1) and show less vertical variability; however, magnitudes do not change appreciably within the cored depth of 14 m. The shear strengths are much higher (20 kPa at 14 m compared with 7 kPa for the plateau sediments) and exhibit an almost linear increase with depth. There is evidence of episodic erosion with furrowing, and mass wasting on some of the slopes.

Only a few, fairly short (2–4 m), cores from the GSOR were available for geotechnical analysis. The water content of these surficial sediments averages around 100%, which is lower than that of comparable materials on the NBR, and the shear strength increases to a fairly high value of over 20 kPa at 0.5 m depth. The lower water contents and higher strengths are attributed to erosion of overburden and resulting exposure of more consolidated sediments (Laine and Hollister, 1980; Tucholke and Laine, in press).

A suite of four long (17.1–30.5 m) GPC cores were recovered from the BBOR area (Fig. 1), three (GPC-7, 8, 9) being in non-eroded laminated sediments and one (GPC-11) in a large erosional furrow. For GPC-7, 8 and 9 the average water content is higher than on the NBR (Table 1), and there is only a slight gradual overall, though erratic, decrease with depth. The shear strengths are relatively low, with a gradual, fairly steady increase with depth to a value of 20 kPa at 20–25 m. The profiles for GPC-11 are quite different, with thick zones of high and low water contents and the presence of a silty sand layer at 17 m. The strengths for GPC-11 are much higher, with values well above 10 kPa near the surface and gradual increase with depth.

The nine cores used in the analysis for the ESAP include two box cores, four LGC cores and three standard piston cores. Water contents were almost constant, although differing in magnitude, in the upper 4 m. The shear strength of these surficial materials was very low (in some cases below the sensitivity of the miniature vane apparatus). The water contents appear to be controlled primarily by the grain size distribution, with very low values (22%) in the 'sand' layer at about 10 m depth.

Consolidation Properties and Stress Histories

Presented here is a general description of compressibility properties of the four North Western Atlantic regimens (Table 1), followed by a discussion of related stress history analyses. Many of the void ratio–log stress curves have been previously published and it is not possible to present all of the more recent results here. Most of the GPC (and LPC), LGC and box core results indicated samples of very good quality.

Three plots of more recent data from the laminated 'plateau' sediments of the NBR (Fig. 4) illustrate the

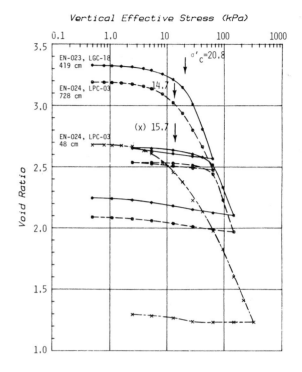

Fig. 4. Typical e–log σ' curves of an overconsolidated (σ'_c = 15.7 kPa) and an underconsolidated (σ'_c = 14.7 kPa) sediments from the North Bermuda Rise Plateau.

behavior of these relatively soft sediments. Two samples are from a large-diameter piston core (LPC) and one is from a nearby LGC. The sample at 48 cm (EN-024, LPC-03) exhibits high apparent overconsolidation (OCR = 6.5), which is fairly typical of these surficial sediments. The sample at 419 cm depth (EN-023, LGC-18) shows only slight overconsolidation (OCR = 1.2) and is probably near the lower limit of the apparent overconsolidation zone. The deeper sample at 728 cm (EN-024, LPC-03) indicates a significant amount of underconsolidation (OCR = 0.5), also typical of the NBR plateau sediments; this phenomenon will be discussed later.

Summaries of consolidation properties for the four regions of the North Western Atlantic are given in Table 1. The average values of C_c and compression ratio, $C_c/1 + e_0$, for the NBR (both plateau and slope) fall within a fairly narrow range of 0.88–1.31 and 0.22–0.33, respectively, but the overall range of all tests is quite high (0.51–2.87 and 0.21–0.40). This variability is a reflection of mineralogical and grain size distribution differences, with the larger values attributable primarily to increased smectite content. The sediments of the acoustically transparent outcrop area of the NBR have compression indices similar to those of the plateau but the compression ratio is considerably

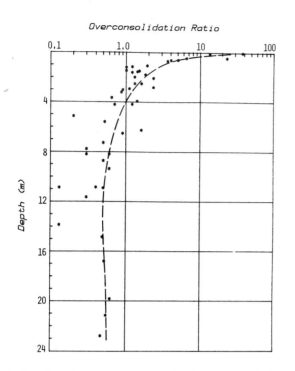

Fig. 5. Profile of OCR versus depth of Bermuda Rise plateau sediments displaying apparent overconsolidation (OCR > 1) in the upper 3–4 m and underconsolidation (OCR < 1) below 6–8 m.

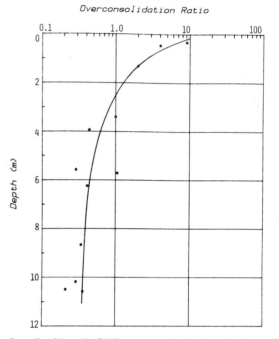

Fig. 6. Profile of OCR versus depth in the Eastern Sohm Abyssal Plain showing apparent overconsolidation (OCR > 1) in the upper 2–3 m and underconsolidation (OCR < 1) below 3–4 m.

greater. The overconsolidation ratios at depths below 4 m indicate that this sediment column is essentially normally consolidated.

A total of nine tests on surficial (< 4.5 m) samples from the GSOR region are summarized in Table 1. The limited results for the GSOR suggest that compressibility characteristics are similar to those on the NBR.

The compressibility characteristics of the BBOR sediments are considerably different from those of the other three North Western Atlantic regions studied (Table 1). The compression indices and compression ratios tend to be much higher, with (except for two samples) a narrower range of values. Lower C_c values of 0.45 and 0.57 in the deeper layers correspond to zones of higher silt and sand content (Silva and Hollister, 1979).

Compared with the nearby NBR, the compressibilities in the upper several metres of the ESAP sediments are only somewhat lower, but the deeper layers do appear to have much lower compression ratios. It is possible that the surficial materials of these abyssal plain areas are derived primarily from the NBR through erosion and mass wasting processes coupled with transport by bottom currents (Silva et al., 1976; Laine et al., 1980).

A composite plot of overconsolidation ratios for all samples from the NBR plateau sediments is shown in

Fig. 5. It should be noted that these results are for a very broad region, with probably significant localized differences in mineralogy, grain size distribution and sedimentation rate; therefore a smooth curve is not expected. Also, because the OCR scale is logarithmic, small differences in the lower values appear accentuated in this plot. The trend line is a best estimate of an 'upper bound' average for a typical sediment section of the NBR. There is a definite trend of high OCRs in the upper few meters, decreasing in a logarithmic fashion with depth and showing appreciable amounts of underconsolidation (OCR < 0.6) below 8 m. The apparent overconsolidation in the upper 3–5 m is attributed to intrinsic strength due to interparticle bonding and/or cementation effects. The existence of underconsolidation (< 8 m) has been previously reported (Silva et al., 1976). The more recent data, which are included with previous results in Fig. 5, confirm that underconsolidation does indeed exist in the laminated sediments of the NBR.

One of the cores on the GSOR (EN-023, LGC-20) was taken to the western side of this regimen, where Tucholke and Laine (in press) hypothesized that considerable erosion has occurred. The OCR values for two samples (OCR = 5.9 at 202 cm and 6.2 at 94 cm) were substantially higher than for any comparable

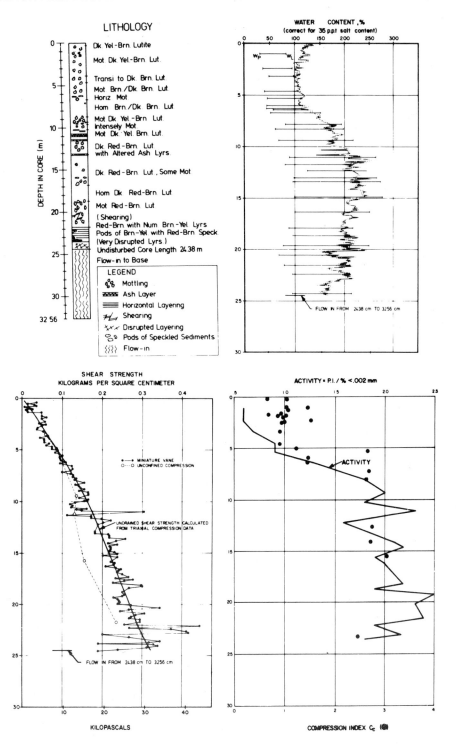

Fig. 7.   Profiles of some physical properties of 'red clay' sediments in MPG-I study area of North Central Pacific.

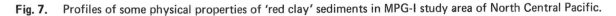

depths in the North Western Atlantic and are considered to be due to removal of overburden. The possible presence of remanent apparent overconsolidation effects complicates the analysis somewhat; however, calculations based on these limited tests indicate that 6–7 m of overburden removal would be required to explain these high OCR values. It is interesting that many of the properties of the GSOR (Table 1) are more in line with the deeper zones of similar laminated sediments on the NBR plateau, which suggests that the source of sediments may be similar.

The OCR data for the BBOR region are not as extensive as for the NBR but a few observations are reviewed here. Exclusive of one core taken in a large erosional furrow, the low values of OCR at depth suggest slight underconsolidation. The high OCR values in one core (KN-31, GPC-11) of 7.0 at 2.3 m and slightly over unity at 17 m are concluded to be due to overconsolidation caused by erosion. There is considerable evidence that this core was taken in the base of a large erosional furrow (Hollister *et al.*, 1974) and calculations indicate that removal of 10 m of sediment is necessary to account for the difference between preconsolidation and the existing overburden stress.

The composite plot of OCR for the ESAP (Fig. 6) shows a trend very similar to that of the NBR. Although the sedimentation history in this regimen is more complex than that of the NBR, there is a clear trend of high OCR near the surface and appreciable underconsolidation below 4–6 cm (OCR < 0.4 below 8 m).

North Central Pacific results

Index Properties

Profiles of lithology and some geotechnical properties are shown in Fig. 7 for the longest (24.4 m) core (GPC-3) retrieved to date from the North Central Pacific in MPG-I. The clays are illite-rich above 4 m (I zone), transitional (illite to smectite, T zone) between 4 and 10 m, and smectite-rich (S zone) below 10 m. The variability of both water content and activity supports the identification of the three zones. Above 4 m (illites) and below 10 m (smectites) water contents and activities are relatively constant, with the deeper material having much higher magnitudes, typical for smectites in comparison with illites. Mitchell (1976) listed a range of activities, with 0.5–1.0 for illites and 1.0–7.0 for smectites, which agree quite well with the I and S zones shown. However, in the T zone activity and water content profiles indicate a gradual transition from illite-rich to smectite-rich material.

The profile of undrained shear strength (determined from a miniature vane apparatus on board ship) does not show the same sensitivity to mineralogical changes as do water content and activity. However, a gradual increase in strength, S_u (kg/cm^2), with depth, z (m), is evident and can be approximated by the following equation.

$$S_u(z) = 0.2 \ln (1.0 + 0.1149z)$$

In addition, a continuous curve which represents undrained shear strengths determined from triaxial test results has been superimposed over the vane measurements (Fig. 7). The excellent agreement suggests that the vane measurements are representative of the sediment strength and that the core quality was very good.

Summaries of North Central Pacific sediment index properties are listed in Table 1. Water contents, initial void ratios, liquid limit and plasticity indices all increase with depth, which is typical for clays increasing in smectite mineral concentration. Mitchell (1976) lists typical ranges of liquid limit and plasticity index for both illites (60–120 for w_L and 25–60 for I_p) and smectites (100–900 for w_L and 50–800 for I_p). Averaged values listed in the table fall within the range of the dominant clay type identified for the particular zone. Water contents are generally above the liquid limit in the I zone, which results in liquidity indices greater than unity, and are usually thought to imply a viscous material behaviour when sheared (Holtz and Kovacs, 1981). However, triaxial test results showed behavior representative of an elastoplastic material (Silva *et al.*, in press). This discrepancy has been previously recognized (Jordan *et al.*, 1982) and is attributed to the phenomenon of apparent overconsolidation which typically occurs in the upper few meters.

Clay-size particles dominate the material composition in all three zones and are relatively constant in concentration with depth (Table 1). Therefore it appears that any appreciable changes in the geotechnical properties and/or behavior with depth could not be explained by the variability in grain size distribution.

Consolidation Properties and Stress History

Results of typical consolidation tests conducted on samples from each of the three zones are shown in Fig. 8. The smooth parabolic transition and sharp curvature between the overconsolidated and normally consolidated states are indicative of good-quality samples with little disturbance. The more compressible nature of smectites (in comparison with illites) is illustrated in Fig. 8 and shown in the values of compression index (Table 1), with magnitudes decreasing in the order smectite > transitional > illite. The results of many consolidation tests on terrestrial illites and smectites with differing ionic compositions were included in a report by Cornell University (1951). The results identified a range of compression indices for each clay type:

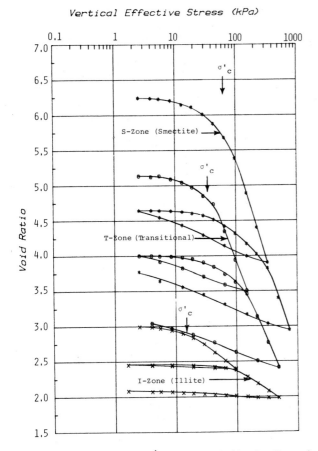

Fig. 8. Typical $e-\log\sigma'$ curves of North Central Pacific sediments from the I, T and S zones.

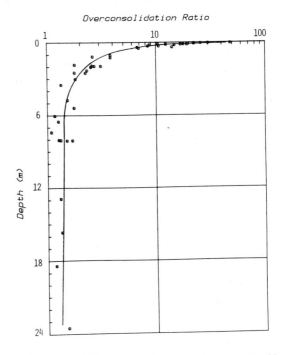

Fig. 9. Variability of OCR versus depth in the North Central Pacific indicating apparent overconsolidation above 4 m and normal consolidation below 4–5 m.

0.5–1.1 for illite and 1.0–2.6 for smectite. The compression indices determined for the North Central Pacific sediments agree quite well with those found for these terrestrial soils. Compression ratio values listed in Table 1 significantly increase in magnitude with depth, again reflecting the more compressible nature of smectite.

Coefficients of consolidation, c_v, represent the mean of determinations after stress application exceeded preconsolidation (Table 1). A comparison of values for each zone suggests that c_v is insensitive to mineralogical changes in these sediments. It should be noted, however, that values were significantly higher for specimens from the upper 3 m and were excluded from the data set to determine the upper zone average of $3.2 \times 10^{-4} \, \text{cm}^2\,\text{s}^{-1}$.

Figure 9 shows a profile of the variability of overconsolidation ratio (OCR) with depth for the North Central Pacific sediments. Very high (25–50) OCR values are observed near the surface, decreasing in an exponential fashion down to approximately 5 m. An independent analysis by Leinen (1980) concluded that there has been continuous sedimentation through the last 60–70 million years within the cored depth. Therefore the high OCRs displayed here are not brought about by erosion. These sediments (above 4 m) are apparently overconsolidated, which is attributed to the development of very high interparticle bonding and possibly cementation effects (Silva *et al.,* 1981, 1982). Mitchell (1976) pointed out that iron oxides (which have been found in these sediments) are capable of precipitating out and cementing the interparticle contacts. Below 4 m, OCR values are slightly higher than unity. The state of stress in these deeper strata may be affected by factors such as ageing (Bjerrum, 1973) and/or cementation. It is concluded that this portion of the sediment column is essentially normally consolidated.

Summary and conclusions

The consolidation and related properties presented here represent a wide range of deep sea sedimentary environments and sediment types. Sediments range from calcareous ooze to very fine-grained smectite-rich clays. Sedimentation rates differ by a factor of over 2000, with the higher magnitudes (of up to 240 cm/ 1000 years) in the North Western Atlantic and very low values (< 1 mm/1000 years) in the North Central Pacific. Depositional and post-depositional processes also vary, with those for the North Central Pacific being primarily

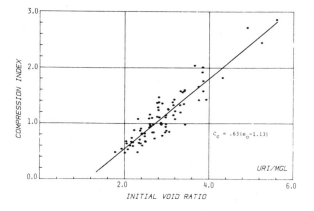

Fig. 10. Correlation between compression index and initial void ratio for Bermuda Rise and Blake Bahama Outer Ridge sediments: correlation coefficient $r = 0.90$.

TABLE 2   Equations for compression index

Equation	Reference	Applicability
$C_c = 1.15 (e_0 - 0.35)$	Nishida (1956)	All clays
$C_c = 0.30 (e_0 - 0.27)$	Hough (1957)	Inorganic cohesive soils (M, MC, CL)
$C_c = 0.75 (e_0 - 0.50)$	Azzouz et al. (1976)	Clays of low plasticity
$C_c = 0.007 (w_L - 10)$	Skempton (1944)	Remoulded clays
$C_c = 0.009 (w_L - 10)$	Terzaghi and Peck (1948)	Normally consolidated clays
$C_c = 0.63 (e_0 - 1.13)$	This study	Northwest Atlantic clays
$C_c = 0.70 (e_0 - 1.65)$	This study	Pacific red clays
$C_c = 0.011 (w_L - 6.4)$	This study	Pacific red clays

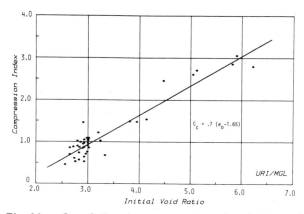

Fig. 11. Correlation between compression index and initial void ratio for North Central Pacific sediments: correlation coefficient $r = 0.92$.

continuous pelagic deposition and those for the North Western Atlantic study regions ranging from pelagic or hemi-pelagic to turbidity current transport, with some areas exhibiting general or localized erosion and mass wasting. Following are summaries and conclusions of the main points presented in this paper.

(1) *Compressibility characteristics and correlations with index properties.* Compression indices and compression ratios for the North Western Atlantic sediments compare quite well with those from the North Central Pacific, where illite is the dominant clay type. However, values from both areas are significantly higher than magnitudes previously reported for terrestrial and bay clays and silty clays (Holtz and Kovacs, 1981). Mitchell (1976) indicated that C_c values for most natural clays are rarely above 1 and are generally less than 0.5. Averaged values for our deep sea clay samples ranged between 0.7 and 2.8. It appears that the compositional and environmental factors affect deep sea sediments such that high compressibilities are the rule rather than the exception.

Correlations between consolidation characteristics and index properties are commonly found for terrestrial soils and are used for preliminary calculations to predict settlement. Some of the better-known equations which relate compression index to initial void ratio and liquid limit are presented in Table 2. The equation derived by Terzaghi and Peck (1948) is most widely used for normally consolidated clays of low sensitivity (<4). Included in Table 2 are equations relating the compression index to the initial void ratio for both regimens and to liquid limit for the North Central Pacific. Corresponding plots of the compression index against initial void ratio are shown in Figs. 10 and 11. The slopes of the correlation lines (0.63 and 0.70) compare reasonably

well with the range (0.30—1.15) for the terrestrial soils; however, void ratio intercepts (1.13 and 1.65) are noticeably greater. Deep sea clays typically develop a highly open, very stable flocculated structure, with initial void ratios much higher than those of their terrestrial counterparts.

The slope and intercept of the compression index—liquid limit relationship agree quite well with those in the terrestrial equations. Unlike the initial void ratio, the liquid limit is not dependent on the original structure of the sediment, and it appears that the agreement in intercept is a reflection of this.

(2) *Areas of continuous deposition.* Apparent overconsolidation exists in the upper 2—4 m of all non-eroded areas studied. In the few locations where detailed data exist, it appears that the preconsolidation stress in this upper zone is constant. There is a depth at which the sediment column is normally consolidated, but in some situations (particularly in the North Western Atlantic) this is simply a transition from an apparently overcon-

solidated condition to a case of underconsolidation in the deeper layers. Apparent overconsolidation is the result of high interparticle bonding and/or cementation effects which result in an intrinsic strength sufficient to prevent consolidation. That is, the 'normal' consolidation process is suppressed until the overburden stress overcomes this intrinsic strength. This phenomenon, coupled with high sedimentation rates (such as exist in some areas of the North Western Atlantic), creates an underconsolidated condition in the deeper strata due to the longer drainage path imposed by apparent overconsolidation. The results of our tests and stress calculations show that large portions of the NBR plateau areas, the ESAP and some areas of the BBOR have underconsolidation stress states below 8 m depth. In contrast, the sediment column in the North Central Pacific is more nearly normally consolidated below 4 m. In this case the sedimentation rates are so low that there is no accumulation of excess pore pressures in the deeper strata.

(3) *Erosional areas.* To analyze possible erosional sites, samples must be taken below the apparent overconsolidation zone. On the basis of somewhat limited consolidation data, we estimate that approximately 6–7 m of sediment has been eroded from one area of the western flank of the GSOR. Additional deeper (> 4 m) samples are needed to investigate further this regimen.

One core on the BBOR evidently was located in the base of a large erosional furrow. Calculations based on one consolidation test indicate that at least 10 m of sediment has been removed at this location. This result is consistent with other independent studies in which furrows of up to 20 m depth were discovered in this area.

(4) *Mass wasting areas.* The slope sediments on the eastern edge of the NBR are normally consolidated. A previous study indicates that intermittent slumping has occurred on some of these slopes such that only a thin drape of recent material is now present. Evidently, there has not been an accumulation of thick overburden, since this would have been reflected in the consolidation test results.

(5) *Analytical models for stress history.* Some preliminary numerical modeling of continuous sedimentation with varying permeability properties has been completed (Sage, 1974). However, on the basis of our understanding of the environmental conditions and geotechnical behavior of these deep sea clays, development of a more realistic model is required to predict the stress history observed in these regimens. For the situation of variable deposition and erosion, the model should incorporate the following aspects: (a) variable rates of sedimentation

and erosion; (b) apparent overconsolidation (c) variable permeability and consolidation properties; (d) the process of ageing; and (e) cementation effects.

(6) *Final notes.* Geotechnical analyses related to stress history can provide valuable insights into geological processes such as erosion, furrowing and slumping, as well as diagenetic processes in the sediment column.

The existence of an apparent overconsolidation zone has important implications. As indicated above, this zone can have a significant influence on the state of stress such that underconsolidation could exist in deeper strata. In addition, settlement analyses of bottom-supported structures should take into consideration the presence of apparent overconsolidation. The predicted settlement will be less than if this phenomenon were not recognized.

The state of stress, and, hence, the long-term rate of water migration, is a very important factor in evaluating sites for possible containment of solidified high-level nuclear wastes within the sediment column. Thus, if a condition of underconsolidation exists, the natural convection would have to be factored into models for prediction of rates of pore fluid migration away from canisters. Although this in itself would not preclude these sites, it would seem preferable to have sites in which the sediment column is normally consolidated.

Notation

The following symbols are used

A	= activity
C_c	= compression index
$C_c/1 + e_0$	= compression ratio
c_v	= coefficient of consolidation
e	= void ratio
I_p	= plasticity index
I_L	= liquidity index
OCR	= overconsolidation ratio
S_u	= undrained shear strength
w	= water content
w_L	= liquid limit
z	= depth
σ'	= vertical stress
$\sigma'_c =$	= preconsolidation stress

Acknowledgments

The results reported have been accumulated over several years as parts of several research contracts. The research was supported primarily by the US Office of Naval Research (most recent contract number 14-76C0226) and also by the Department of Energy through Sandia National Laboratories (most recent contract number 37-2235).

References

American Society for Testing and Materials, 1982. *Annual Book of ASTM Standards,* Part 19, Natural building stones; soil and rock; peats, mosses and humis. ASTM, Philadelphia.

Anderson, D. R., Boyer, D. G., Deese, D., Herrmann, H., Kelly, J. and Talbert, D. M., 1980. *The Strategy for Assessing of Subseabed Disposal.* Sandia National Laboratories Report SAND79-2245, Alburquerque.

Azzouz, A. S., Krizek, R. J. and Corotis, R. B., 1976. Regression analysis of soil compressibility. *Soils Fndns,* 16 (2), 19–29.

Bjerrum, L., 1974. Problems of soil mechanics and construction on soft clays and structurally unstable soils: General Report, Session 4. *Proc. 8th Int. Conf. Soil Mechanics and Foundation Engineering,* Vol. 3, pp. 111–159.

Bryant, W. R., Deflache, A. P. and Trabant, P. K., 1974. Consolidation of clays and carbonates in deep-sea sediments. In: A. L. Inderbitzen (ed.), *Physical and Mechanical Properties.* Plenum Press, New York, pp. 209–244.

Casagrande, A., 1936. The determination of the pre-consolidation load and its practical significance, Discussion D-34. *Proc. 1st Int. Conf. Soil Mechanics and Foundation Engineering, Cambridge,* Vol. III, pp. 60–64.

Casagrande, A., 1938. *Notes on Soil Mechanics – First Semester.* Harvard University (unpublished).

Cornell University, 1951. Final Report, *Soil Solidification Research.* Ithaca, New York.

Ewing, M., Carpenter, G., Winisch, C. and Ewing, J., 1973. Sediment distribution in the oceans: the Atlantic. *Geol. Soc. Am. Bull.,* 84, 71–88.

Heath, G. R., Hayes, D., Silva, A. J., Hollister, D. C. and Leinen, M., 1980. Size characterization–North Pacific. *Proc. Marine Technology,* Washington, D.C.

Hollister, C. D. and Heezen, B. C., 1972. Geologic effects of ocean bottom currents: Western North Atlantic. In: *Studies in Physical Oceanography, 2.* Gordon and Breach, New York, pp. 3–66.

Hollister, C. D., Flood, R. D., Johnson, D. A., Lonsdale, P. L. and Southard, J. B., 1974. Abyssal furrows and hyperbolic echo traces on the Bahama Outer Ridge. *Geology* 2, 395–400.

Holtz, R. D. and Kovacs, W. D., 1981. *An Introduction to Geotechnical Engineering.* Prentice-Hall, Englewood Cliffs.

Hough, B. K., 1957. *Basic Soils Engineering,* 1st edn, McRonald Press, New York.

Jordan, S. A., Silva, A. J. and Levy, W. P., 1982. Consolidation, permeability and classification properties of deep sea sediments: North Central Pacific. 1982 ASCE Meeting, New Orleans (unpublished report).

Laine, E. P. and Hollister, C. D., 1980. Geological effects of the Gulf Stream on the Northern Bermuda Rise. *Mar. Geol.,* 39, 277–310.

Laine, E. P., Heath, G. R., Silva, A. J., Ayer, E. and Kominz, M., 1980. Evaluation of the Northern Bermuda Rise for the subseabed disposal of nuclear waste. *Proc. Marine Technology '80.*

Leinen, J., 1980. Paleochemical Signatures on Cenozoic Pacific Sediments. Ph.D. Thesis, University of Rhode Island, Graduate School of Oceanography.

Mitchell, J. K., 1976. *Fundamentals of Soil Behavior.* Wiley, New York.

Nishida, P., 1956. A brief note on compression index of soil. *J. Soil Mech. Fndn Engng Div., ASCE,* 82 (SM3), Proceedings Paper 1027, 1027–1, 1027–14.

Noorany, I. 1984. Classification of marine sediments. Submitted to ASCE.

Richards, A. F. and Hamilton, E. L., 1967. Investigation of Deep-Sea Sediment Cores, III. In: A. F. Richards, (ed.), *Consolidation Marine Geotechnique.* University of Illinois Press, Urbana, pp. 93–117.

Richards, A. F., 1976. Marine geotechnics of the Oslofjorden region. In: N. Janbu, F. Jorsted and B. Kjoernsli, (eds.), *Laurits Bjerrum Memorial Volume.* NGI, pp. 41–63.

Sage, J. D., 1974. Preliminary study of underconsolidation ratios in marine sediments. Worcester Polytechnic Institute (unpublished report).

Silva, A. J., Hollister, C. D., Laine, E. P. and Beverly, S., 1976. Geotechnical properties of deep sea sediments: Bermuda Rise. *Mar. Geotech.,* 1 (3).

Silva, A. J., 1979. Geotechnical properties of deep-sea clays – a brief discussion. *1st Canadian Conf. Marine Geotechnical Engineering.*

Silva, A. J. and Hollister, C. D., 1979. Geotechnical properties of ocean sediments recovered with the giant piston corer: Blake-Bahama Outer Ridge. *Mar. Geol.,* 29, 1–2.

Silva, A. J., Laine, E. P., Lipkin, J., Heath, G. R. and Akers, S. A., 1980. Geotechnical properties of sediments from North Pacific and Northern Bermuda Rise. Marine Technology Society Paper, *Proc. 16th Annual Meeting,* pp. 491–499.

Silva, A. J., Hetherman, J. R. and Calnan, D. L., 1981a. Low-gradient permeability testing of fine-grained marine sediments. In: T. F. Zimmie and C. O. Riggs (eds), *Permeability and Groundwater Contaminant Transport,* ASTM STP 746. American Society for Testing and Materials, Philadelphia, pp. 121–136.

Silva, A. J., Jordan, S. A. and Levy, W. P. 1981b. Geotechnical aspects of subseabed disposal of high level radioactive wastes. DOE/SLA Contract Nos. 13–9927 and 74–1098, March.

Silva, A. J., Jordan, S. A. and Levy, W. P., 1982. Geotechnical studies for subseabed disposal of high level radioactive wastes. DOE/SLA Contract No. 16-3310, December.

Silva, A. J., Moran, K. and Akers, S. A., 1983. Stress–strain–time behavior of deep sea clays. *Can. Geotech. J.,* 20, (3).

Skempton, A. W., 1944. Notes on the compressibility of clays. *Q. Jl Geol. Soc. Lond.* 100, 119–135.

Terzaghi, K. and Peck, R. B., 1948. *Soil Mechanics in Engineering Practice.* Wiley, New York.

Tucholke, B. F. and Laine, E. P. (in press). Neogene and Quaternary development of the Lower Continental Rise off the Central U.S. East Coast. Submitted to American Association of Petroleum Geologists, *Hedberg Research Conf. Continental Margin Processes.*

4

Settlement and Consolidation in the Laboratory of Steadily-deposited Sediment

G. C. Sills and R. C. Thomas

Department of Engineering Science, University of Oxford, UK

Large amounts of sediment are moved in suspension in water – in rivers and estuaries, for example – and then deposited onto the bed as the available energy reduces. An understanding of the settling behaviour and subsequent consolidation of the sediment is important for the analysis of siltation and dredging problems and for the disposal of slurried waste.

In the initial stages the sediment is mainly fluid-supported and behaves as a non-Newtonian fluid. Once the flocs become sufficiently close together, a framework develops along with increasing shear strength, providing a resistance to erosion. The consolidation process in which this occurs is affected by a number of factors, including the rate of deposition of sediment and the total mass of sediment as well as the sediment characteristics, pore-water chemistry, and so on.

This paper examines the specific influences of the conditions of deposition in the well-controlled environment of the laboratory. Sediment is introduced into 2 m high, 100 m diameter columns at different rates, and the subsequent consolidation is monitored using X-rays for accurate, non-destructive measurements of density and pressure transducers for pore-water pressures. The results confirm earlier conclusions that, for any given soil, the density of sediment for which a framework can exist will depend on the stress history, particularly the deposition rate. It was observed that a given mass of sediment will form a bed of greater thickness if it is deposited slowly than if it is deposited fast.

The quantitative results described in the paper cannot be applied directly to specific field situations, since the processes observed will be affected by the field conditions, but they do demonstrate the complexity of the sediment behaviour and indicate the significant processes.

Introduction

An understanding of the behaviour of sediment deposited under water and allowed to consolidate under its own weight is of importance in a number of areas. These include harbour maintenance, where a knowledge of the expected magnitude and time scale of settlement could influence dredging programmes, and disposal of waste material in settling ponds after mineral extraction. The field conditions are inevitably complex and behaviour at a specific site will be affected by the energy available in the water, lateral currents, and so on. Nevertheless an insight into the important physical parameters can be obtained by examining a one-dimensional process of deposition and consolidation in well-controlled laboratory conditions.

The simplest laboratory experiments are those in which a quantity of sediment is introduced into a settling column as a slurry of specified initial concentration and is then allowed to settle and consolidate while measurements are made of density, settlement, pore-water pressure and total stress. The results and analysis of such experiments have been reported by Been (1981), Been and Sills (1980), and Sills and Been (1981). However, in many field conditions the sediment will accumulate over a period of time, and it is proposed to examine here the effect on the subsequent consolidation of a steady input of sediment into the settling column.

Experimental details

The settling columns are acrylic, 2 m high and with an internal diameter of 101.6 mm.

The sediment used in the experiments was from Combwich in Somerset, and is an estuarine silty clay (termed Combwich 6), wet-sieved to remove any fine sand or particles coarser than $63\,\mu m$. In contains 40–45% clay sizes, and has a plastic limit of about 32% and a plasticity index of about 32%. It was mixed to a uniform slurry with tap-water at an approximate density of $1.08\,g\,cm^{-3}$ and was circulated by pumping from a 25 l container through a length of 12 mm diameter tubing back to the container. At any point along the tubing an outlet to a peristaltic pump, driven by a stepper motor, can be attached. This pump then feeds the sediment slurry at a predetermined rate into the settling column just below the top water level. Both inlet and outlet of the circulating system are below the slurry level in the container of mud, so no air becomes entrained with the circulating mud. The large container acts as a reservoir which can be kept filled with mud of the correct density. The rate of input can be varied by varying the diameter of the tubing in the peristaltic pump or by changing the stepper motor speed.

The most important measurement in this series is density, achieved by projecting a collimated beam of X-rays through the settling column and using a sodium iodide crystal and photomultiplier assembly to 'count' the X-rays passing through. The measurement count rate is converted to density, assuming an exponential relationship

$$N = N_0 \exp{(-k\gamma)}$$

where N is the count rate, N_0, k are constants, and γ is the bulk density. The constants N_0 and k are obtained for each density measurement from two or more calibration samples of known density placed beneath the settling column. Thus a continuous density profile can be obtained by traversing the system up and down the column automatically. This is shown schematically in Fig. 1, which also shows pore pressure and total stress measurement techniques. Pore pressure is measured by using pressure transducers mounted behind porous sand/Araldite filters in the walls of the column. For total stress measurements pressure transducers are mounted in the wall (for horizontal stress) and in the base (for vertical total stress).

The technique is described in detail in Been (1981). The accuracy of the density measurement achieved in these experiments is of the order of $\pm 0.02\,g\,cm^{-3}$, with a spatial resolution of better than 5 mm. This is less accurate than the method is capable of, and results from experimental problems with this series.

Results

Previous measurements (Been, 1981; Been and Sills, 1980) of density and pore-water pressure in settling

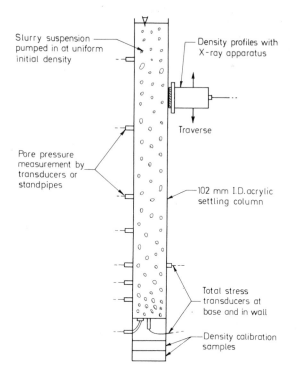

Fig. 1.   Schematic density measurement by X-rays.

columns had shown that there are two fundamentally different stages in the settling and consolidation process. While the concentration, or density, of the suspension is sufficiently low, the sediment is entirely fluid-supported. The surface settles quite fast (at a rate which is dependent to some extent on the concentration). Once the sediment particles, or flocs, become sufficiently close together, they form a structure which provides some support.

This behaviour is demonstrated in Fig. 2, which shows a typical density profile for a slurry placed all at once in the settling column. The left-hand curve shows a typical density profile at an early stage (in this case, 4.75 h after the start) of the settling experiments. The upper part of the curve indicates that the sediment there is still at the original suspension density. Lower in the column the sediment has settled sufficiently to produce a significantly higher density and it can be seen that there is a fairly abrupt change from the suspension density. At the base of the column there is a thin layer of still higher density, caused partly by some segregation of silt-size particles. The right-hand curve shows the change in total vertical stress through the column, calculated by integrating the density profile, and the pore pressures, measured by pressure transducers. Both these measurements are shown as excess pressures above hydrostatic, so that eventually the pore-water pressures will dissipate back to the y-axis,

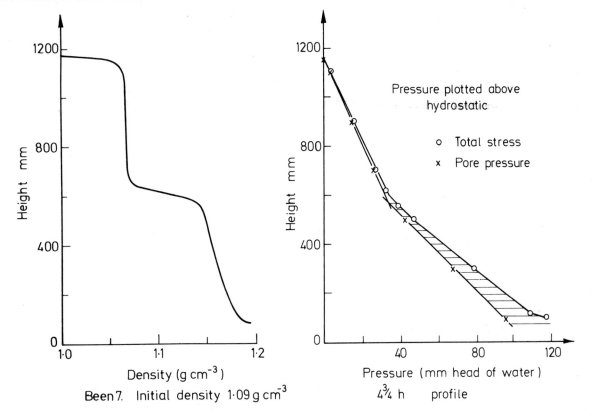

Been 7.   Initial density $1.09 \, \text{g cm}^{-3}$

$4\frac{3}{4}$ h   profile

Fig. 2.   Typical density profile.

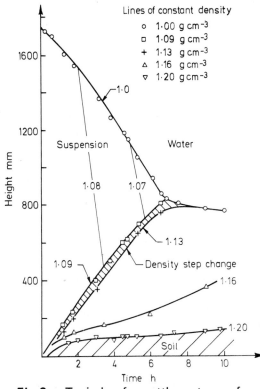

Fig. 3.   Typical surface settlement curve for instantaneous deposition.

corresponding to the end of consolidation settlement. At the level of the suspension the pore-water pressures and total stresses are equal, which shows that the sediment is entirely fluid-supported. Once the higher densities are achieved, the pore pressures are less than the total stresses, which indicates that a framework of sediment is now providing some support. This support increases as the sediment compresses under its own weight until the pore pressures have dissipated to hydrostatic pressure. During the whole of this stage the sediment will be referred to as 'settling mud'.

The suspension stage disappears quite quickly in these experiments — typically, 7–8 h after the start — but the settling mud may continue to consolidate for months. The disappearance of the suspension stage may therefore also be seen by an examination of the interface between water and sediment. Initially this interface separates water and suspension and settles quickly: once the sediment in suspension has all reached the settling bed, the interface represents the surface of the settling bed and compresses much more slowly. Figure 3 shows the change of the various sediment regions with time for the same experiment as shown in Fig. 2. The surface of the suspension settles fast, while the settling mud increases in height as sediment is deposited onto it.

TABLE 1 Details of continuous sediment experiments

Experiment	Mean rate of output (g h⁻¹)	Period of input (h)	Total mass (g)	Height of column (mm)
XRYCI1	362	2.5	906	1694
XRYCI2	6.90	105	724	1726
XRYCI3	7.89	70	552	1703
XRYCI4	12.61	183	2308	1699
XRYCI5	4.24	215	916	1699
KB7	Instantaneous		2040	1742
KB12	Instantaneous		1190	1800

Accuracies in this table are of the order of 3%

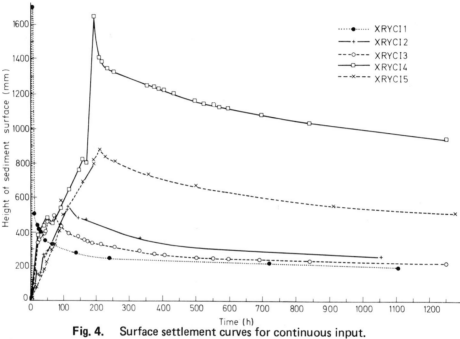

Fig. 4. Surface settlement curves for continuous input.

The step change in density between suspension and settling bed is shown between the density lines $1.09\,\mathrm{g\,cm^{-3}}$ and $1.12\,\mathrm{g\,cm^{-3}}$. The slow growth of the layer of higher density at the base of the column is also shown. Once the suspension surface meets the settling mud surface, all the sediment has fallen onto the settling mud surface and the subsequent settlement is much slower. In the absence of density measurements this change in the slope of the sediment—water interface identifies the disappearance of the suspension phase. The density at which a transition occurs from a suspension to settling mud depends on the soil itself and the pore-water chemistry, but also on the initial concentration of the suspension. For Combwich mud the transition density lies between 1.14 and $1.18\,\mathrm{g\,cm^{-3}}$.

In the present series of tests results for five different rates of input are compared with one another and with earlier experiments in which the sediment slurry was placed in the column all at once.

These earlier experiments (Been, 1981; Been and Sills,

1980) have shown that both the total mass and the initial slurry density affect the subsequent settling behaviour when the sediment is introduced all at once. It is therefore to be expected that the rate of introduction, measured in terms of sediment mass added in unit time, and the length of time during which the sediment is added will affect the behaviour during and following deposition. Table 1 shows the conditions for the five experiments: there is one comparatively fast input and four very much slower. Figure 4 indicates the position of the sediment surface for each of the experiments. This is not usually clearly defined, as the water throughout the column contains mud during the input stage, but, typically, a denser layer grows upward from the bottom and will be visible unless a stage is reached where the overlying suspension has become sufficiently thick to obscure the interface. The pattern is the same in each case: an increasing height initially during the period of deposition, followed by settlement. Some difficulties were experienced

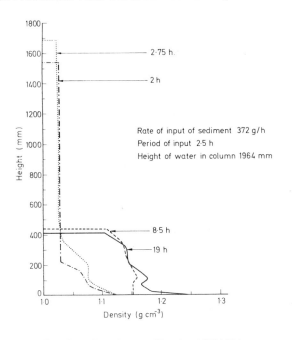

Rate of input of sediment 372 g/h
Period of input 2·5 h
Height of water in column 1964 mm

Fig. 5.   Density profiles for XRYCI1.

with the control mechanism during the experiments, so that the rate of introduction of sediment was not consistently uniform, and this can be seen as small peaks in the input stage of the slower four experiments.

The two stages of suspension and settling mud would be expected to occur if the conditions are suitable. Examining the change in slope of the surface settlement in Fig. 4, it is apparent that experiments XRYCI1 and XRYCI4 show the existence of some suspension phase that has disappeared by 10 h and 240 h, respectively. This is consistent with the surface observation during input, that over a short period of time the overlying water acquired sufficient sediment in suspension for the interface to appear to rise suddenly. There is some evidence also that similar behaviour, although less marked, occurred in each of the other three experiments. In experiment XRYCI1 the sediment had little time to consolidate during deposition and the surface moved very quickly up the column. Had this sediment been placed all at once rather than over a period of 2.5 h, it would have had an initial density of 1.04 g cm^{-3}, well below the suspension/settling bed demarcation, so that the presence of a suspension phase is not surprising.

Figure 5 shows density profiles for experiment XRYCI1 during and after input, and confirms that both suspension and settling mud are present. The profile marked '2 h' shows a small amount of settling mud at

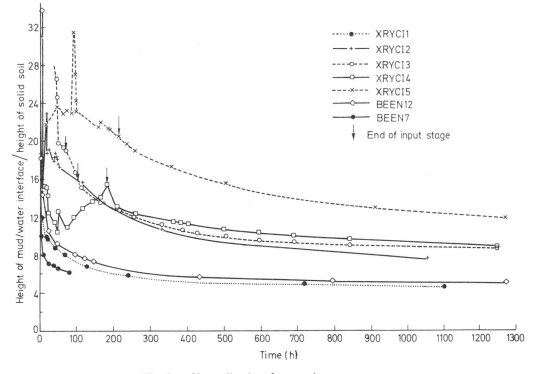

Fig. 6.   Normalized surface settlement curves.

the base of the column with a substantial layer of suspended sediment above. The 2.75 h curve (15 min after sediment input has ceased) also shows the suspension quite clearly, although it has disappeared 6 h later.

The settlement curves for all the experiments are examined in a different way in Fig. 6. During the input stage the mass of sediment in the column is changing. A better indication of the behaviour will therefore be obtained by normalizing the results with respect to this mass of sediment, so the surface settlement has been divided by the height that the corresponding sediment mass would occupy in the column if it contained no voids. This normalizing value (i.e. the sediment mass) increases steadily during the input stage. If no consolidation were occurring, then the mud surface would also increase steadily, so that the ratio would be expected to hold some constant value, indicative of an initial concentration, to be called the equivalent initial density. However, consolidation occurs, thereby increasing the initial density, and the ratio will tend to fall during the input stage. Once the input is complete, the curve maintains a constant ratio to the corresponding mud surface settlement curve.

If the height of the mud–water interface is represented by h and the solid sediment height by h_s, then the equivalent density can be obtained from

$$\gamma = 1 + \frac{h_s}{h}(G_s - 1)$$

where G_s is the specific gravity of the solid, taken to be 2.64 for Combwich mud.

For experiment XRYCI1 this gives a density at the end of the input of 1.05 g cm^{-3}. Since little consolidation has occurred in such a short time, this should be close to the equivalent initial density, which has already been calculated to be 1.04 g cm^{-3} on the assumption of instantaneous deposition. This curve is then compared with one from the earlier series of experiments, Been 12, where a slurry of initial density 1.05 g cm^{-3} was introduced all at once into the column. (The sediment used in Been's experiment was Combwich 3, very similar to Combwich 6, although with a slightly lower proportion of clay-size particles.) It may be seen that the curves are very similar.

The choice of initial h/h_s value (height ratio) for experiment XRYCI2 of around 19 (neglecting the initial spikes) leads to an equivalent initial density of 1.09 g cm^{-3}. A similar density experiment (initial slurry density 1.09 g cm^{-3}) using Combwich 3, deposited instantaneously, gives a significantly lower settled bed thickness. The experiment lasted only 79 h, so that a long-term comparison is not possible. This curve is shown also in Fig. 5.

None of the curves conform exactly to the idealization

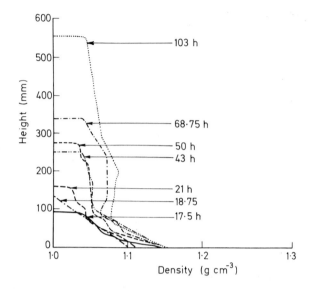

Fig. 7.   Density profiles for XRYCI2 during input.

of height ratio dropping slowly but steadily during the input state, although XRYCI1, XRYCI2 and XRYCI5 are close enough to make this interpretation worth while. With the other two experiments, at least some of the variation in the height ratio may be attributed to non-uniform deposition rates due to input control problems. The other factor that makes the height ratio a somewhat unreliable parameter is the fact that some of the sediment introduced into the column, and therefore included in the sediment mass height, will be in suspension in the water above the settled bed, and is therefore not part of the bed. This will cause the height ratio to be unrealistically low or the corresponding density too high.

After input has ceased, the water above the mud clears as the final flocs settle and the height ratio then provides a reasonable method for comparing behaviour for different sediment masses. It appears that the general trend is for increasing height ratio with decreasing rate of sedimentation. Figure 6 shows that after 1300 h, the height ratios seem to fall into three categories: the highest value is shown by the slowest input rate, corresponding to 4.24 g h^{-1} for experiment XRYCI5, while the lowest value occurs in experiment XRYCI1, with an almost instantaneous deposition at a rate of 372 g h^{-1}. Between these extremes the intermediate rates (6.87, 7.89 and 12.61 g h^{-1}) show broadly similar height ratios. It should be noted, however, that the trend in settlement ratios with time suggests that the settlement may not yet be complete, so that these observations cannot yet be substantiated in the long term.

Turning in more detail to one of the slower input rate experiments, Fig. 7 shows the density profiles obtained in XRCYI2 during the 105 h period of input sediment.

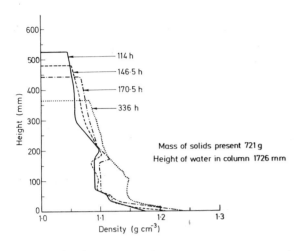

Mass of solids present 721 g

Height of water in column 1726 mm

Fig. 8. Density profiles for XRYCI2 after input.

Without pore-water pressure measurements it is not possible to determine exactly at what stage a framework develops but the increase in density between 50 h and 68.75 h could indicate a change from suspension to settled mud. If so, the transition density would be somewhere between 1.05 and 1.07 g cm^{-3}. This is considerably lower than that associated with the earlier instantaneous input experiments (with a range of 1.14–1.18 g cm^{-3}) or with XRYCI1, the fast input experiment of this series, where the settling mud shows a density of around 1.14 g cm^{-3} during the early stages of its development. Nevertheless the subsequent profiles, both during and, in Fig. 8, after the input stage, reinforce this interpretation, since they all show the general characteristics of shape associated with the settled bed from the earlier experiments.

At the conclusion of each experiment the settled bed was subsampled to determine the particle size variation

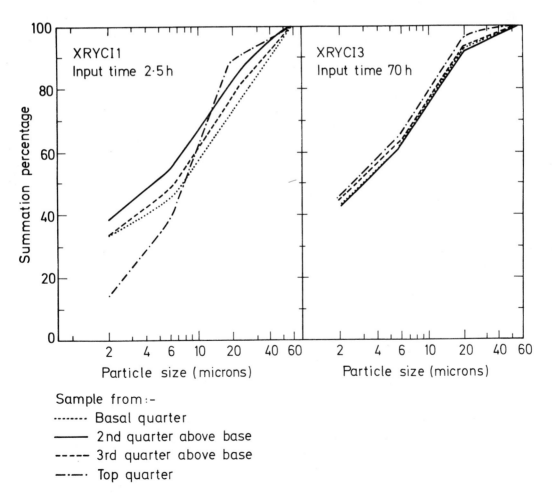

Sample from :-

.... Basal quarter

———— 2nd quarter above base

– – – – 3rd quarter above base

– · – · Top quarter

Fig. 9. Particle size distribution for XRYCI1 and XRYCI2.

at different heights. Figure 9 shows the results from two experiments, dividing the sediment into four equal regions and taking a sample from each region. Earlier experiments with instantaneously deposited sediment had shown that some small amount of particle segregation occurs, despite the sediment clearly having flocculated. This behaviour is not seen with the continuous input experiments: in XRYCI1 and XRYCI2 little segregation has resulted, presumably because the larger particles do not fall through the settled mud building up from the bottom. The apparently low clay proportion at the top of the settled bed in XRYCI1 is surprising and difficult to explain. It is possible that the subsample here may not be representative.

Discussion of results

In order to determine accurately the effect of sedimentation over a period of time, a much wider range of experiments than the present series would be necessary, especially to discover the separate consequences of variable sediment mass and different rates of input.

The problems of interpretation stem from the fact that these experiments suggest that a comparatively low settled bed thickness in the field may be due, obviously, to a low mass of sediment or, alternatively, to a larger mass of sediment with a short period of deposition at a fast rate. In these experiments an attempt has been made to isolate the effect of deposition rate by normalizing each of the surface settlement curves by dividing by the height that the mass of sediment alone would occupy in the column. This approach allows an extension of the conclusions of the earlier work, on slurries introduced all at once into the settling column. Previously it was noted that the initial slurry density had an important effect on the subsequent settled bed formation: now the present work has demonstrated that the slower the rate of sedimentation and therefore the more isolated the individual flocs before reaching the settled mud, the more open the framework that they are able to establish and the thicker the initial settled mud layer. In order to identify this behaviour in the field, *in situ* density measurements would be required.

These differences are quite clear over a period of the first 50 days and are illustrated particularly well by the two extreme rates of deposition: XRYCI1 and XRYCI5. Apart from their deposition rates, these two experiments are closely comparable and they contain very similar total masses of sediment. The bed thickness for the slower input is of the order of twice the thickness for the fast rate. It is not clear from these data whether the layers will be able to achieve the

same thickness by the time that the excess pore-water pressures have dissipated and consolidation is complete. There is, however, evidence from other experiments at Oxford that the volumetric strain behaviour of this sediment is strongly dependent on time and stress history, and it is unlikely that, even given a period of years, the two layers will ever become similar.

Conclusions

The results obtained in controlled laboratory conditions will not translate directly to the field, since the quantitative behaviour will be influenced by the sediment properties and the water chemistry. Nevertheless the general conclusion would be expected to be applicable in most circumstances. Thus it appears that the higher the rate of deposition of a given mass of sediment, the more compacted is the resulting settled mud layer. Once deposition is complete, the layer continues to consolidate, with the rate of settlement influenced by two factors: the permeability (which will be higher in the more open, thicker layer associated with low deposition rates) and the drainage path length (greater in the thicker layer). These two factors have opposing effects. Although none of these experiments lasted long enough for pore-water pressures to dissipate completely and for consolidation to finish, the rate of compression of all the layers appears comparable, which suggests that the two factors of permeability and drainage path lengths more or less balance.

Two major questions remain unanswered by the present experiments. The first relates to the behaviour of the settled mud bed under additional load — for example, a sand layer. It might be expected that the layer with a more open framework would undergo collapse to a denser state, while the already denser mud would exhibit smaller compression. The second question is the resistance to erosion of the settled mud. In all the layers the existence of a framework would be coupled with some resistance to erosion: it is not, however, possible to determine at this stage whether the denser mud dumped instantaneously would be more resistant than the slowly deposited sediment.

These experiments have demonstrated the need for measurements of density, total stress and pore-water pressures both in laboratory conditions and in the field. Field measurements are of particular importance, since the density differences due to the rate of input in the laboratory columns are sufficiently marked for them to be measurable in the field. Further insight would then be provided into the mechanisms operating in the field and the feasibility of modelling them in the laboratory.

References

Been, K., 1981. Non-destructive soil bulk density measurement using X-ray attenuation. *Geotech. Test. J.* December, 169–176.

Been, K. and Sill, S. G. C., 1980. Self weight consolidation of soft soils and experimental and theoretical study. *Geotechnique,* **31** (4).

solidation of soft soils – an experimental and theoretical study. *Geotechnique,* **31** (4).

Sills, G. C. and Been, K. 1981. Escape of pore fluid from consolidating sediment. In: *Transfer Processes in Cohesive Sediment Systems,* Plenum Press, New York.

Section 2 — Submarine Slopes and Sediment Testing

Introduction
Submarine slope failures

A. B. Hawkins (Chairman)
Geology Department, University of Bristol, UK

Although major advances have been made in seabed mechanics during the last 20 years, it is still a Cinderella subject with many problems. The biggest difficulty is assessing the data or parameters needed to establish mathematical solutions, assuming we have determined what are the reliable parameters and the appropriate equations to quantify the seabed sediment properties.

This paper gives examples of slope failure modes in many parts of the world. It is clear that while sediment loading may be important, it is probable that most fialures are due to sudden changes in pore-water pressure associated with either earthquake shock or, in shallow water, to wave loading. Only in certain cases, such as failures in fiords or in such examples as that off Nice, can the cause be explained simply by undrained loading.

Introduction

Although submarine cables have been placed across the ocean floor since early this century, it was the exploration and exploitation of hydrocarbons that initiated significant research into submarine slopes. Partly spurred by this industrial/commercial need, research resulted in major advances in marine geology and the development of seabed mechanics.

While it could be said that since Terzaghi (1956) the study of seabed mechanics has become more widespread and scientific, it must also be pointed out that there are even more problems than with on-land geotechnics. For instance, there is a natural difficulty in acquiring the data, determining accurately the position and size of seabed features, and obtaining realistic test data in the laboratory under very different conditions from those on the seabed. Although there are now some methods of acquiring *in situ* test data, these are basically restricted to shallow depths and the marine geologist/technologist is left with seismic profiling supported by grab and core sampling to obtain field data.

It is not intended here simply to review the papers which follow, but instead to give some details from a number of papers that have appeared mainly since 1980, in order to illustrate a number of seabed characteristics and failures. In addition to the papers in this volume, the reader's attention is drawn to two excellent publications, the NATO workshop on *Marine Slides and other Mass Movements* edited by Saxov and Nievwenhuis (1982) and the Geomarine Letters (1982) on *Seafloor Stability of Continental Margins.* A conscious decision was made not to include the slumping of large blocks or the effects of liquefaction (see Taylor, 1984).

No attempt has been made fully to explain all the failures discussed briefly in this paper. As an introduciion to this section it was considered more appropriate to highlight some of the main types of failure and give enough details to encourage others to examine subjects of their interest in more detail.

Effect of sea-level changes

Most seabed surfaces are covered with geologically recent (Flandrian) sediment. In a few places, however, where sea cliffs were cut into the bedrock during periods of low sea level, these steep features still remain as rock outcrops. In 1948 Cooper described a 10 m high marine cliff with a base at 40 m below sea level offshore of Plymouth (Fig. 1). In 1970, 12 echo traces were taken across the cliff and it was found to be a prominent feature for a distance of about 8 km.

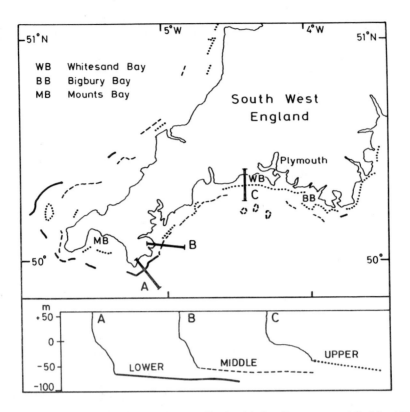

Fig. 1.   Marine cliffs off south-west England (after Donovan and Stride, 1975).

Donovan and Stride (1975), having examined the Admiralty Charts around south-west England, found that there are a number of such features (Fig. 1). They note that such features have not been identified in areas near sandy beaches or seawards of major rivers where the bedrock would be covered by recent sediments. Three main base of slope levels have been recognized, 38–49 m, 49–58 m, 58–69 m; but although the 'cliffs' may be steeper, in some cases they are still of only shallow grade. Upslope of the cliff-like features the seabed is irregular and not yet fully covered with sediment. Offshore, however, the seabed off southwest England is remarkable for its gentle gradient (1:200–1:1000).

As a result of the Pleistocene glacial advances/retreats, changes in sea level were very pronounced. During the last glacial advance (Devensian in Britain) the ocean level dropped to about − 120 m below present (Fig. 2). For another 80 m below, the sediments would have been within the range of wave influence. As a consequence the sediments which have accreted in the upper 120 m either have accumulated by movement away from the coasts, have been supplied by rivers, etc., in the normal manner, have moved landwards with the transgressive front, or have been deposited from suspension directly onto an old land surface.

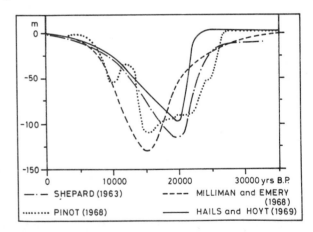

Fig. 2.   Sea level changes during the last 30 000 years.

Exposure and desiccation would have changed the normally consolidated marine sediments such that at least the upper surface became overconsolidated. As a result there would be a natural break in the sediment with a smooth defined base. This is likely to have a lower permeability than both the geology below and the later sediments deposited above.

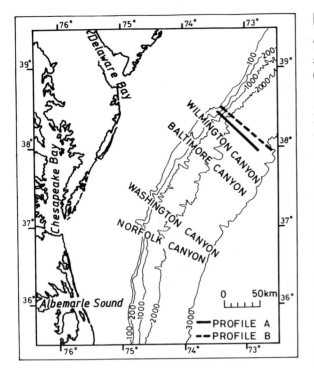

Fig. 3. Map of eastern USA offshore zone to locate two profiles shown in Fig. 4.

Unconformities — eastern USA

The importance of unconformities beneath a rapidly accumulating sediment pile has been noted by McGregor (1981). He, and others such as Almagor *et al.* (1982, 1984) and Booth *et al.* (1984), discuss the continental slope off the eastern United States coast (Fig. 3). McGregor's studies confirm the conclusion by Vail *et al.* (1977) that an important unconformity exists between the Pliocene and Pleistocene.

As determined from the seismic profile, this unconformity is generally smooth, dips seawards and has a dip of the order of 7° (Fig. 4). The smooth unconformity has not been located near Baltimore Canyon, where the sub-bottom horizons, although still sloping seaward, do not contain the continuous planar surface. The overlying sediments are up to 600 m thick.

Almagor *et al.* (1982, 1984) indicate the type of surface sediment in the area. The clay fraction forms about 20–55% of the material, while the clay minerals are dominated by illite, with subsidiary kaolinite, smectite and chlorite. In general, to 2400 m the water content is below 60% while below that depth the sediments were wetter, with values of the order of 90%. The plasticity index, generally between 10 and 30 per cent, is fairly consistent throughout the sampling depth. Typically the overconsolidation ratio (OCR) is

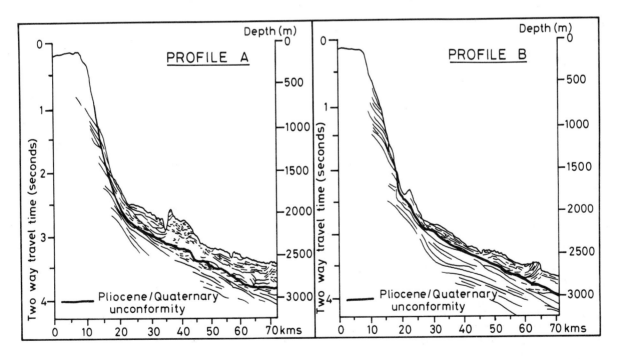

Fig. 4. Seismic profiles to show the unconformity beneath the recent sediments (after McGregor, 1981).

1–2 (i.e. normal to slightly overconsolidated). By use of the undrained shear strengths obtained, the calculated C_u/P_o (ratio between undrained cohesion and effective sediment overburden stress) can be divided into three groups; one-third were lower than 0.2 (underconsolidated), half lay between 0.2 and 0.4, and the remainder were greater than 0.4 (overconsolidated).

With drained strength measurements indicating $\phi' = 28$–$32°$, figures higher than expected for a silty clay soil, most of the natural slopes would be stable if a drained analysis were applicable. In the case of the undrained analysis, however, the frictional parameter, ϕ_u, has been measured as 14–18°, angles similar to those of the steeper slopes in the area. In both Almagor *et al.* (1982) and Almagor *et al.* (1984) the natural slopes are presented against the calculated stable angles based on a C_u/P_o relationship. The authors indicate that in only 5 of the 34 analyses is the calculated maximum slope less than the measured seabed slope.

Either the basic assumptions are wrong, the strengths measured in the laboratory are inaccurate, or the cause of the instabilities recorded off the mid-US eastern coastline must be due to other factors. An obvious possibility is that of earthquake acceleration. Morgenstern (1967) has shown that with a C_u/P_o ratio of 0.25 on a 7° slope an acceleration of 5% of gravity is sufficient to cause failure. McGregor (1981) points out, however, that in the case of the Wilmington and Albemarle areas the failures took place above a smooth continuous seaward-dipping horizon. In a situation where the sediment pile has accumulated quickly and high water contents are present, under shock conditions a high pore-water pressure is likely to be reached, yet transmissivity is impeded where a sudden change in permeability is present. Thus, where underconsolidated deposits overlie overconsolidated sediments, especially if the two are separated by a smooth surface, a natural potential failure surface will exist.

Cyclic loading by earthquakes off Israel

Almagor and Wiseman (1982) discuss the seabed off Israel, in which again it is believed small earthquake shocks must occur in order to reduce the shear strength and cause failure in the sediment pile. Seismic and sonar mapping off the coast of Israel shows that a number of disturbed areas exist (Fig. 5) in a region where the slope angle varies from 3–6° off the Nile delta region north of Sinai to 6–18° off Lebanon. Slump scars up to 45 m deep, 3 km wide and 4 km long and with side/rear angles of up to 20° have been located in water 400–500 m deep; notably where the continental slope attains its steepest gradient. Mass creep and small rotation

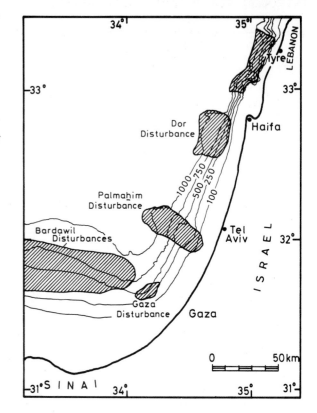

Fig. 5. Seabed topography and disturbance areas off Israel (after Almagor and Wiseman, 1982).

slides are ubiquitous in the 0.5–2.5° slopes of the shelf and uppermost continental slope where the water depths are 80–150 m and 200–325 m, respectively.

The sediments have been studied from 80 mm diameter, 3–5 m long cores (Almagor and Wiseman, 1982). The silty clays had a clay fraction which increases from 30% at 30 m to 60% at 400 m, but remains generally constant at greater depths. Analysis of the clay fraction indicated 60–80% montmorillonite, 20–40% kaolinite, but less than 15% illite and a maximum of 9% carbonate. Quartz sand is common at depths less than 350 m.

The sediments can be conveniently divided into two types on the basis of water depth. It is of note that the measured geotechnical parameters support this subdivision (Table 1). The shear strength recorded by Almagor and Wiseman (1982) is again quite high for such a clay-rich material, ϕ' being 24–25° and ϕ_u 15–17°. The measured permeabilities, however, ranging between 1.4×10^{-9} and 2.5×10^{-10} m s^{-1}, support the sediment grade as being clay-rich.

Some of the individual cores show pronounced changes in sediment character. For instance, the data from Core CA-41, taken at 860 m depth, shows a sudden change of water content, unit weight, void ratio

TABLE 1   Geotechnical properties of sediments off the Israel coastline

Property	Near shore	Base of continental slope
Water content (%)	75–95	128–143
Porosity (%)	68–72	78–80
Void ratio	2.1–2.6	3.5–4.0
Liquid limit (%)	60–80	85–115
Plasticity index (%)	35–50	65–80
Unit weight (g cm^{-3})	1.41–1.61	1.25–1.37
C_u/P_o	0.24	0.76–0.91

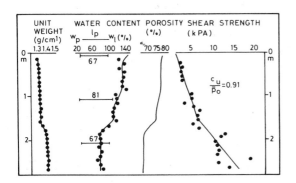

Fig. 6.   Geotechnical data from core CA-41 (after Almagor and Wiseman, 1982).

Cyclic loading by waves in the Mississippi delta

In some ways the cyclic effect of earthquake shocks is similar to that resulting from wave pressure. Using data from the Mississippi area, Fisk and McClelland (1959), McClelland (1967) and Henkel (1970) point out that the C_u/P_o ratio is 0.3–0.5. Calculations indicate that the maximum stable slope would therefore be about five times the measured seabed angle; hence, the slopes should not have failed. Using $\phi' = 20°$ again, the drained slope conditions should be capable of maintaining a stable grade about five times higher than the actual seabed angle. On this evidence Henkel (1970) concludes that it is difficult to explain how the failures occurred, if they result simply from overloading and gravity forces. Henkel does, however, examine the effects of waves, which are a frequent meteorological occurrence in the Mississippi delta area, especially at the time of hurricanes. Henkel's calculations show that in water 14 m deep a wave 9 m high and 180 m in amplitude exerts a maximum pressure of more than 30 kPa and that for over a depth range of 3–60 m the overall increase in pressure is 10 kPa. Such a wave-imposed loading would be transient and approximately sinusoidal in space and time.

While Henkel appreciates that underwater instability associated with wave forces is complex, his wave tank experiments confirmed that if the sediment surface had an initial slope and waves were large enough to overcome the undrained shear strength, there was a slow downslope mass movement. Consequently, within the wave zone (approximately 50 m) the stability of marine slopes in environments of rapid sediment accretion and subject to hurricanes may be influenced mainly by the pore-water pressures associated with the temporary stress induced by wave loading. Below, say, 100 m, however, it is likely that gravity sliding, possibly assisted by earthquake shock waves, is the main mechanism of slope failure.

It is important to relate the work of Henkel (1970) on waves in the Mississippi area and that of McGregor (1981), who considers the presence of smooth discontinuity surfaces to be important in stability considerations off the mid-USA to the comments made earlier on Flandrian sea level rise. Flandrian sediments within 125 m depth will overlie a surface which would have been exposed and hardened during the last period of low sea level (Late Devensian glaciation). Thus, again there will be a problem where normally or underconsolidated sediments overlie a hardened desiccated surface: a situation prone to instability if local wave-induced pressures have the effect of producing sudden increases in load, so raising pore pressures as well as intensifying the undrained loading.

and shear strength at a depth 1.6 m below the seabed (Fig. 6). Unfortunately, Almagor and Wiseman (1982) do not explain this, but record that the proportion of clay fraction is below 20% at the top of many of the cores taken from deeper sections of the continental slope. This sedimentological change in the succession may be the most satisfactory explanation for the sudden 5 kPa increase in the C_u strength.

Despite the fact that an analysis using C_u/P_o data indicates that the slope should be stable, ubiquitous mass creep phenomena exist along the entire shelf edge zones of Isreal. Unable to explain the slope failure by measured parameters, Almagor and Wiseman (1982) again suggest earthquake shocks as the cause of the long-term deterioration in shear strength. Such cyclic loading of deposits of low permeability will induce a rise in pore-water pressure, increase strain and subsequently decrease the shear strength. Although it is easy to suggest a cause that cannot be checked, it should be remembered that the eastern part of the Mediterranean is a well-established earthquake zone. Indeed, it could be considered that much of the eastern coast of the United States was geotechnically stable, yet failures take place and earthquakes occur.

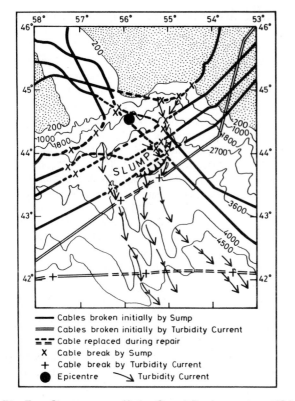

Fig. 7. Slump area off the Grand Banks, eastern USA (after Heezen and Drake, 1964).

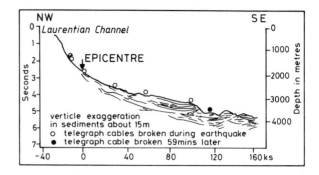

Fig. 8. Seismic reflection profile through Grand Banks slump mass (after Heezen and Drake, 1964).

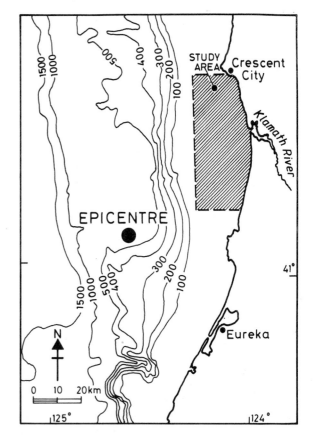

Fig. 9. Location of the Klamath disturbance area off northern California (after Field and Hall, 1982).

Earthquakes

One of the best-known submarine slope failures was that south of the Grand Banks on 19 November 1929. Although much has been written on this slope failure, it was Heezen and Drake (1964) who drew attention to the difficulty, not covered by previous workers, of explaining why simultaneous cable breaks in the large epicentral area were followed, after 59 min, by a break in a nearby cable (Fig. 7). In 1961, however, a seismic trace profile indicated a prominent feature beneath the sediments upslope of the later cable break (Fig. 8). This feature has been interpreted by Heezen and Drake as being a large gravity slump, which could satisfactorily explain the distribution of the cable breaks. The simultaneous breaks would have been caused by the slump, and the later one by a turbidity current resulting from the retrogressive failure of the steep back scar of the original slump.

Reflection profiling showed the upper zone of sediments to have a significant seismic penetration and a layer different from that of the normal adjacent continental slope. It appears therefore that the turbidity current which was responsible for the later cable failures was triggered somewhere near the epicentre and that the density currents mainly flowed around the slump block before converging towards the south along a broad front (Fig. 7).

Field and Hall (1982) describe the results of an earthquake which occurred on 8 November 1980 75 km off the Californian coast. Following the quake, ridges appeared off the mouth of the Klamath River (Fig. 9).

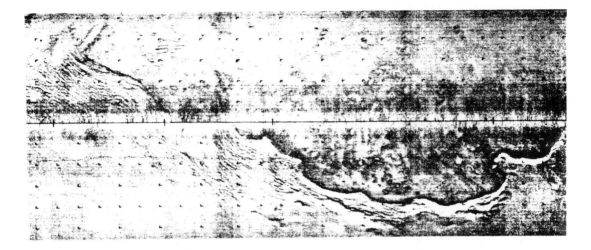

Fig. 10.   Sonograph of seaward edge of failure zone (after Field and Hall, 1982).

Seismic reflection profiling showed a zone of disturbance 20 km long and 1.5 km wide in water 30–70 m deep in sediments attain a thickness of 20–50 m. In this area the sediments in the upper levels are dominantly a muddy sand, while sandy clayey silt is typical of the deeper water. The seaward boundary of the failure zone coincides with the change of line of the sediment types.

Side scan records indicate that along the southern end of the slide there is a 1–2 m high toe ridge, which appears as a prominent feature on a slope of less than 1/4°. Seaward of the toe lies a zone of small scarp-like subparallel pressure ridges (Fig. 10). Side-scan records also showed clear evidence of discrete rings measuring up to 25 m in diameter. These have been intepreted as large sand boils.

The boundary of the failure, being close to the level of sediment change, suggests that the latter was a major controlling factor. It is possible that with such a wide continuous failure lateral spreading may be a cause, developing as a shallow-seated liquefaction. This is supported by the presence of sand boils, while the overlapping of sediment lobes indicates some fluidization as part of the failure process.

Although not mentioned by Field and Hall (1982), the depth of failure below sea level is such that the sediment would undoubtedly have accumulated on an old overconsolidated surface. Unfortunately they do not provide any seismic profile records, but it is hoped that in their subsequent research they will confirm whether or not a clear 'unconformity' exists and, hence, whether the sudden changes in permeability at such a surface influenced the depth of the slide. The broad movement shown in Fig. 10 is typical of what might be anticipated from a failure on a wide disconformity surface.

Organic content

In the last few years various authors have noted the importance of the organic carbon content of submarine sediment. Keller (1982) describes in some detail the occurrence of such material in the upwelling zone of the Humbolt current off the coast of Peru.

Figure 11 indicates that in parts of the area, notably off Pisco, the organic carbon content may be more than 20% of the sediment mass. Most of the material is in the form of faecal pellets which descend into water with little or no oxygen. In the case of the Californian current off Oregon, the upwelling is more essential and the organic carbon contents are only up to 3.2 per cent. Off the Peru coast the sediments are silty clays, 50–70% being clay grade, 25–40% silt and only 5–10% sand. The prominent clay minerals are illite (45–60%) and chlorite (30–40%), but only a little smectite (5–20%). This is the opposite of the general marginal sediments off Peru.

Geotechnical examination of these sediments indicates water contents often in excess of 200% off the Callao–Pisco coast (Fig. 12), the maximum recorded being 835% measured at specific horizons. Busch and Keller (1982) comment on the two isolated 150–200% areas and record that in these areas the sediment was slightly more consolidated as a result of erosion of some of the recently deposited sediment.

Figure 12 indicates that there is a very clear relationship between the organic carbon and water contents of the sediments. McDonald (1983) confirmed a similar relationship in the sediments of Oregon.

Figure 13 shows that the organic sediments have distinctly higher liquid limits and plasticity indices than the other sediments along the Peru coastline. Again

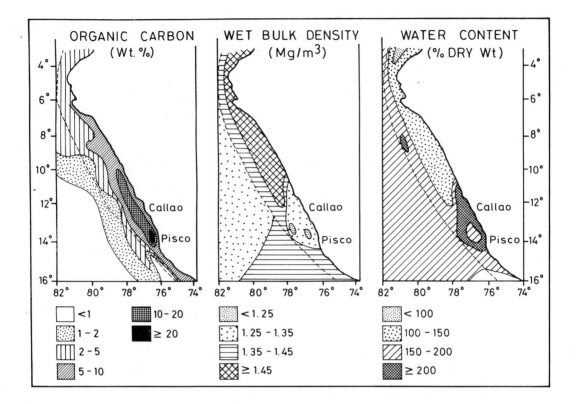

Fig. 11. Organic content, water content and bulk density of the marine deposits off the coast of Peru (after Keller, 1982).

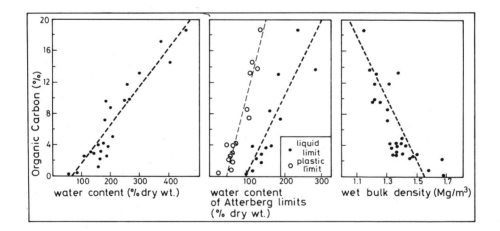

Fig. 12. Relationship of water content, Atterberg limits, and wet bulk density to the organic content of the sediments off the coast of Peru (after Keller, 1982).

there is a clear relationship between the Atterberg limits and the organic carbon content of the host sediments. Typically, geotechnical engineers expect the smectite-rich (finer) clays to have higher liquid limits but, as indicated in Fig. 11, while these may be high (up to 200%) the organic-rich clays have liquid limits exceeding 300%. As the organic-rich deposits have been found to have a water content greater than the liquid limit, it is clear that in its depositional state there must be sufficient strength between the particles to inhibit flow, which

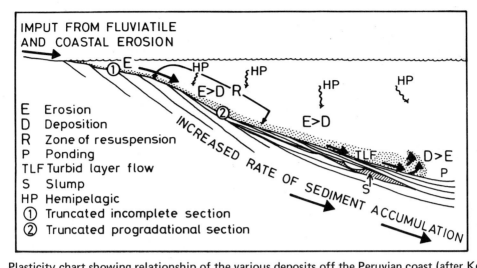

IMPUT FROM FLUVIATILE
AND COASTAL EROSION

E   Erosion
D   Deposition
R   Zone of resuspension
P   Ponding
TLF Turbid layer flow
S   Slump
HP Hemipelagic
① Truncated incomplete section
② Truncated progradational section

INCREASED RATE OF SEDIMENT ACCUMULATION

Fig. 13. Plasticity chart showing relationship of the various deposits off the Peruvian coast (after Keller, 1982).

would happen if the soil became disturbed or was in a remoulded state.

The undrained shear strengths vary between 0 and 34 kPa. In general, there is little difference from the other coastal sediments except that in the area influenced by upwelling there is a higher proportion of low values. In the remoulded state it was found that the strength dropped markedly as the organic content increased. It is considered that the presence of cellulose and polyvalent cations such as Mg^{2+} and Ca^{2+} gel complexes will form which, in the undisturbed state, will impart a bonding effect. In the disturbed remoulded state, however, the bonding is broken. From the work of Pusch (1973), Rashid and Brown (1975) and McDonald (1983) is is likely that organic contents less than 5% do not have a significant influence on shear strengths.

The sensitivity of the sediment is the relationship between the disturbed and undisturbed shear strengths. As such the general marine clays off Peru have sensitivities of 3–4, indicating a 50–75% reduction in strength, but the organic-rich sediments have a mean of 9, within the range 1–21. This indicates a loss of strength of up to 90%.

The organic-rich deposits have distinctly lower wet bulk densities ($1.2–1.3\,Mg\,m^{-3}$) than the other deposits along the coastal zone ($1.4–1.7\,Mg\,m^{-3}$; Fig. 12). Specific horizons have given values as low as $1.09\,Mg\,m^{-3}$, akin to those of peat ($0.5–1.2\,Mg\,m^{-3}$).

For most submarine deposits the OCR is between 1 and 4. In the case of the Peruvian organic-rich sediments, however, the highest recorded OCR is 17 and the mean 13. It is difficult to explain the reason for this. It is not likely to be due to past loading but is probably associated with bonding which, in turn, de-

velops organic sheaths suggested by Keller (1982) to be a significant factor in increasing the strength of sediments.

With the properties enumerated above it is surprising that there has not been a major slumping or sliding of the sediments, especially as this is a seismic area with shocks up to 7.75 on the Richter scale. Keller notes that the cores show clear evidence of post-flow features such as contorted bedding. It may be that these failures are quite frequent but the scientific interest in the area has not been over a sufficient period. What is required is a resurvey after an earthquake to determine the relationship between the high pore-water pressures and the overburden load. If stress is transferred from the particles to the water and there is a failure of the bonds, then a fluidized failure will result, similar to those of the quick clay areas of the northern hemisphere.

Unifites

Records of high-resolution shallow-penetrating seismic profiles show a remarkable continuity of acoustic reflectors across flat basin plains to produce what Hersey (1965) described as sediment ponding. The material in the ponded deposits is referred to as 'unifites' — a descriptive term applied to a fine layer of sediment which may show some faint laminations but may also be structureless.

Stanley (1981) notes that this structureless mud has been recorded from a great variety of depths, up to 4000 m in the Hellenic Trench basin. Unifites are almost totally free of sand-sized material and yet contain reworked benthic and planktonic tests. Detailed analyses show that the unifites display a degree of

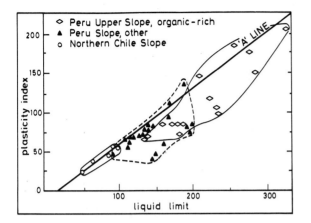

Fig. 14. Sedimentation characteristics to produce unifite deposits (after Stanley, 1981).

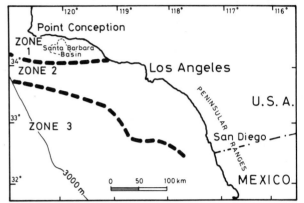

Fig. 16. Main sedimentation rate zones in the coastal area off Los Angeles (after Hein and Gorsline, 1981).

fining upwards, with subtle organic and carbonate changes.

The origin of the unifites is not certain. Stanley believes, however, that they are distal end products of a gravity flow related to mud-charged turbidity currents in the final water transport phase, probably across near-horizontal planes. Figure 14 shows, in a simplified form, the main phases in the model.

The sediment failure leading to unifite deposition may be triggered by quick deposition in the upper slopes, seismic activity, tsunamis or the influence of bottom current activity. High-resolution seismic profiles and the study of core samples from the area of unifite deposits (Fig. 15) indicate that, in general, sediments tend to reside only briefly on the steep Mediterranean basin slopes. On the basis of radiocarbon dating, it appears that unifite deposition has

been pronounced during the last 7000 years, which indicates a large amount of submarine slope failures at shallower depths.

Mass wasting

A typical example of mass wasting has been given by Hein and Gorsline (1981), who studied the geotechnical aspects associated with an area of rapid sedimentation off southern California (Fig. 16). In Zone I (Santa Barbara Basin) the sediment accumulation is $(50 \, \text{mg} \, \text{cm}^{-2})$ /year and mass wasting is prevalent even on slopes as gentle as $1°$ or less. In Zone II the sedimentation rate is $(30 \, \text{mg} \, \text{cm}^{-2})$/year, which it appears in this environment is about the critical rate to foster major mass wasting. In Zone III, where the slope angle is $7-10°$, there is

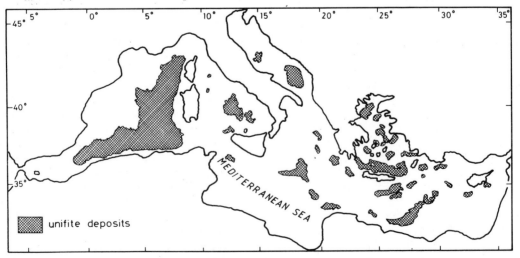

Fig. 15. Location of the main unifite deposits in the Mediterranean (after Stanley, 1981).

TABLE 2 Sedimentological and geotechnical data off the Southern California coast (after Hein and Gorsline, 1984)

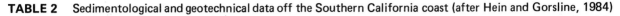

Flow type	Sedimentological					Geotechnical						
	No. of flows	No. of samples	% Sand	% Silt	% Clay	No. of samples	Wet water content	Liquid limit	Plastic limit	Plasticity index	Activity	Wet bulk density
Rotational slump	12	81	9.94	61.71	28.25	31	45.75	94.29	48.47	45.82	1.62	1.555
Debris flow	3	27	25.73	55.04	18.66	8	27.53	34.14	29.33	4.81	0.26	1.843
Mud flow	9	114	3.94	57.29	40.78	34	62.94	115.98	63.92	52.06	1.28	1.318
Thick turbidite	3	28	0.62	55.04	44.32	10	61.71	136.50	78.33	57.57	1.30	1.317
Thin turbidite and hemipelagic	6	59	1.73	56.05	42.24	66	62.71	102.17	53.06	40.11	1.16	1.166
Liquefied	23	167	2.27	62.06	35.32	61	53.87	117.44	63.62	53.82	1.52	1.431

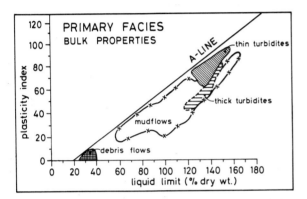

Fig. 17. Main zones on the Casagrande chart occupied by the main facies types (after Hein and Gorsline, 1981).

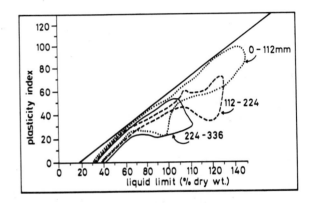

Fig. 18. Main zone occupied by samples from different depths when plotted on the Casagrande chart (after (Hein and Gorsline, 1981).

little evidence of any major slumping where the sediment rate is only (10 mg cm^{-2})/year.

Samples collected as box cores and piston cores have been analysed. The sediments were considered too weak to produce satisfactory vane shear strengths. Six facies have been defined and the sedimentological and geotechnical properties are grouped together as Table 2.

Rotational slumps are composed mainly of poorly sorted clayey silts. Debris flows consist of very poorly sorted sandy silts which may be ungraded or even have upward coarsening. Mudflows are the finer-grain equivalents of debris flows; they are ungraded, poorly sorted clayey silts. The thick turbidites are poorly sorted sandy clayey silts, while poorly sorted clayey silts grading into hemipelagic clays are typical of thin turbidites. The deposits termed liquefied are poorly-sorted clayey silts with well-developed fluid escape features.

As no change was discernible in plasticity index in a vertical distribution within the upper 0.5 m of the sediment, it is possible to plot this on a typical Casagrande classification chart. It can be seen (Fig. 17) that there is a clear difference between some of the deposit types, although they all fall below the 'A' line. In the case of the slump deposits and those of liquefaction it is noted that these span the plasticity chart, agaain below the 'A' line.

The effect of bioturbation within the sediment is marked and tends to separate the sediment types on a depth basis (Fig. 18). It is assumed that this reflects the breakdown of original sedimentary fabric and structure due to burrowing, creating a more open structure and providing horizons with increased porosity and permeability which would be most susceptible to consolidation.

Auffret *et al.* (1982) summarized an example of mass wasting which took place off Nice on 19 October 1979. At 14.00 h there was a lowering of the sea level followed by a tsunami of several metres amplitude. At the same time an embankment area 300 m long collapsed. Between 14.00 and 17.45 several other sea level oscillations were observed along 100 km of the Mediterranean coastline. At 17.45 h a telephone cable was broken 80 km offshore and at 22.00 h another at 110 km.

Offshore of Nice the marine topography consists of a narrow shelf and a steep (5°) continental slope. Typically the fans in this area are 1000 m thick and consist of Pliocene and Quaternary sediments overlying an erosion surface of Upper Miocene age. As no seismic shocks were recorded it is obvious that such a large mass movement of sediment was not associated with an earthquake and is therefore likely to have developed simply as a response to rapid undrained loading.

Fiords

Fiord instability is different from the general submarine instability in two main ways. First, recent fiord sediment will accrete only as a result of direct river inflow and, hence, occur as localized deltas in a rock-eroded basin; second, fiords are found only in regions which have suffered from isostatic uplift.

Two examples of recent fiord instability are worthy of some note. Prior *et al.* (1982) describe the instability which occurred in the Kitimiat area of British Columbia in 1975. Here there were several slides between 1952 and 1968 and again in the early 1970s. For instance, in 1971 a slide caused damage to the Eurocan dock, while in 1974, probably resulting from a sea floor disturbance, a 2.4 m high sea wave was observed at the head of the inlet. Typical of such an area, however, constructional activity continued. In January 1975 a wharf was built with 6–7.5 m of fill placed behind it and in April a breakwater was constructed of fill and dredged material. The result of this localized loading was a major failure of the fiord on 27 April 1975, sufficient to destroy a 75 × 20 m piled jetty and place it below water level. At the time of the slide, waves over 8 m high caused considerable flooding of the surrounding land. The prograding delta sediments are sand and silt, producing a depositional angle of 8°. The fiord floor, however, consists of marine clays which are generally soft and are normal to underconsolidated. In these areas seasonal snow melt associated with accretion of the delta produces an episodic loading.

In the Kitimiat area the tidal range is 6.2 m. It is of note that two of the documented failures occurred about half an hour after low water – i.e. at a time when the upper delta would have been exposed and drawdown would have increased the loading of the delta front sediments to a maximum. Indeed it is of interest that the main instabilities in the muds on the River Avon banks at Bristol occur at low water periods at times of equinoctial tides.

Karlsrud and Edgers (1982), on the basis of a paper by Bjerrum have summarized the studies on a number of slides in Norwegian fiords. All the slides occurred in Flandrian river delta sediments in the area referred to as 'middle' Norway.

Bjerrum recorded that a number of slides took place on 2 May 1930 and several of them affected

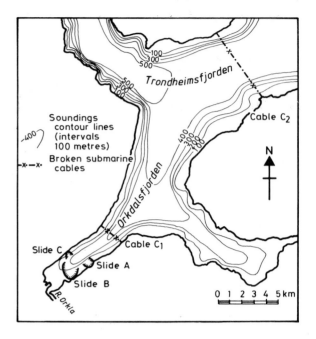

Fig. 19.   Slide at Orkdalsfjord (after Bjerrum).

structures along the side of the fiord. In particular, Karlsrud and Edgers (1982) describe the slide at Orkdalsfjord. The slide was 8 m deep and 1–2 km wide, widening northwards as it met the Trondheimsfjord 25 km away (Fig. 19). From the inlet of the River Orkla to the fiord base the slope is 3–6°.

The failure was initiated by slide A (Fig. 19) when over 1000 m³ of recent fill failed, which created a scar 500 m along the fiord edge. When slide B occurred a few minutes later, some piers and harbour works were destroyed. Slide C took place 7 min after slide A. The moving sediment-charged water broke the telephone cable running across the fiord. Back-calculating, it can be shown that these slides were moving at about 25 km h⁻¹. Another cable 18 km downfiord in 500 m of water broke 112 min after slide A, which indicated a slowing down of the water/sediment movement. It has been calculated that 25 × 10⁶ m³ of material was moved by slides A, B and C.

Borings in the on-land failure area indicated very loose (soft) non-plastic silt with a water content up to 33%. The quantity of material moved is high and indicates that slide A must have eroded deeply into bottom sediments, with the maximum erosion probably being where there was a change in flow grade at the bottom of the slope. In soft sediment such sudden bed lowering would have caused side failures, and it is believed that it was retrogressive flow sides which, in turn, resulted in slides B and C.

Of the eight slides described by Bjerrum, four were caused primarily by man-made fill being placed on the shoreline. Typically four of the initial slides occurred at about low tide, when the natural loading effects are at their maximum. In the case of the four slides no seismic effects were recorded, but in the case of the Sandnessjøen slide the instability was at a time when there was blasting within a shipwreck in the fiord; hence, shock wave disturbance may well have triggered the failure.

It is clear from such writers as Bjerrum and Karlsrud and Edgers that large-dimension fiord failures can occur, especially in sand/silt soils which often overlie finer fiord bottom sediments. They suggest that there is a complete loss of strength after shear failure typical of sand and coarse silt and that the slide mass assumes the character of a viscous liquid. The particular character of silt/sand flows is their exceptionally high erodability due to the lack of cohesion. Consequently any such slide is likely to cut a deep channel and undermine any obstruction that is in the path of the flow. It is a characteristic therefore that oversteep side slopes will occur which, in turn, will fail, the result being a retrogressive upslope failure.

Conclusion

Using examples from several parts of the world, this paper draws attention to the factors which affect marine slope instability, but does not discuss in detail the problem of liquefaction of seabed sediments, as this is considered by Taylor (1984). Although recently there has been an increased interest in determining the stability of marine slopes, it is unfortunate that there are very few substantiated case references compared with on-land geotechnical studies. The failure of a marine slope is often not recognized at the time of its occurrence, and those that are appreciated are often identified from such features as tidal waves, or the failure of marine cables and pipelines.

Two continuing difficulties are obtaining on-site dimensions and appropriate parameters. Many of the slopes are of such low grade that under either drained or undrained conditions they should remain stable, yet they have obviously failed. These failures cannot be attributed to any single factor, except possibly undrained loading in regions of either very high sediment input or fiords. Elsewhere it is likely that vibrations caused by earthquakes or waves could create such high pore pressure that instability occurs. There is no difficulty in obtaining disturbed, albeit core, samples from the seabed, but the reliability of undrained strength measurements in these samples is questionable. Undoubtedly in the future we shall see an emphasis on *in situ* measurements of drained and undrained parameters which can then be better related to such classifications as the Atterberg limits obtained on disturbed samples.

Acknowledgements

The author thanks Jean Bees for preparing the text figures.

References

Almagor, G., Bennett, R. H. and McGregor, B. A., 1984. Analysis of slope stability, Wilmington to Lindenkohl Canyons, US mid-Atlantic margin. This volume.

Almagor, G., Bennett, R. H., McGregor, B. A. and Shephard, L. E., 1982. Stability studies of superficial sediments in the Wilmington-Lindenkohl Canyons Area, Eastern US Margin. *Geo Marine. Lett.* 2, 129–134.

Almagor, G. and Wiseman, G., 1982. Submarine slumping and mass movements on the continental slope of Israel. In: S. Saxov and J. K. Nieuwenhuis (eds), *Marine Slides and other Mass Movements*, pp. 95–128. Nato Conf. Series IV– Marine Sciences, Vol. 6.

Auffret, G. A., Auzende, J. M., Gennesseaux, M., Monti, S., Pastouret, L., Pautot, G. and Vanney, J. R., 1982. Recent mass wasting processes on the Provencal margin (Western Mediterranean). In: S. Saxov and J. K. Nieuwenhuis (eds), *Marine Slides and Other Mass Movements*, pp. 53–58. Nato Conf. Series IV– Marine Sciences Vol. 6.

Busch, W. H. and Keller, G. H., 1982. Consolidation characteristics of sediments from the Peru-Chile continental margin and implications for past sediment instability. *Mar. Geol.,* **45,** 17–39.

Cooper, L. H. N., 1948. A submerged ancient cliff near Plymouth. *Nature, Lond.,* **161,** 280.

Donovan, D. T. and Stride, A. H., 1975. Three drowned coastlines of probable Late Tertiary age around Devon and Cornwall. *Mar. Geol.,* 19, 35–40.

Field, M. E. and Hall, R. K., 1982. Sonographs of submarine sediment failure caused by the 1980 earthquake off Northern California. *Geo Mar. Lett.* **2,** 135–144.

Fisk, H. N. and McClelland, B., 1959. Geology of the continental shelf off Louisiana; its influence on offshore foundation design. *Bull. Geol. Soc. Am.,* **70,** 1369–1394.

Geo Marine Letters, 1982. Seafloor stability of continental margins research conference, October 1982 2 (3/4) 115–248.

Heezen, B. C. and Drake, C. L., 1964. Grand Banks slump. *Bull. Am. Ass. Petrol. Geol.* 48 (2), 221–233.

Hein, F. J. and Gorsline, D. S., 1981. Geotechnical aspects of fine-grained mass flow deposits: California Continental Borderland. *Geo Mar. Lett.* **1,** 1–5.

Henkel, D. J., 1970. The role of waves in causing submarine landslides. *Geotechnique,* **20** (1), 75–80.

Hersey, J. B., 1965. Sediment ponding in the deep sea. *Geol. Soc. Am. Bull.* **76,** 1251–1260.

Karlsrud, K. and Edgers, L., 1982. Some aspects of submarine slope stability. In: S. Saxov and J. K.

Nieuwenhuis (eds), *Marine Slides and Other Mass Movements,* pp. 63–82. Nato Conf. Series IV—Marine Sciences, Vol. 6.

Keller, G. H., 1982. Organic matter and the geotechnical properties of submarine sediments. *Geo Mar. Lett.* **2**, 191–198.

McClelland, B., 1967. Progress of consolidation in delta front and prodelta clays of the Mississippi River. *Mar. Geotech.,* 22–33.

McDonald, W., 1983. Influence of Organic Matter on the Geotechnical Properties and Consolidation Characteristics of Northern Oregon Continental Slope Sediments. Unpublished M.S.C. Thesis, Oregon State University, 69 pp.

McGregor, B. A., 1981. Smooth seaward-dipping horizons — an important factor in sea-floor stability. *Mar. Geol.,* **39**, 89–98.

Morgenstern, N. W., 1967. Submarine slumping and initiation of turbidity currents. In: Richards, A. F. (ed.), *Marine Geotechnique.* University of Illinois Press, Urbana, Ill., pp. 189–220.

Prior, D. B., Coleman, J. M. and Bornhold, B. D., 1982. Results of a known seafloor instability event. *Geo Mar. Lett.* **2**, 117–122.

Pusch, R., 1973. Influence of organic matter on the geotechnical properties of clays. Statens Institut for Byggnads Forskning, Stockhol., Document D11, 64pp.

Rashid, M. A. and Brown, J. D., 1975. Influence of marine organic compounds on the engineering properties of a remoulded sediment. *Engng Geol.,* **9**, 141–154.

Saxov, S. and Nieuwenhuis, J. K. (eds), 1982. *Marine Slides and Other Mass Movements.* Nato Conf. Series IV—Marine Sciences Vol. 6.

Stanley, D. J. 1981. Unifites: structureless muds of gravity flow origin in Mediterranean Basins. *Geo Mar. Lett.* **1**, 77–83.

Taylor, R. K., 1984. Liquefaction of seabed sediments: Triaxial test simulations. This volume, pp. 131–138.

Terzaghi, K., 1956. Varieties of submarine slope failures. *Proc. 8th Texas Oil Mech. and Engr Conf.,* pp. 1–41.

Vail, P. R., Mithum, R. M. Jr. and Thompson, S. III, 1977. Seismic statigraphy and global changes of sea level. Part 4: global cycles of relative changes of sea level. In: C. E. Payton (ed.), *Seismic Stratigraphy — Applications to Hydrocarbon Exploration. Am. Ass. Petrol. Geol. Mem.,* **26**, 83–97.

5

Slope-stability Analysis and Creep Susceptibility of Quaternary Sediments on the Northeastern United States Continental Slope

James S. Booth*, Armand J. Silva† and Stephen A. Jordan†

*US Geological Survey, Woods Hole, USA
†Marine Geomechanics Laboratory, University of Rhode Island, Narragansett, USA

The continental slope off the northeastern United States is a relatively steep, morphologically complex surface which shows abundant evidence of submarine slides and related processes. Because this area may be developed by the petroleum industry, questions arise concerning the potential for further slope failures or unacceptable deformations and the conditions necessary to cause such instabilities. Accordingly, a generalized analysis of slope stability and the stress–strain–time-dependent behavior of the sediments is being conducted.

Piston cores provided samples for general geotechnical analysis as well as for drained triaxial creep tests. Samples were composed primarily of inorganic silt and clay of medium to high plasticity (CH). Although the direct application of results is limited to the upper 10 m of sediment, deep borings have shown that this same type of sediment may predominate to a sub-seafloor depth of as much as a few hundred meters over a large portion of the area of investigation.

Analysis of regional stability based on infinite-slope methods (static case, peak strengths) indicates that the continental slope is inherently stable: no drained factor of safety values and only 15% of the undrained values were less than 2. The minimum value was 1.3. However, strength reduction through cyclic loading or strain softening (e.g. due to creep) could lower the static factor of safety values significantly. Calculations of the dynamic factor of safety show that 18% of the core sites would require only modest horizontal acceleration ($\leqslant 5\% \, g$) to reduce the static factor of safety (undrained) to unity.

Results of finite-slope stability analyses based on both a method of slices (for uniform sediment section) and a wedge method (for planes of weakness) show that canyon headwalls and sidewalls may be metastable. Inasmuch as unloading due to canyon erosion may result in elastic rebound and, hence, strength degradation, potential failure planes in such locations may be in a progressive failure mode.

The stress–strain–time-dependent behavior of the sediment was investigated by performing drained triaxial creep tests on a normally consolidated, representative core sample. One indicator of creep potential, the Singh–Mitchell creep-parameter, m, was derived from these tests. Values for m varied between 1.5 and 2 for various consolidation stresses. The lowest values of m, which indicate the greatest creep potential, were associated with conditions close to preconsolidation stress. Sample calculations of surface creep for 20 m and 30 m thick sections on a 5° slope yielded displacements of 0.023 m and 0.193 m, respectively, for a hypothetical first year. The derived m values suggest that for typical slope conditions creep rupture is not a likely cause of discrete slope failure in the study area.

Introduction

Of the numerous geologic processes which may impose constraints on the commercial development of the seabed, mass movement (the downslope unit movement of a portion of a slope surface) is one of the more conspicuous. Accordingly, as the petroleum industry has focused its attention on frontier areas on the continental slope off the northeastern United States, research attention has focused on mass movement and related processes in this area. These studies (for example, Embley

and Jacobi, 1977; O'Leary and Twichell, 1981; Robb et al., 1981) have shown, chiefly through high-resolution seismic-reflection profiles and sidescan-sonar images, that evidence of past slope instabilities is present throughout much of the region and is abundant in specific locales. The variety of features that appear to have been derived from mass movement processes includes rotational and translational slide scarps, Toreva blocks, debris flows, rubble fields and scree, allogenic blocks and terracettes, as well as other features. However,

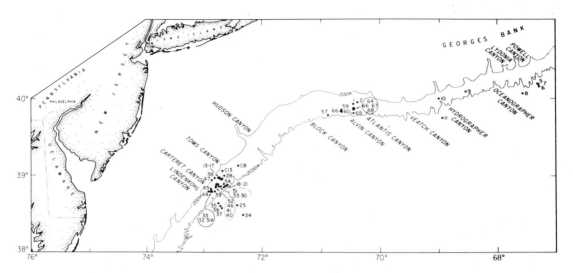

Fig. 1. General study area and piston core sites. Core 11 (right center) was used in creep tests.

the geophysical evidence, although pertinent for identifying features, establishing geometry and classifying failure types, does not indicate the potential for future slope failures. This report addresses questions concerning the present conditions of the slope with respect to mass movement.

Emphasis has been placed on the most prominent manifestation of mass movement — discrete slope failure — and a related but more subtle process — creep. Taken separately, each of these processes is capable of doing significant amounts of geologic work and each may impose formidable constraints on, or even block, offshore development. Knowing the potential for either is an important part of an overall geologic characterization of a region. Taken together, as a cause and effect couplet, the role of creep as a precursor to discrete slope failure may be examined: strength degradation through creep deformation or localized creep rupture may lead indirectly or directly to discrete slope failure. Our purpose, therefore, was to present the initial results of slope-stability analyses and creep susceptibility studies on samples collected from the continental slope off the northeastern United States and to establish the relationship between the two processes in the area.

The study area (Fig. 1) is morphologically complex and dominated by the presence of numerous submarine canyon systems (Emery and Uchupi, 1972) which in themselves are testimony to large-scale slope degradation. Regional slope gradients may approach 7–10°; locally slopes exceed 30°. Segments of some canyon sidewalls approach the vertical. The Quaternary sediment section, which has been the most affected by mass-movement processes, is as much as 450 m thick (Robb *et al.,* 1981) and is dominated texturally by clastic silt and clay, and mineralogically by illite and chlorite (clay minerals) and

quartz and feldspars (nonclay minerals). Previous geotechnical work has shown that, on average, the sediments in the upper 10 m of the area are primarily inorganic silts and clays of high plasticity and have a Unified Soil Classification System designation of CH (Keller *et al.,* 1979; Booth *et al.,* 1981a, b), and that the plasticity of the lower parts of the Quaternary section generally ranges from medium to high plasticity (Richards, 1978). However, considerable variability exists within these Quaternary sediments.

Methods

Sampling

The piston cores used for the geotechnical aspect of the study were collected by the US Geological Survey during two cruises aboard R.V. *Endeavor.* Locations are shown in Fig. 1. The coring system was designed to obtain cores with minimal mechanical disturbance (Booth *et al.,* 1981a). More than 50 cores (diameter, 89 mm; maximum length, approximately 10 m) were recovered.

Laboratory

Undrained shear strength was measured in split core sections by means of a four-bladed, 12.7 mm square laboratory vane at intervals of 0.50 m and at lithologic changes. The blade was inserted normal to the long direction of the core and buried at least 20 mm into the section. A rotation rate of 90°/min was used. Remoulded strength was also determined with the vane apparatus.

The suite of geotechnical index property tests (water

TABLE 1   Piston-core geotechnical data summary

Property	Units	No. of measurements	Minimum	Maximum	Average
Vane shear strength					
natural (S_u)	kPa	414	1.7	90	9.8
sensitivity (S_t)[a]		320	1.0	15.0	5.4
Index properties					
water content (w)	%	508	26	113	64
bulk density (γ_t)	g/cm^3	469	1.41	2.05	1.71
porosity (n)[b]	%	470	41	78	62
liquid limit (w_L)	%	497	21	110	58
plastic limit (w_p)	%	501	15	41	24
plasticity index (I_p)	%	495	6	76	34
liquidity index (I_L)		491	0.15	4.80	1.23
grain specific gravity (G_s)		327	2.59	2.89	2.71
Strength parameters					
cohesion (c')	kPa	34	0	10	4
angle of internal friction (ϕ')	°	34	12	33	25
strength/overburden (S_u/α'_{v0})		27	0.18	1.00	0.44
Consolidation properties					
overconsolidation ratio (OCR)[c]		27	1.5	372	3.2
compression index (c_c)		35	0.12	1.02	0.38
coefficient of consolidation (C_v)	cm^2/s	28	4.3E−4	1.0E−1	1.3E−3
coefficient of permeability (k)	cm/s	28	7.0E−8	2.0E−2	6.2E−7

Note: All data related to water content are corrected for salt content assuming $S^0/_{00} = 35^0/_{00}$.

[a] Because many samples were too weak to measure after remolding, average and maximum values shown are minima.

[b] Void ratio (e) average = 1.63.

[c] Data from obvious erosional surfaces omitted from average.

content, liquid and plastic limits, and grain specific gravity) was conducted according to procedures recommended by the American Society for Testing and Materials (1982), with two exceptions: grain specific gravity was measured by means of an air comparison pycnometer and all water content data were corrected for a salt content of $35^0/_{00}$.

Consolidated undrained triaxial tests with pore pressure measurements were conducted in accordance with procedures given by Bishop and Henkel (1957). The constant rate of strain (CRS) method (Wissa *et al.*, 1971) was used for consolidation testing.

Isotropically consolidated, drained creep tests were performed on core 11 in special cells having linear ball bushing piston guides sealed with a rolling diaphragm (Moran, 1981). salt water ($35^0/_{00}$) was used as the cell fluid. The deviator stress required to achieve a specified stress level was attained by gradually adding loads to a dead-weight load frame and adjusted daily to account for volume reductions. Loads were added as a percentage of the current load so as to maintain a load increment ratio (LIR) of less than 0.25. Stress level is here defined as the ratio between the applied deviator stress and the failure strength at the end of isotropic consolidation.

Axial deformation was recorded every 45 min, and volumetric readings were taken daily on a calibrated burette system. Test durations were as long as 50 000 min (35 days), including the time required to achieve the desired deviator stress.

Results and analysis

Geotechnical characterization

Although they are not pertinent to specific slope-stability or creep analyses, the ranges and mean values of the geotechnical properties set bounds and guidelines for general evaluations and are useful in cases where direct analytical results are unavailable. Accordingly, these data are shown in Table 1. The data indicate that the upper 10 m of sediment has a soft consistency, is very sensitive, is highly plastic and tends to be overconsolidated. Further, the sediments have most of the geotechnical properties which are normally associated with a fine-grained sediment dominated by illite and less active minerals.

Discrete slope failure

Infinite slopes

The infinite-slope model of stability analysis is appropri-

TABLE 2   Infinite slopes: static and dynamic slope-stability analysis

Core	Static factors of safety		Horizontal ground acceleration required to reduce factor of safety to unity (% g)	
	F_d	F_u	d	u
PC01	9.9	18.4	19	29
PC02	22.6	54	30	44
PC59	8.0	16.6	20	43
PC66	3.1	14.9	10	60
P05	12.9	6.2	10	5
P06	12.9	6.2	17	14
P07	12.9	9.4	23	15
P08	9.2	7.5	13	10
P09	2.0	1.4	9	3
P10	2.2	5.2	10	33
P11	45.8	57.3	22	6
PC39	3.3	2.0	13	22
PC40	2.5	1.3	12	3
PC41	2.5	3.7	14	25
PC43	6.6	3.7	20	20
PC44	3.4	1.9	17	26
PC46	2.8	2.9	14	13
PC52	2.3	1.4	12	11
PC53	2.3	1.5	9	2
PC54	3.3	2.1	10	10
P14	6.4	4.5	14	9
P15	4.1	4.4	11	12
P16	4.2	3.6	11	9
P17	11.9	—	17	—
P18	8.4	7.5	16	14
P19	5.8	10.9	9	19
P20	2.2	3.6	13	29
P28	5.3	4.1	17	12
P31A	3.6	5.1	6	9
P33	5.3	4.0	16	11
P34	5.8	6.9	8	10
P35	3.7	2.1	9	4
P36	7.8	4.8	19	10

F_d = static factor of safety − drained condition.
F_u = static factor of safety − undrained condition.
g = acceleration due to gravity.
d = drained condition.
u = undrained condition.

ate for much of the area of investigation. Assumptions attendant to the application of this model are presented in Morgenstern and Sangrey (1978). In addition, specific results derived from the method are somewhat compromised by limitations related to sampling, extrapolation and the geologic environment itself (see, for example, Booth *et al.,* in press, a or b). Nevertheless, this method provides an acceptable preliminary assessment of regional slope stability.

Results of the factors of safety calculations for both drained and undrained conditions are shown in Table 2. For the static case, and assuming no excess pore pressure or progressive strain softening such as from creep or other time-dependent deformational processes, the slope and rise are apparently stable. All factor of safety values (FS) are > 1 and only five (15% of total) have values < 2 (each of these is in the undrained class only). If creep or an analogous process is active in certain areas, the factors of safety may be decreasing. The widespread presence of apparently overconsolidated sediment, as reflected by the higher average overconsolidation ratio (OCR) and S_u/σ'_{v0} values in Table 1, skews the results of the undrained analysis toward higher FS values. Because most of the core sites are not associated with identified erosional surfaces and because the OCR values appear to decrease with subbottom depth (Booth *et al.*, in press, a, b), we suggest that on average a state of normal consolidation persists down the sediment column (i.e. $S_u/\sigma'_{v0} \sim 0.25$). If this is so, the undrained FS values would be considerably reduced: approximately one-third of the sites would be considered unstable. However, there is a general absence of evidence for rapid regional undercutting or rapid oversteepening from depositional or tectonic processes. Thus, we consider undrained failure of an 'infinite' slope on a regional scale to be an unlikely phenomenon in the study area unless dynamic loading is involved.

Dynamic loading can occur through a variety of processes, but because the study area is below wave base for major storms and data are lacking on internal wave forces in the area, we have restricted our analysis to earthquake-induced effects. Earthquakes may affect slope stability in two ways: by increasing shear stress through ground accelerations and by reducing shear resistance because of possible elevated pore pressure resulting from the cyclic loading. Only the former effect was considered in this analysis. Equations for evaluating the effects of ground accelerations have been published by Morgenstern (1967), Hampton *et al.* (1978) and Sangrey and Marks (1981). In general, both drained and undrained stability equations may be streamlined by accounting only for horizontal accelerations (for discussion see Booth *et al.*, in press, c).

The results of the calculations are shown in Table 2. The values shown represent the horizontal ground accelerations in per cent gravity (% g) necessary to reduce the static factors of safety to unity. Seed *et al.* (1975) imply that accelerations from a $6.5 \, m_b$ earthquake would probably not exceed 5% g at a distance of 100 km from the energy source and would probably not exceed 10% g at a distance greater than 50 km from the source. Recent earthquake epicenter locations for the northeastern United States (including offshore locations) have been published by Yang and Aggarwal (1981). Some epicenters appear to be within or near 100 km of the study area, including sites near Hudson and Lydonia Canyons. Magnitudes determined thus far have been small (m_b 2–4), however. Nonetheless, 15% of the sites would require accelerations of $\leqslant 5\% \, g$ to reach limit equilibrium, and thus would seem somewhat vulnerable to the effects of a proximal earthquake. An even higher percentage of sites ($> 40\%$) would be vulnerable if acceleration of 10% g were experienced.

Finite slopes

Although applicable to a significant portion of the area of investigation, the infinite-slope model alone cannot represent the entire area. Local slopes (e.g. canyon walls) which lend themselves to finite-slope stability analysis are also important. Submarine-canyon systems and other features of pronounced relief are so abundant, in fact, that as much as 80% of certain parts of the slope and more than 50% of the area as a whole lends itself to a finite-slope analysis. Clearly, an analysis of finite-slope stability must be included in any real investigation of discrete slope failures. The importance of this type of analysis is underscored by sidescan-sonar images, which show that most recognized slope failures in the study area are local in nature (O'Leary and Twichell, 1981; Robb *et al.*, 1981). Finite slopes were analyzed for cases in which (a) strength parameters are isotropic throughout the sediment section, (b) undrained strength increases linearly with depth and (c) planes of weakness exist within the sediment column. Previous work within the area (e.g. Swanson and Brown, 1978; Hathaway *et al.*, 1979; O'Leary and Twichell, 1981) indicates that any of these conditions may be present.

Both a method of slices and a wedge method were used to estimate slope stability for the finite slopes. The data used, the rationales and the sources are shown in Table 3. The results of such an analysis are hypothetical; however, the values used are in accord with existing knowledge of the area. Further, although our results do not provide definitive solutions, they do offer a first approximation of general local slope stability.

Results of the analysis by the method of slices are summarized in Fig. 2(A). The upper and lower curves for each set of analyses were calculated by use of maximum and minimum strength parameters, respectively. Regarding the lower bounds, Lo and Lee (1973) demonstrated the effect of stress changes arising from excavation (which is analogous to erosion) on a resultant slope. They imply that an ultimate or residual factor of safety should be determined for any slope that could be in a state of progressive failure due to these stress changes. In addition, strength degradation due to creep could also drive the factors of safety toward the lower bounds.

For the undrained case, slopes greater than about $15°$ are metastable. Because rapid undercutting is possible in canyons, undrained failure is more likely than it is for regional slopes. The residual factor of safety for the drained example also reaches limit equilibrium at about $15°$, but the peak FS indicates that these slopes are stable until the slope attains an angle between 25 and $30°$. Nonetheless, and without

TABLE 3   Values for finite slope-stability analysis

Property	Assumed value	Explanation	Source(s)
ϕ	25°	Average value	This study
ϕ_r	13°	Based on dominance of illite and on plasticity	Mitchell (1976)
c'	4 kPa	Average value	This study
c_r'	0 kPa	Minimum residual value	
S_u/σ_{vo}'	0.25	From CIU tests and plasticity	This study Skempton (1954)
S_u/σ_{vo}' (weak planes)	0.18	S_u reduction of > 25% common in profiles	This study (< 10 m) Richards (1978) (< 300 m)
S_u/σ_{vo}' (weak planes, ultimate)	0.12	S_u reduction of > 50% observed occasionally in profile	This study (< 10 m) Richards (1978) (< 300 m)
γ_b	9.5 kN/m³	Typical value for surface to 300 m	This study (< 10 m) Richards (1978) (< 300 m)
U_e	0 kPa	No direct evidence of excess pore pressure	

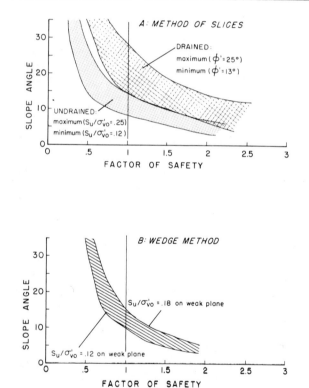

Fig. 2. Factor of safety versus slope angle for finite slopes. Calculations based on assumed values shown in Table 3. (A) method of slices: maxima and minima for drained and undrained conditions; (B) wedge method. Symbols: ϕ' = effective stress internal friction angle; S_u = undrained shear strength; σ_{vo}' = assumed *in situ* effective vertical stress.

establishing the effects of dynamic loading, it appears that many local slopes are vulnerable to failure by a slip-circle mechanism.

Wedge analysis seeks to predict the possible effect of natural changes in sedimentation and related geologic phenomena through time. Our estimate of the possible effect on undrained shear strength is shown in Table 3. The derived factors of safety are plotted against the angle of slope in Fig. 2(B). When derived by this method, the slope angles required to achieve instability are similar to those predicted by the undrained analysis of the method of slices.

Many slopes associated with submarine canyons may be metastable. If progressive failure is taking place, whether due to induced stress changes or to strength reduction because of creep, the presence of unstable sidewalls and headwalls may be even more widespread. Our results, although limited and hypothetical, imply that a dynamic geologic environment exists on a local scale.

Creep behavior and processes

Creep is here defined as the slow downslope movement of sediment under the influence of gravity. Creep processes may be important on submarine slopes where relatively low stresses exist over long periods of time such that large deformations can result. Creep may also be significant in progressive failures or as a precursor to catastrophic mass-wasting processes.

Several investigators (Bennett *et al.*, 1977; O'Leary and Twichell, 1981; Silva *et al.*, 1983; K. Moran, unpublished data) have reported evidence of creep processes in submarine slopes, and others (Moran, 1981; Silva *et al.*, 1982) have shown the creep susceptibility of fine-grained deep sea sediments. However, the data are quite sparse and usually site specific, and most of

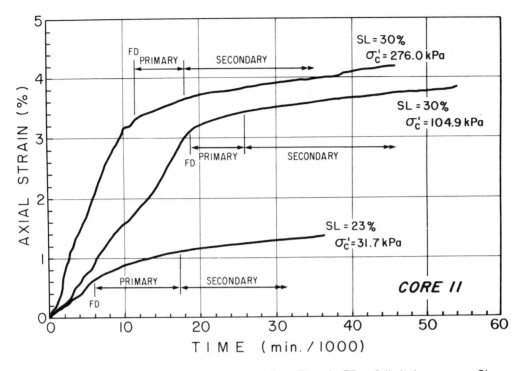

Fig. 3. Triaxial drained creep test results from core 11 (see Fig. 1). FD = full deviator stress; SL = stress level (ratio between applied deviator stress and failure strength at the end of isotropic consolidation, expressed as a percentage); σ_c' = confining stress. Note different consolidation stresses and stress levels. Line plots are based on cubic spline fit to data points. The lowest stress (31.7 kPa) is approximately preconsolidation stress.

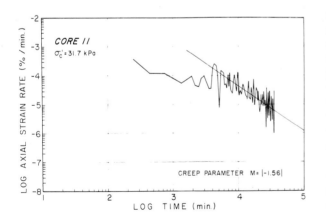

Fig. 4. Singh–Mitchell m parameter at preconsolidation stress. Slope of regression line based on data from cubic spline fit.

the previous experimental investigations have been for undrained conditions. As the processes are by definition very slow, it would seem that the drained conditions used in this study are more representative of the field conditions.

The samples used in the creep testing represent as

nearly as possible the average sediment in the field; that is, the index properties of the samples are very close to the mean values shown in Table 1 ($w = 85$; $w_L = 61$; $w_P = 27$; $G_s = 2.74$). Accordingly, although just one set of samples was used in the test program, we believe that the results may be roughly extrapolated to much of the area. Inasmuch as the primary use of the creep data herein is to analyze geologic processes, we defer detailed comment on many of the test implications and theoretical considerations to the future.

Pertinent creep test results are plotted in Figs. 3 and 4 and listed in Table 4. A composite plot of axial strain as a function of time is shown in Fig. 3. Consolidation stress (σ_c'), stress level (SL), and times to achieve full deviator (FD) stress for each test are noted on this figure and/or listed in Table 4. Careful examination of Fig. 3 shows that each curve displays a trend of primary creep (decreasing strain rate) followed by secondary creep (constant strain rate) after the desired stress level (or full deviator) was attained and that the accumulated strains are much less at the lower consolidation stress (31.7 kPa) than at the higher stresses. In addition, a comparison of the averaged axial strain rates during secondary creep (Table 4) at the same stress level (30%) indicates a decrease in strain rates as effective stress increases.

TABLE 4   Creep tests data summary

Consolidation stress (kPa)	Stress level (%)	Average axial secondary stage strain rate ($\times 10^{-5}$ % min)	m parameter
31.7	23	1.14	1.56
104.9	30	1.85	1.69
276.0	30	1.51	2.06

Singh and Mitchell (1968) developed a phenomenological creep equation in which they utilized an empirical material parameter (m), which is the absolute value of slope on a plot of logarithm of strain rate ($\dot{\epsilon}$) against the logarithm of time (t). An example of the derivation of the m parameter from our data is shown in Figure 4. Singh and Mitchell showed that this parameter is a key factor in assessing creep potential; the larger the absolute value of m, the lower the potential of creep. A comparison of the m values listed in Table 4 shows a trend of decreasing creep potential (increasing m) as consolidation stress increases. The data suggest that the highest creep potential of a normally consolidated sediment (in this case 31.7 kPa) is at approximately preconsolidation stress.

By use of the Singh–Mitchell model, the drained creep results were incorporated into the basic equations of stability analysis to calculate downslope displacements. For the problem of an infinite slope of homogeneous isotropic sediment, Yen (1969) reduced the gradient of shear strain to a first-order derivative wherein the downslope velocity is dependent only on y: $\dot{\epsilon}_{xy} = \mathrm{d}u/\mathrm{d}y$ (the geometry and notations are shown in Fig. 5). The applied shearing stress is obtained by substituting the equations of static equilibrium into the equations of principal stress. Substituting these basic equations into the Singh–Mitchell relationship and integrating over some finite depth and time, a relationship for predicting displacement is obtained. An outline of the mathematical treatment of this derivation is given in the Appendix.

Table 5 lists predicted creep displacements after one year for declivities ranging from 5° to 20° and sediment thicknesses from 20 to 50 meters using Equations 7 and 10 in the Appendix. For slopes 5° or less, predicted one-year surface displacements generated by gravity alone are insignificant, being on the order of 3 cm or less for sediment depths up to 50 m. However, for a sediment thickness of 50 m and a slope angle near 10°, surface displacements are over one-half meter which would most likely result in localized slope failure. Slopes of 20° or more experience surficial displacements such that localized failure is probable even for relatively shallow sediment thicknesses. It must be remembered that the analysis presented is very

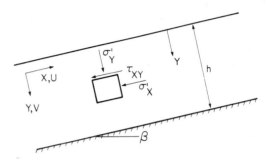

Fig. 5.   Geometry and nomenclature definitions for infinite-slope model of creep deformation (Yen, 1969).

TABLE 5   Predictions of creep displacements at the surface for one year[a]

Sediment thickness (m)	Sea-floor slope angle (β)		
	5°	10°	20°
20	< 1 cm	1 cm	6cm
30	1 cm	3 cm	95 cm
50	3 cm	53 cm	$\simeq \infty$

[a]Calculations are based on the following laboratory-derived creep parameter values: $A = 4 \times 10^{-5}$/min; $\alpha = 0.065$/kPa; $\gamma_b = 6.6$ kN/m^3; $m = 1.56$.

preliminary. Inasmuch as other sediment properties vary considerably, it is likely that the value of m will vary also. Therefore, some cumulative displacements may be near nil, while others will be greater as shown in Table 5. In general, the results do indicate that the slopes are creep susceptible, and that the problem is a tractable one. In addition, because the case where $m > 1$ indicates an eventual cessation of creep (Singh and Mitchell, 1968), creep rupture does not appear to be a candidate for causing discrete slope failure. Additional studies and future refinements of the analysis techniques are planned to assess more accurately the role of creep on submarine slope analysis.

Summary

The likelihood of mass movements on the continental slope off the northeastern USA is circumstantial. Given a state of normal consolidation for the Quaternary sediments, static conditions and the criteria of the infinite-slope model of stability analysis, the slope is essentially stable. Factors of safety associated with undrained conditions are typically lower than those representing drained conditions for this model. No mechanism is known at present, however, that would cause widespread undercutting, oversteepening or other process that may induce undrained failure. Thus, large-scale slope failure under static conditions seems unlikely. However, because creep may be an active process on parts of the slope, concomitant strength degradation may lead to slope failure. Further, if segments of the slope are subjected to occasional, modest ($\leqslant 5\%$ g) ground accelerations from earthquakes, slope failures may also result. Given the m values derived herein and typical conditions, creep rupture appears to be an unlikely cause of slope failure.

For local areas, such as those near or a part of the numerous canyon systems, rapid undercutting or oversteepening is more likely; hence, so is undrained failure. Canyon cutting and resultant release of lateral stress may also affect slope stability as would strain softening associated with creep. For these areas, general application of finite-slope stability analyses indicates that slopes greater than 15° may be metastable. This applies to both homogeneous sediment sections and sections characterized by planes of weakness (e.g. some bedding planes). Discrete slope failure and creep have occurred in the past. Our preliminary research suggests that they will continue to occur.

Acknowledgments

We are grateful to Ronald Circé, Alfred Dahl and William Levy for their contributions. We also acknowledge the support of the US Bureau of Land Management (AA851-MU0-18).

References

American Society for Testing and Materials, 1982. *Annual Book of ASTM Standards,* Part 19, Natural building stones; soil and rock; peats, mosses, and humis. ASTM, Philadelphia.

Bennett, R. H., Lambert, D. N. and Hulbert, M. H., 1977. Geotechnical properties of a submarine slide area on the U.S. Continental Slope northeast of Wilmington Canyon. *Mar. Geotech.* 2, 245–262.

Bishop, A. W. and Henkel, D. J., 1957. *The Measurement of Soil Properties in the Triaxial Test.* Edward Arnold, London.

Booth, J. S., Circé, R. C. and Dahl, A. G., in press (a). Geotechnical characterization and mass-movement potential of the United States Mid-Atlantic Continental Slope and Rise. *US Geol. Surv. Open-File Rep. 83.*

Booth, J. S., Circé, R. C. and Dahl, A. G., in press (b). Geotechnical characterization and mass-movement potential of the United States North Atlantic Continental Slope and Rise. *US Geol. Surv. Open-File Rep. 83.*

Booth, J. S., Farrow, R. A. and Rice, T. L., 1981a. Geotechnical properties and slope stability analysis of surficial sediments on the Baltimore Canyon

Continental Slope. *US Geol. Surv. Open-File Rep. 81-733.*

Booth, J. S., Farrow, R. A. and Rice, T. L., 1981b. Geotechnical properties and slope stability analysis of surficial sediments on the Georges Bank Continental Slope. *US Geol. Surv. Open-File Rep. 81-566.*

Booth, J. S., Sangrey, D. A. and Fugate, J. K., in press (c). A nomogram for interpreting slope stability in modern and ancient marine environments. *J. Sediment. Petrol.*

Embley, R. W. and Jacobi, R. D., 1977. Distribution and morphology of large sediment slides and slumps on Atlantic Continental margins. *Mar. Geotech.,* **2,** 205–228.

Emery, K. O. and Uchupi, E., 1972. Western North Atlantic Ocean: topography, rocks, structure, water, life, and sediments. *Am. Ass. Petrol. Geol. Mem.,* **17.**

Hampton, M. A., Bouma, A. H., Carlson, P. R., Molnia, B. F., Clukey, E. D. and Sangrey, D. A., 1978. Quantitative study of slope instability in the Gulf of Alaska. *Proc., Offshore Technology Conf., 10th, Houston, Texas.* OTC No. 3314, pp. 2307–2318.

Hathaway, J. C., Poag, C. W., Valentine, P. C., Miller, R. E., Schultz, D. M., Manheim, F. T., Kohout, F. A., Bothner, M. H. and Sangrey, D. W., 1979. U.S. Geological Survey core drilling on the Atlantic Shelf. *Science, N.Y.,* **206,** 514–527.

Keller, G. H., Lambert, D. N. and Bennett, R. H., 1979. Geotechnical properties of continental slope sediments – Cape Hatteras to Hydrographer Canyon. In: L. J. Doyle and O. H. Pilkey, (eds.), *Geology of the Continental Slopes.* Society of Economic Paleontologists and Mineralogists Special Publication No. 27, pp. 131–151.

Lo, K. V. and Lee, C. F., 1973. Stress analysis and slope stability in strain-softening materials. *Geotechnique,* **23,** 1–11.

Mitchell, J. K., 1976. *Fundamentals of Soil Behaviour.* Wiley, New York.

Moran, K., 1981. Drained Creep of Deep Sea Sediments. M.S. Thesis, University of Rhode Island.

Morgenstern, N. R., 1967. Submarine slumping and the initiation of turbidity currents. In: A. F. Richards, (ed.), *Marine Geotechnique.* University of Illinois Press, Urbana, Ill., pp. 189–220.

Morgenstern, N. R. and Sangrey, D. A., 1978. Methods of stability analysis, In: *Landslides: Analysis and Control.* Transportation Research Board, National Research Council, Washington, D.C., pp. 155–171.

O'Leary, D. W. and Twichell, D. C. 1981. Potential geologic hazards in the vicinity of Georges Bank Basin. In: J. A. Grow (ed.), Summary report of the sediments, structural framework, petroleum potential and environmental conditions of the United States middle and northern Continental Margin in area of proposed oil and gas lease sale no. 76. *US Geol. Surv. Open-File Rep. 81-765.* pp. 48–68.

Richards, A. F., 1978. Atlantic Margin Coring Project, 1976: Preliminary report on shipboard and some laboratory geotechnical data. *US Geol. Surv. Open-File Rep. 78-123.*

Robb, J. M., Hampson, J. C. and Twichell, D. C., 1981. Geomorphology and sediment stability of a segment of the U.S. Continental Slope off New Jersey, *Science, N.Y.*, **211**, 935–937.

Sangrey, D. A. and Marks, D. L., 1981. Hindcasting evaluation of slope stability in the Baltimore Canyon Trough Area. *Proc., Offshore Technology Conf., 13th, Houston, Texas*, pp. 241–252.

Seed, H. B., Murasaka, R., Lysmer, J. and Idriss, I. M., 1975. Relationships between maximum acceleration, maximum velocity, distance from source and local site conditions for moderately strong earthquakes. University of California, Berkeley, Earthquake Engineering Research Centre Report No. 71-17.

Silva, A. J., *et al.*, 1982. Geotechnical studies for sub-seabed disposal of high level radioactive wastes: Annual Progress Report No. 9, Department of Energy/SNLA, Contract No. 16-3110.

Silva, A. J., *et al.*, 1983. Stress–strain-time behavior of deep sea clays. *Can. Geotech. J.*, **20**, (3), 517–531.

Singh, A. and Mitchell, J. K., 1968. General stress–strain-time function for soils. *J. Soil Mech. Foundn Engng. Div., ASCE*, **94**, (SM1), Proceedings Paper 5728, 21–46.

Singh, A. and Mitchell, J. K., 1969. Creep potential and creep rupture of soils. *Proc., Int. Conf. Soil Mechancis and Foundation Engineering, 7th, Mexico City*, pp. 379–384.

Skempton, A. W., 1954. Discussion of the structure of inorganic soil. *J. Soil Mech. Foundn Engng. Div., Proc., ASCE*, **80**, Report no. 478, 19–22.

Swanson, P. G. and Brown, R. E., 1978. Triaxial and consolidation testing of cores from the 1976 Atlantic Margin Coring Project of the United States Geological Survey. *US Geol. Surv. Open-File Rep. 78-124*.

Wissa, A. E. Z., Christian, J. T., Davis, E. H. and Hieberg, S., 1971. Consolidation at constant rate of strain. *J. Soil Mech. Foundn Engng. Div., ASCE*, **97** (SM10), Proceedings Paper 8447, 1393–1413.

Yang, J. and Aggarwal, Y. P., 1981. Seismotectonics of northeastern United States and adjacent Canada. *J. Geophys. Res.*, **86**, (B6), 4981–4998.

Yen, B. C., 1969. Stability of slopes undergoing creep deformations. *J. Soil Mech. Foundn Engng. Div., ASCE*, **95**, (SM4), 1075–1095.

Appendix: Analysis of potential creep displacements

Model (Singh and Mitchell, 1968)

$$\dot{\epsilon}_1 = A e^{\alpha D}(t_r/t)^m \qquad (1)$$

where $\dot{\epsilon}_1$ = principal axial strain rate; A = empirical constant coefficient determined at $t_r = 1$ min; α = empirical constant coefficient (slope of $\log \dot{\epsilon}$ against D); D = deviator stress; t = time elapsed after initiation of creep; t_r = reference time (usually 1 min); m = empirical tant exponent (|slope| of $\log \dot{\epsilon}$ against $\log t$).

Basic equations for stability analysis

Continuity in plane strain, no volume change (Yen, 1969): $(\partial u/\partial x) + (\partial v/\partial y) = 0$, where u and v are constant exponent (|slope| of $\log \dot{\epsilon}$ against $\log t$). velocities in x and y directions, or $\dot{\epsilon}_x + \dot{\epsilon}_y = 0$; if all variables are independent of x, then

$$\dot{\epsilon}_x = 0 \quad \text{and} \quad \dot{\epsilon}_y = 0 \qquad (2)$$

Gradient of shear strain, $\dot{\epsilon}_{xy}$ (Yen, 1969):

$$\dot{\epsilon}_{xy} = \frac{\partial u}{\partial y} + \frac{\partial v}{\partial x} = \frac{\partial u}{\partial y} \quad \text{or} \quad \frac{du}{dy} \qquad (3)$$

Equilibrium, (from geometry):

$$\sigma_y = \gamma_b y \cos \beta \qquad (4a)$$

$$\sigma_x = K_0 \gamma_b y \cos \beta \qquad (4b)$$

$$\tau_{xy} = \gamma_b y \sin \beta \qquad (4c)$$

where γ_b = bouyant unit weight; y = depth; β = angle of infinite slope; K_0 = coefficient of lateral earth pressure at rest.

Principal strain

$$\dot{\epsilon}_1 = \frac{\dot{\epsilon}_x + \dot{\epsilon}_y}{2} + \left[\left(\frac{\dot{\epsilon}_x - \dot{\epsilon}_y}{2}\right)^2 + \dot{\epsilon}_{xy}^2\right]^{1/2}$$

from equations (2)

$$\dot{\epsilon}_1 = \dot{\epsilon}_{xy} \qquad (5)$$

Principal stress:

$$D = \sigma_1' - \sigma_3' = 2\left[\left(\frac{\sigma_y' - \sigma_x'}{2}\right)^2 + \tau_{xy}^2\right]^{1/2}$$

where σ_1' = maximum principal effective stress; σ_3' = minimum principal effective stress; σ_y' = vertical effective stress; σ_x' = lateral effective stress; τ_{xy} = shear stress.

From equation (4)

$$D = Fy \qquad (6)$$

where:

$$F = 2\gamma_b \left[\left(\frac{1-K_0}{2}\right)^2 \cos^2\beta + \sin^2\beta\right]^{1/2}$$

Final equation for long-term stability

Through substitution of equations (3), (5) and (6) into equation (1):

$$\frac{du}{dy} = A e^{\alpha F y}(t_r/t)^m \qquad (7)$$

Integrating with respect to y and boundary conditions $u = 0$ at $y = h$; $t_r = 1$:

$$u = \frac{dx}{dt} = \frac{A}{\alpha F}(t)^{-m}(e^{\alpha F y} - e^{\alpha F h}) \qquad (8)$$

Integrating with respect to time, t, where $t = 1$ min at which creep is initiated; i.e. $x = 0$ at $t = 1$:

$$x = \frac{A}{\alpha F(1-m)}(e^{\alpha F y} - e^{\alpha F h})[(t)^{1-m} - 1] \quad \text{for } m \neq 1 \qquad (9)$$

Applying creep data

Equation (9):

$$x = \frac{A}{\alpha F(1-m)}(e^{\alpha F y} - e^{\alpha F h})[(t)^{1-m} - 1] \text{ for } m \neq 1$$

$$F = 2\gamma_b\left[\left(\frac{1-K_0}{2}\right)^2 \cos^2\beta + \sin^2\beta\right]^{1/2}$$

Data: $A = 4 \times 10^{-5}$/min; $\alpha = 0.0605$/kPa; $\gamma_b = 6.6$ kN/m^3; $K_0 = 0.5$; $\beta = 5°$; $m = 1.56$.

For $y = 0$ – displacement at surface:

Condition 1: $t = 525\,600$ min (1 year) and $h = 20$ m
$x = 2.3$ cm

Condition 2: $t = 525\,600$ min and $h = 30$ m
$x = 19.3$ cm

6

Analysis of Slope Stability, Wilmington to Lindenkohl Canyons, US Mid-Atlantic Margin

Gideon Almagor*, Richard H. Bennett†, Douglas N. Lambert†, Evan B. Forde‡
and Les S. Shepard§

*Marine Geology and Geomathematics Division, Geological Survey of Israel,
Jerusalem, Israel
†Seafloor Geosciences Division, Naval Ocean Research and Development Activity,
NSTL, Mississippi, USA
‡Atlantic Oceanographic and Meteorological Laboratories, National Oceanic and
Atmospheric Administration, Miami, Florida, USA
§Sandia National Laboratories, Albuquerque, New Mexico, USA

The continental slope gradient in the study area averages $7-8°$. Many valleys, canyons and occasionally large sediment slumped masses occur. Moderate to steep slopes ($19-27°$) as well as very steep to precipitous slopes ($> 27°$) are abundant and occupy about 7% of the investigated area.

The surficial sediments are predominantly terrigenous silty clays of medium to high plasticity ($I_p = 10-35\%$, $w_L = 30-70\%$), but contain varying quantities of sands. Angles of internal friction are $\bar{\phi}_d = 27-32°$, $\phi_{cu} = 30-33°$ and $\phi_{cu} = 14-17°$. The sediments are normally to slightly overconsolidated, but some unconsolidated sediments also were identified. $c_u/\bar{p}_0$ values range from 0.12 to 0.78.

An analysis of force equilibrium within the sediments reveals (a) that the gentle slopes in the study area are mostly stable; (b) that the stability of some steep slopes ($19-27°$) is marginal; and (c) that on precipitous slopes ($> 27°$) only a thin veneer of sediments can exist. Observations of these slopes during steep dives support these results. The analysis shows that additional accumulation of sediments and small shocks caused by earthquakes or internal waves can cause the slopes to fail. Collapse resulting from liquefaction in the uppermost slope along the canyons and valleys axes, where fine sands and silt accumulate, also is likely.

Introduction

Many gently steep to precipitous slopes were created by the formation of numerous valleys and canyons that incise the continental slope, and by downslope slumping of large sediment blocks.

A full-fledged geotechnical study, based on data from tests made on a suite of 48 cores, provides the basis for a quantitative analysis of slope stability in the study area. This is a simplified pseudo-static analysis, first discussed by Morgenstern (1967).

The continental margin

The US mid-Atlantic continental margin consists of a Tertiary–Quaternary sedimentary sequence overlying Cretaceous strata. The continental slope in the study area (Fig. 1) has an average gradient of $7-8°$. Three large canyons – Wilmington, Spencer and Lindenkohl – cut the outer shelf. Many submarine valleys and ridges trending perpendicular to the slope extend to

the upper rise, forming a highly scalloped topography. The large valleys are 300–1000 m wide and 50–180 m deep, and begin seaward of the shelfbreak. Mass gravity processes during the Pleistocene led to the formation of these valleys (McGregor, 1979) and to re-excavation of the major canyons that were originally incised in Cretaceous and Cenozoic times (Kelling and Stanley, 1970; Schlee et al., 1976). Repeated slumping of very large blocks of Tertiary–Pleistocene sediments to the lower portion of the continental slope has further modified the morphology of the area by diverting the original paths of the canyons and by creating hummocky ground (McGregor, 1977; McGregor and Bennett, 1981).

Scarps formed by the canyons and valleys, and those associated with the detached slump blocks, are high and steep, attaining angles of $19-27°$. A considerable number of these scarps are very steep to precipitous, attaining slopes of over $30°$. A few are nearly vertical. These slopes occupy about 7% of the area (Bennett et al., 1978).

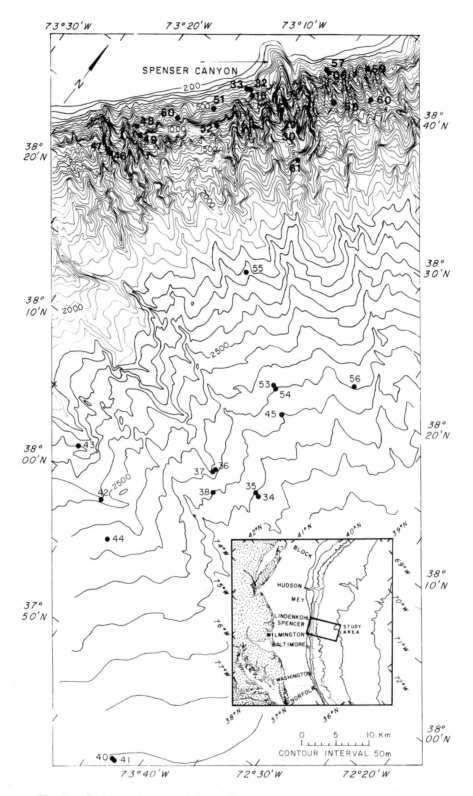

Fig. 1. Bathymetric map of the study area and location of coring stations.

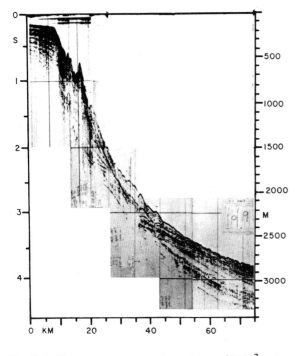

Fig. 2. Mass movement downslope (40 in³ airgun profile, vertical exaggeration is x 7.5) (from McGregor and Bennett, 1977).

Mass movements

On a grand scale the recent sediments of the continental margin of the US mid-Atlantic bight appear to be textrally graded (Doyle *et al.*, 1979). Sands and some gravels accumulate on the continental shelf and grade progressively to silty and clayey sediments on the slope and rise. The veneer of the fine-grained sediments on the relict Late Pleistocene slope and in the canyons suggest generally quiescent conditions and filling up of canyons (Doyle *et al.*, 1979; Robb *et al.*, 1982). Identifiable slump and slide features occupy only a small part of the slope, which suggests a non-vigorous or intermittent present-day activity (Robb *et al.*, 1982). Seismic and sidescan-sonar records reveal slump scars, disrupted exotic blocks of varying sizes that slid downslope, and acoustically fuzzy-looking bottom and seismically transparent sub-bottom sediment layers that resulted from mass movements (McGregor and Bennett, 1977; Knebel and Carson, 1979; Robb *et al.*, 1982) (Fig. 2). Observations during submersible dives of step-like depressions, notches and ledges carved into canyon walls and their outer banks, and of fields of blocky debris, upturned clay beds and disaggregated gravels provide additional evidence of slumping activity (Robb *et al.*, 1982; Stubblefield *et al.*, 1982).

Spill-over of shelf sands across the shelfbreak and their accumulation on the uppermost continental slope have formed a highly sandy bottom (Bennett *et al.*, 1977; Doyle *et al.*, 1979). Finer-grained sediments are winnowed out or by-pass the shelfbreak through complex processes of resuspension, cascading downslope (McCave, 1972). Excessive accumulation on the inter-canyon slope portions and on the valley walls, combined with local erosional activity associated with down-canyon mass flow, lead to oversteepening of the slopes and eventual failure. Redistribution and mixing of the downslope-flowing sediments highly modifies the simple regional sedimentation pattern, and the result is the deposition of highly heterogeneous sediments on the slope and inside the canyons.

Mass physical properties

Illite is the dominant clay mineral, with lesser quantities of kaolinite, smectite and chlorite. The silt-sized minerals are predominantly quartz and there are some feldspars (Bennett *et al.*, 1977; Demars *et al.*, 1979). Variable amounts of sands, up to 15% of the sediment bulk, are mixed throughout the finer material or occur in the form of thin discrete layers. In a few cores, notably those from the upper slope, sands comprise 30–60% of the upper 0–30 cm. The sand fraction is composed primarily of quartz, with lesser amounts of mica, pyrite and heavy minerals. The $CaCO_3$ content is typically very low (0–5%) (Doyle *et al.*, 1979). The textural composition of the slope and the rise sediments is really quite variable. In this study, sediment properties were averaged over the entire core lengths (Table 1). On the basis of these averages, the bulk of the sediment is defined as inorganic silty clays of medium to high plasticity ($I_p = 10–35\%$, $w_L = 30–70\%$). Grain specific gravity values range from 2.68 to 2.85, and average 2.75. The mean values of water content, plasticity and compressibility indices, although highly variable, gradually increase with water depth, whereas the unit weight of the sediments decreases with depth (Fig. 3). These results support the concept of a texturally graded continental margin.

For most of the cores the unit weight increases with burial depth, whereas the water content and the liquidity index decrease with depth (Fig. 4). The wide scatter of the values of these properties with depth reflects mainly the highly variable and sand content of the cores.

Similar geotechnical data were reported by several investigators who worked in the area between Baltimore and Mey canyons (Bennett *et al.*, 1977; Lambert *et al.*, 1980; Booth *et al.*, 1981; Olsen *et al.*, 1982).

TABLE 1 Physical properties of core samples (average values)

Core	Latitude (N)	Longitude (W)	Water depth (m)	Core length (m)	Topographic feature	Sand:silt:clay (no. of tests)	Specific gravity of solids, G_s	Unit weight, γ_t (mg/cm³)	Water content, W (%)	Porosity, n (%)	Void ratio, e	Liquid limit, w_L (%)	Plastic limit, w_p (%)	Plasticity index, I_p (%)	Liquidity index, I_L	Vane shear strength, s (kPa)	Sensitivity, S_t
8057	38°38.58'	73°05.82'	372	5.01	Upper slope	7:42:51 (13)	2.73	1.72	54	60.0	1.50	43	25	18	1.64	12.3	5.1
8032	38°33.52'	73°11.23'	381	5.50	Upper slope	7:43:51 (11)	2.76	1.78	52	56.4	1.29	42	26	16	1.50	15.5	3.3
8015	38°36.56'	73°08.71'	397	1.13	Ridge-upper slope	40:40:20 (6)	2.74	1.88	34	48.4	0.95	26	18	8	1.97	10.8	3.8
8033	38°33.53'	73°11.24'	400	6.27	Upper slope	11:51:38 (7)	2.75	1.83	42	51.1	1.07	38	25	13	1.53	12.3	3.4
8006	38°38.51'	73°05.54'	421	1.71	Upper slope	14:58:28 (14)	2.73	1.78	45	55.4	1.25	27	19	8	1.89	9.7	6.2
8046	38°22.86'	73°19.55'	458	4.46	Valley slope	7:52:41 (12)	2.70	1.56	45	67.8	2.16	55	29	26	2.04	5.4	5.4
8016	38°33.36'	73°10.84'	489	1.07	Upper slope	41:30:29 (2)	2.75	1.80	45	54.2	1.20	37	24	13	1.43	10.4	3.6
8051	38°30.64'	73°13.24'	495	4.04	Upper slope	—	2.76	1.70	53	58.9	1.44	50	29	21	1.18	16.1	4.0
8059	38°40.52'	73°02.53'	498	4.10	Upper slope	5:54:41 (14)	2.74	1.79	46	55.2	1.25	43	24	19	1.16	29.1	4.1
7045	38°28.30'	73°15.52'	640	5.60	Upper slope	13:38:49 (15)	2.73	1.70	—	—	—	40	25	15	1.94	7.5	4.6
8050	38°28.35'	73°15.60'	652	8.36	Slope	—	2.73	1.68	57	61.2	1.60	49	27	22	1.47	12.4	4.3
8048	38°25.86'	73°18.78'	851	5.31	Slope	—	2.72	1.67	58	61.2	1.59	47	23	24	1.60	8.5	4.2
8049	38°25.19'	73°17.82'	1055	7.91	Slope	5:54:41 (20)	2.70	1.74	51	57.4	1.36	48	27	21	1.07	21.6	4.7
8058	38°36.82'	73°03.38'	1080	10.01	Slope	—	2.72	1.68	57	60.8	1.60	43	28	15	1.86	9.8	5.8
8052	38°29.58'	73°11.81'	1089	5.76	Slope	7:57:36 (13)	2.74	1.74	49	59.3	1.52	41	25	16	1.56	15.9	5.0
8047	38°22.89'	73°19.74'	1144	4.21	Valley	19:51:30 (13)	2.69	1.67	57	60.9	1.61	49	29	20	1.83	16.0	6.0
8060	38°38.78'	73°00.08'	1183	5.46	Slope	4:57:39 (15)	2.72	1.58	38	51.5	1.05	59	32	27	1.78	5.9	4.7
8043	38°02.44'	73°04.71'	1257	9.81	Ridge	4:61:35 (22)	2.73	1.77	47	56.5	1.31	37	22	15	1.98	8.3	4.8
8029	38°33.39'	73°04.87'	1271	5.16	Ridge	19:48:34 (6)	2.75	1.75	50	57.8	1.39	39	25	14	1.90	14.0	4.7
8030	38°33.37'	73°04.86'	1271	5.36	Ridge	14:43:41 (6)	2.73	1.76	48	55.7	1.29	39	25	16	1.55	15.7	4.5
8031	38°33.76'	73°07.15'	1288	4.70	Valley	15:56:29 (12)	2.71	1.68	57	60.6	1.60	54	33	21	1.34	10.0	3.3
7046	38°26.92'	73°13.53'	1298	5.52	Valley	3:50:47 (16)	2.71	1.57	—	—	—	66	42	24	1.61	7.6	5.4
8061	38°31.23'	73°02.96'	1807	3.59	Valley	—	2.72	1.86	38	51.2	1.05	42	24	18	0.86	30.8	1.8
8055	38°21.53'	73°00.56'	2296	2.39	Valley	—	2.73	1.51	92	70.8	2.44	55	31	24	2.54	6.5	5.9
8042	37°59.86'	72°59.83'	2413	10.16	Ridge, steep slope, rise	9:58:33 (24)	2.72	1.77	46	55.8	1.31	36	22	14	1.99	11.0	4.6
8036	38°07.29'	72°51.72'	2530	8.31	Hill	—	2.73	1.76	48	57.4	1.35	46	23	23	1.16	17.8	3.2
8037	38°07.21'	72°51.68'	2537	3.83	Hill	—	2.73	1.76	49	60.2	1.52	47	21	26	1.11	17.5	4.5
8054	38°15.65'	72°51.57'	2590	8.16	Rise	4:42:54 (13)	2.72	1.54	86	70.2	2.42	59	33	26	2.48	10.1	5.0
8053	38°15.67'	72°51.61'	2592	9.51	Rise	—	2.72	1.50	80	70.1	2.38	56	35	21	2.47	11.9	5.7
8056	38°19.58'	72°44.78'	2598	10.51	Rise	11:05:30 (24)	2.74	1.72	52	58.7	1.46	37	24	15	2.13	11.3	6.2
8038	38°05.89'	72°50.42'	2640	2.93	Rise	—	2.71	1.79	47	55.8	1.27	49	29	20	0.96	30.9	2.5
8045	38°14.21'	72°49.31'	2643	11.46	Rise	10:44:46 (5)	2.81	1.54	82	67.2	2.21	61	33	28	2.27	6.9	6.7
8035	38°07.99'	72°46.50'	2700	10.81	Rise	14:50:36 (24)	2.72	1.66	63	61.0	1.64	42	27	15	2.44	7.3	5.0
8034	38°07.97'	72°46.47'	2705	5.51	Rise	13:43:44 (15)	2.72	1.58	79	69.7	2.36	44	25	19	2.89	4.5	6.2
8044	37°57.76'	72°56.67'	2719	10.42	Rise	—	2.73	1.68	59	61.6	1.63	44	24	20	1.76	12.4	6.3
8041	37°43.80'	72°42.45'	2799	5.86	Rise	1:50:49 (23)	2.70	1.57	80	66.6	2.12	58	32	26	2.06	7.5	5.3
8040	37°43.65'	72°46.23'	2800	11.23	Rise	—	2.70	1.54	83	69.1	2.32	55	29	26	2.37	7.3	6.4

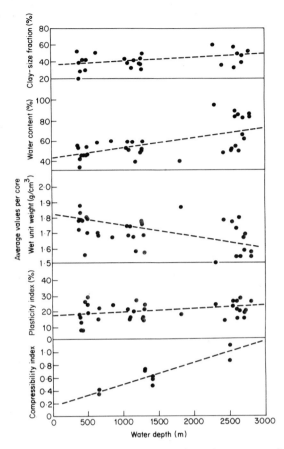

Fig. 3. Average physical properties of core sampling plotted against water depth.

Shear strength

Laboratory Vane Shear Tests

The vane shear strength profiles usually display an increase of strength with burial depth (Fig. 4); however, for most of them the data points are highly variable. The textural variability of the sediments, especially the presence of sand, causes this variability. The areal distribution of the strength values are erratic, and no relationship between shear strength and distance from shore or water depth could be established. The average shear strength of the surficial sediments is 7.3 kPa, and the average strength for all the samples tested is 12.9 kPa; maximum and minimum values range from 1.3 to 36.8 kPa (Table 1). The sensitivity of the sediments is within the range 1.8–6.7, the mean value being 4.8.

Direct Shear Tests

Five direct shear tests under drained and undrained conditions were conducted. Angles of internal friction

under both undrained and drained conditions are in the range $\bar{\phi}_d = 28\text{–}32°$.

Consolidated–undrained Triaxial Compression Tests

Four tests on samples from various depths were conducted. The total angles of internal friction that were measured are in the range $\phi_{cu} = 14\text{–}17°$, and the effective angles are in the range $\phi'_{cu} = 30\text{–}33°$. These results support the assumption that the fine-grained sediments in the study area are geotechnically similar. These effective angles of internal friction are 1.4–4.4° higher than those of the sediments on the continental slope northeast of the study area (Lambert *et al.*, 1980). Except for one core which had a large concentration of sand, the $c_u/\bar{p}_0$ (ratio between the undrained vane shear strength of the sediments and the sediment overburden stress) values calculated from the rupture envelopes in these tests are in agreement with the $c_u/\bar{p}_0$ values calculated from the vane shear tests. The pore pressure parameters at failure are in the range $A_f = 0.56\text{–}1.01$, which is characteristic of normally consolidated to slightly overconsolidated clayey sediments.

Consolidation state of the sediments

The results of the consolidation tests are summarized in Fig. 5. The overconsolidation ratios obtained range from 1.0 to 1.9, which suggests that the sediments tested were normally to slightly overconsolidated. One sample appears to be slightly overconsolidated. Disturbance of the samples caused during coring, and shipboard and laboratory handling, may account for the deviation from closer agreement between pre-consolidation and effective overburden stresses. The results of 32 consolidation tests on samples collected in the adjacent area to the northeast also show normal to slight overconsolidation state of the sediments (Olsen *et al.*, 1982).

A state of normal to slight overconsolidation is indicated by the $c_u/\bar{p}_0$ values computed for the core samples. They range from 0.12 to 0.78, of which seven values are above 0.5; a low value of 0.04 in Core 8046, which is texturally fairly uniform, may be attributed to an underconsolidation state in this particular core (Fig. 6; Table 2). $c_u/\bar{p}_0$ values of 0.12–0.5 are considered typical of slightly underconsolidated to normally consolidated sediments, and those above 0.5 indicate an overconsolidation state (Skempton, 1954).

Plots of $c_u/\bar{p}_0$ against I_p serve as a crude criterion for the state of consolidation of the sediments (Skempton, 1954). When plotted, most of the $c_u/\bar{p}_0$ ratios are grouped on and near the correction line, which also indicates that the sediments are mostly normally to

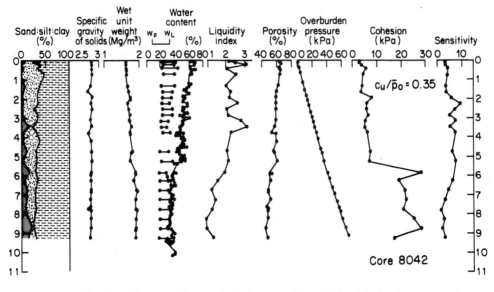

Fig. 4.   Changes of geotechnical properties with burial depth.

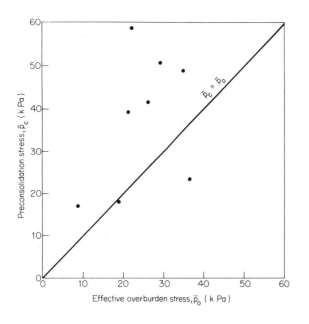

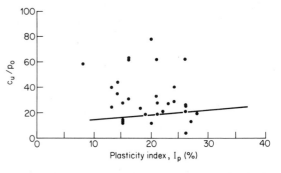

Fig. 6.   Relation of undrained strength $(c_u/\bar{p}_0)$ to plasticity index (I_p).

Stability analysis

Infinite-slope stability analysis is the best method to assess the sediment stability, since the slopes in the study area, including many of the walls of the valleys and canyons, are long and are made of fairly continuous thin lobate sediment sheets that are conformable over great distances. In this method the force equilibrium between the shearing stress exerted by the bouyant weight of the sedimentary column and the shear resistance of the sediments along a potential failure plane is calculated (Figs. 7, 8). When the shearing stress becomes equal to or exceeds the shearing resistance of the sediments along the failure surface, slumping is initiated. Depending on the drainage conditions in the sediments during the shear and on the time required for draining, one of the following approaches may be selected for

Fig. 5.   Relation of preconsolidation stress to effective overburden pressure.

slightly overconsolidated (Fig. 6). Some disturbance of the samples during laboratory treatment, and the fact that the $c_u/\bar{p}_0$ values are average values of considerably scattered c_u data points, may account for some deviation from the correlation line.

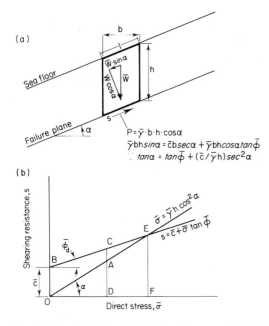

(a)

$$P = \bar{\gamma} \cdot b \cdot h \cdot cos\alpha$$
$$\bar{\gamma}bh sin\alpha = \bar{c}b sec\alpha + \bar{\gamma}bh cos\alpha tan\bar{\phi}$$
$$\therefore tan\alpha = tan\bar{\phi} + (\bar{c}/\bar{\gamma}h) sec^2\alpha$$

(b)

Fig. 7. (a) Equilibrium of infinite slope under drained conditions; (b) Mohr diagram illustrating the limited thickness to which a sedimentary layer can accumulate on a slope when the slope angle is greater than the angle of internal friction of the sediment.

the analysis of the stability of the slope under investigation:

(1) *drained slumping,* which is a long-term process (the fine-grained sediment is stressed commensurate with the slow dissipation of the resultant excess pore pressures); or

(2) *undrained slumping,* which is a short-term process (build-up of excess pore pressure, which cannot be dissipated during shear). A third possible mechanism of failure is

collapse slumping, where the sediments are meta-stable, i.e. the grain packing of the sediment is extra-spacious. Failure may occur initially under drained conditions, but the deformation associated with the failure causes a large increase in the pore pressure, drastically reducing the shear resistance of the sediments to the point where they have the character of a liquid with low viscosity (liquefaction), which results in the flow of sedimentary material downslope. Accumulation of large quantities of fine sands and silts in the US mid-Atlantic margin may give rise to the occurrence of collapse failure, especially in the canyon and valley heads at the shelfbreak and uppermost slope. However, the normal state of consolidation observed in these sediments and their sensitivity suggest that this type of failure is unlikely, except possibly in very confined localities.

TABLE 2 Comparison of natural slopes with calculated maximum stable slopes based on analysis of undrained slumping

Core	Length (m)	Water depth (m)	Natural slope (deg)	$c_u/\bar{p}_0$	Calculated maximum stable slope (deg)
8006	1.71	421	12	0.59[a]	>45
8016	1.07	489	9	0.24[a]	14
8029	5.16	1271	15	0.44	31
8030	5.36	1271	15	0.62	>45
8031	4.70	1288	15	0.19	11
8032	5.50	381	8	0.14	8
8033	6.27	400	8	0.40	27
8034	5.51	2705	1	0.19[a]	11
8035	10.81	2700	1	0.13	8
8036	8.31	2530	4	0.27[a]	16
8037	3.83	2537	4	0.25	15
8038	2.93	2640	2	− 2.09[b]	−
8040	11.23	2800	0.57	0.21	12
8041	5.86	2799	−	0.26	16
8042	10.16	2413	4.5	0.35	22
8043	9.81	1257	2	0.12	7
8044	10.42	2719	0.5	0.12	7
8045	11.46	2643	0.5	0.19	11
8046	4.46	458	22	0.04	2
8047	4.21	1144	18	0.78	>45
8048	5.31	851	22	0.29	18
8049	7.91	1055	35	0.28	17
8050	8.36	652	5	0.21	12
8051	4.04	495	12	0.33	21
8052	5.76	1089	16	0.63	>45
8053	9.51	2592	0.57	0.62	>45
8054	8.16	2590	17	0.62	>45
8055	2.39	2296	5	0.40	27
8056	10.51	2598	0.5	0.14	8
8057	5.01	372	7	0.23	14
8058	10.01	1080	18	0.28	17
8059	4.10	498	18	1.60[b]	−
8060	5.46	1183	8	0.13[a]	8
8061	3.59	1807	14	− 2.00[b]	−

[a] Denotes $c_u/\bar{p}_0$ values calculated for silty and clayey portions of cores only.
[b] Questions $c_u/\bar{p}_0$ values due to high sand content in cores.

Drained Slumping

Figure 7(b) illustrates the conditions for slope failure under drained conditions. As long as the drained angle of internal friction, $\bar{\phi}_d$, remains greater than the angle, α, of a natural slope, sediments may accumulate with no impact on the potential slumping of the slope. In that case line BC, which depicts the maximum shear resistance of the sediments, will never cross line OA, which depicts the normal stress, $\bar{\sigma}$, that acts on the sediments.

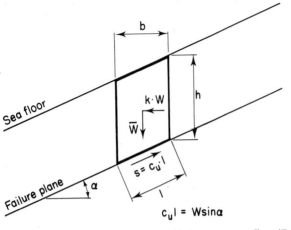

$$c_u l = W \sin \alpha$$

$$1/2 \sin 2\alpha = c_u / \bar{\gamma} h \cong c_u / \bar{p}_o$$

Fig. 8. Equilibrium of infinite slope under undrained conditions.

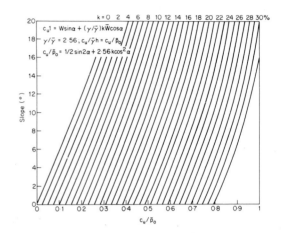

Fig. 9. Relation of slope angle (α) to undrained strength ($c_u/\bar{p}_0$) for an infinite slope at limiting equilibrium and subject to an earthquake acceleration (k % g) (after Morgenstern, 1967).

However, if the slope angle is greater than the drained angle of internal friction, lines OA and BC will cross at point E. As long as the normal stress exerted by the weight of the accumulating sediments is within the area bounded by line BE, the shear stress generated by the weight of the sediments is smaller than the shearing resistance of the sediments. Thus, when a certain sediment column generates a shearing stress AD, which is smaller than the shearing resistance of the sediments along the potential sliding surface (line CD), the slope is stable. If accumulation continues, then at a critical thickness the shearing stress becomes equal to the shearing resistance (line EF) at that depth, and incipient slumping occurs.

The measured effective angles of internal friction (28–32°) designate the maximum possible angles for stable slopes for the sediments in the study area on the basis of infinite-slope analysis under drained conditions. These angles are appreciably larger than the angles of the gentle to moderately steep slopes that predominate in the study area (6–19°) (Bennett et al., 1978), so that drained slumping is unlikely for these slopes. However, many steep slopes (> 27°), consisting of older heavily overconsolidated sub-bottom sediments exposed by erosion (Stubblefield et al., 1982), can support only a thin cover of recent sediments. The recent sediments that accumulate on these slopes at low rates of deposition (thus maintaining zero pore pressures) are prone to drained slumping with continuing deposition when a critical thickness is achieved. The authors predict that accumulation of 3–7 m of the silty–clayey sediments ($\gamma_t = 1.67 \, \text{g/cm}^3$; $\bar{p}_c = 1.8 - 6.0 \, \text{kPa}$) can take place on scarp slopes of 50° before failure occurs, as is indeed confirmed by observations during submersible dives.

Undrained Slumping

Increase of the load on a sediment layer in which the drainage conditions are poor may generate excess pore pressure in the sediments, resulting in reduction of their shear strength. The slope is stable as long as the total angle of internal friction is larger than the inclination of the infinite slope (assuming that the pore pressures developed during laboratory triaxial shear compression tests are identical with the *in situ* pore pressures in the sediments) (Fig. 8). Undrained shear failure can be caused by rapid changes in slope geometry — by erosional undercutting of the slope, such as submarine canyon and valley formation, or its oversteepening due to rapid sediment accumulation.

The total angles of internal friction under undrained conditions ($\phi_{cu} = 14–17°$) in the study area are appreciably larger than the average angle of the continental slope in the study area (7–8°), thus indicating that, on the average, most of the continental slope is stable. However, in numerous localities the slopes are 14–19°, which indicates marginal slope stability. The angles of the scarp walls of the many valleys and canyons in the area, and those of the large slump blocks (McGregor, 1977; Bennett et al., 1978), are large; this clearly makes slope instability very imminent for the sediments that accumulated on them. Many of the cores used for the present study were purposely collected from these steep slopes. Comparison of the maximum stable angles determined for these cored sediments from the undrained slope stability analysis (Fig. 8) with the actual slopes at the coring sites clearly reveals that stability in many localities is marginal, and that in a few places the slope is actually unstable (Table 2).

Fluctuations in pore pressures which are associated

with earthquakes and passage of internal waves may also lead to undrained shear failure in deep water. Morgenstern (1967) discussed a pseudo-static approach to assess the effects of earthquake-induced accelerations on sediment slopes. Assuming that the suddenness of an earthquake shock renders the drainage of pore water from the sediments impossible, even if the sediments are very poorly permeable, and that the vertical accelerations induced by the earthquake are negligible compared with the horizontal accelerations (if the earthquakes originate far away), then the horizontal body force, k, introduced by the earthquake, which acts on the whole sediment mass, will add to the static shearing force along the potential failure plane (Fig. 8). Figure 9 illustrates the relationship between the undrained strength of the sediments in the investigated area $(c_u/\bar{p}_0 = 0.1-0.8)$ at their limiting equilibrium at slope angles of $0-20°$ when they are subjected to earthquake-induced accelerations which are up to 20% of gravity (% g). Figure 9 demonstrates that even small earthquake-induced accelerations are sufficient to trigger undrained slumping of the surficial sediments in the moderate slopes of the study area. Thus, an acceleration of 5% g is sufficient to disrupt sediments with $c_u/\bar{p}_0 = 0.25$ which rest on a $7°$ slope. Although earthquake activity in this region is rare, shock waves that originate far away may cause undrained slumping on these slopes, despite the fact that the greatest part of their original energy had dissipated over the long distance they travelled. Likewise, it was shown that internal gravity waves propagating over the continental slope may be of such proportions as to cause sediment failure (Wunch and Hendry, 1972; Cacchione and Southard, 1974).

Summary and conclusions

Geotechnical tests on a large number of relatively long cores suggest that the vast majority of the sediments are normally to slightly overconsolidated (Table 1; Figs 5, 6). The rates of sediment accumulation in the area (0.08–0.44 cm/1000 years) (Doyle *et al.*, 1979; T. A. Nelsen, personal communication) corroborates this conclusion. Results of shear tests indicate that the surficial sediments are generally sufficiently strong to form stable slopes as steep as $14-18°$ (Table 2). However, a few of the slopes were found to be marginally stable, and on a few occasions unstable (Table 2), and therefore highly susceptible to slumping. Only a thin cover of sediments, a few metres in thickness, can accumulate on steep slopes ($> 27°$) before drained failure and subsequent slumping occur. Observations of scarp slopes and their sediment cover made during deep submersible dives support these conclusions (Stubblefield *et al.*, 1982). Small additional overloading

of these sediments by horizontal earthquake-induced accelerations or by pressures transmitted within the sedimentary mass by internal waves propagating over the slopes may readily lead to undrained slumping of surficial sediments.

The slight overconsolidation state observed in many of the cores (Figs 5, 6) is attributed to exposure of older sediments after slumping. The underconsolidation state observed in a few cores (Figs 5, 6) most probably resulted from rapid deposition of the slumping sediments.

Collapse failure may not be excluded either. Metastable granular sediments that are highly prone to collapse are likely to exist in various localities, especially in the uppermost continental slope and along the axes of submarine canyons, where accumulation of fine sands and silts occurs. Collapse of these sediments can be triggered by internal waves or earthquake-induced accelerations. Highly sensitive fine-grained sediments, however, have not been detected in the cores.

The potential role of bioturbation and gases in the slope stability analysis was not studied.

Acknowledgements

This study was carried out within the framework of the National Research Council — National Oceanographic and Atmospheric Administration (NRC–NOAA) associateship program, which enabled the first author to work in the Atlantic Oceanographic and Meteorological Laboratories (AOML) of NOAA in Miami, Florida, Much of the laboratory geotechnical testing was done by T. A. Nelsen, F. L. Nastav and G. Romero of NOAA/AOML. The assistance of T. J. Burns, S. A. Bush, A. Meyer and G. Santos in carrying our direct and triaxial compression shear tests is acknowledged with thanks. We are indebted to G. H. Keller, T. A. Nelsen and B. A. McGregor for critical reading of the manuscript.

References

Bennett, R. H., Lambert, D. N. and Hulbert, M. H., 1977. Geotechnical properties of a submarine slide area on the US continental slope northeast of Wilmington Canyon, *Mar. Geotech.*, 2, 245–261.

Bennett, R. H., Lambert, D. N., McGregor, B. A., Forde, E. B. and Merrill, G. F., 1978. Slope map: a major submarine slide on the US Atlantic continental slope east of Cap Mey. U.S. Dept Commerce, NOAA, A-5758, USCOMM-NOAA-DC.

Booth, J. S., Farrow, R. A. and Rice, T. A., 1981. Geotechnical properties and slope stability analysis of surficial sediments on the Baltimore Canyon continental slope. *US Geol. Surv., Open-File Rep. 81-733.*

Cacchione, D. A. and Southard, 1974. Incipient sediment movement by shoaling internal gravity waves. *J. Geophys. Res.*, 79 (15) 2237–2242.

Demars, K. R., Charles, R. D. and Richter, J. A., 1979. Geology and geotechnical features of the mid-Atlantic shelf. *OTC, Paper 3397*, 343–347.

Doyle, L. J., Pilkey, O. H. and Woo, C. C., 1979. Sedimentation on the eastern US continental slope. In: L. J. Doyle and O. H. Pikley (eds.), *Geology of the Continental Slopes*. SEPM Spec. Pub. 27, 119–129.

Kelling, G. and Stanley, D. J., 1970. Morphology and structure of Wilmington and Baltimore submarine canyons, eastern US. *J. Geol.*, **78** (6), 637–660.

Knebel, H. J. and Carson, B., 1979. Small-scale slump deposits, middle Atlantic continental slope off eastern US. *Mar. Geol.*, **29**, 221–236.

Lambert, D. N., Bennett, R. H., Sawyer, W. B. and Keller, G. H., 1980. Geotechnical properties of continental upper rise sediments, Veatch Canyon to Cape Hatteras. *Mar. Geotech.*, **1**, (4), 281–306.

McCave, I. N., 1972. Transport and escape of fine-grained sediment from shelf areas. In: D. J. P. Swift, D. B. Duane and O. H. Pilkey, (eds.), *Shelf Sediment Transport*. Dowden, Hutchinson and Ross, Stroudsburg, pp. 225–248.

McGregor, B. A., 1977. Geophysical assessment of submarine slide northeast of Wilmington Canyon. *Mar. Geotech.*, **2**, 229–244.

McGregor, B. A., 1979. Variations of bottom processes along the US Atlantic continental margin. *Am. Assoc. Petrol. Geol. Mem.*, **29**, 139–149.

McGregor, B. A. and Bennett, R. H., 1977. Continental slope sediment instability northeast of Wilmington Canyon. *Am. Ass. Petrol. Geol. Bull.*, **6** (6), 918–928.

McGregor, B. A. and Bennett, R. H., 1979. Mass movement of sediment on the continental slope and rise seaward of the Baltimore Canyon Trough. *Mar. Geol.*, **33**, 163–174.

McGregor, B. A. and Bennett, R. H., 1981. Sediment failure and sedimentary framework of the Wilmington geotechnical corridor, US Atlantic continental margin. *Sediment. Geol.*, **30**, 213–234.

Morgenstern, N. M., 1967. Submarine slumping and initiation of turbidity currents. In: A. F. Richards (ed.), *Marine Geotechnique*, University of Illinois Press Urbana, Ill., pp. 189–220.

Olsen, H. W., McGregor, B. A., Booth, J. S., Cardinell, A. P. and Rice, T. L., 1982. Stability of near-surface sediment on the mid-Atlantic upper continental slope. *OTC, Paper 4303*, 21–27.

Robb, J. M., Hampson, J. C., Jr. and M. Kirby, J. R., 1982. Surficial geologic studies of the continental slope in the northern Baltimore Canyon Trough area – techniques and findings. *OTC, Paper 4170*, 39–43.

Schlee, J. S., Behrendt, J. C., Grow, J. A., Robb, J. M., Mattick, R. E., Taylor, P. T. and Lawson, B. J., 1976. Regional geologic framework off northeastern United States. *Am. Ass. Petrol. Geol. Bull.*, **60**, (6), 926–951.

Skempton, A. W., 1954. The structure of inorganic soil (discussion). *Proc. Am. Soc. Civil Engrs*, **80**, separate **478**, 19–22.

Stubblefield, W. L., McGregor, B. A., Forde, E. B., Lambert, D. N. and Merill, G. F., 1982. Reconnaissance in DSRV ALVIN of a 'fluvial-like' meander system in Wilmington Canyon and slump features in South Wilmington Canyon. *Geology*, **10**, 31–36.

Wunch, C. and Hendry, R., 1972. Array measurements of the bottom boundary layer and the internal wave field on the continental slope. *Geophys. Fluid Dynam.* **4**, 101–145.

7

A Device to Facilitate the Sampling and *in situ* Testing of Sediments

Colin Nunn
Zetelogic Ltd., High Spen, Rowlands Gill, UK

A device is described which can aid the accurate assessment and sampling of sediment properties *in situ*. This probe consists of two units, uphole and downhole, joined by a logging cable, and a multiprocessor electronic control system. The downhole probe consists of both Dutch cone penetrometer and a standard penetration test tool, with individual power units. The simplicity of the device allows easy operation on site, and data may be displayed, stored and manipulated readily.

Introduction

An essential part of all major construction projects is a successful site investigation. This provides the basis for the informed design of foundations, excavations and other specialized groundworks by illuminating the mechanical characteristics of the natural materials beneath the site, as they can vary immensely across an individual site and to an even greater degree with depth.

A common foundation — especially beneath dam sites, in flood plains, in estuaries and at sea — involves layered soft sedimentary rocks and soils of generally low strength. Such an assemblage often provides the main design constraint and can necessitate expensive foundations to spread superimposed loads or take them down to firmer ground below. A typical investigation of such a site includes sampling and probing in combination to determine the *in situ* properties of the strata in order to optimize the design.

Samples can be taken in a wide range of core barrels. The object is to secure a specimen as undisturbed as possible with a view to determining its *in situ* properties, especially shear strength, by laboratory testing. The success of this endeavour often relates as much to the care taken of the sample during and after its recovery as to the intricacy of the core barrel and installation technique. In any case the procedure is fraught with opportunities to introduce sample disturbance. This

has led to the increasing use of devices to determine the strength of engineering soils *in situ*.

To assess the strength of a soil in the ground has predominantly required the use of a penetrometer. Here again the investigator is faced with a range of options. Both on land and at sea devices in common use can be operated either from the surface of the ground or seabed, or down a borehole. Naturally, deeper penetration is possible from a borehole but, of course, it requires the installation of the hole and the interruption of drilling while the penetration test is performed; nevertheless it is frequently considered to make a worthwhile contribution to the suite of geotechnical data assembled.

In recognition of the value of *in situ* measurement, considerable efforts have been made, over the few decades since it became commonplace, to improve both the interpretation of penetration resistance and the design of penetrometers themselves. The many devices on the market divide themselves essentially into two groups, static and dynamic. Static penetrometers are championed by the Dutch cone model, (CPT) which is pushed into the ground at speeds up to 2 cm/s and measures cone resistance and shaft friction separately. Dynamic devices centre around the split spoon sampler which performs the standard penetration test (SPT) and also retrieves a sample. Both these methods are applicable to soft or loose strata, with the SPT enjoying applications to stiffer media as well. Each can be calibrated to permit the mechanical and geological interpretation of a wide range of typical sedimentary environments.

While the use of both a CPT and SPT is possible and indeed desirable on many sites, or down boreholes, their application is often limited to one or the other instead of both in concert because of the expense and inconvenience necessitated by their separate operation and support facilities. However, their combined use would not only extend the range of useful data recovered, but also provide a check on the consistency of calibration between the two devices and thereby increase confidence in their interpretation for engineering design. It is primarily towards that end that the new downhole system described here has been devised. At the same time, with little significant deviation from the standard test specifications, the new facility offers a far more comprehensive, self-contained *in situ* test package than the application of both the CPT and the SPT together. It is sophisticated in its testing capability, instrumentation and control, yet simple to use. Furthermore, despite its current voracious appetite for data acquisition, its modular construction and unified software render it suitable for ready extension to incorporate other sensors. These characteristics also contribute significantly to easy servicing and replacement in the event of damage.

Background

Sounding by static probe first proved popular in Sweden and The Netherlands about 50 years ago. Its exposure to the international site investigation community was extended by such milestone texts as those by Barentsen (1936) and Terzaghi (1953). Since that time the literature has bourgeoned with the descriptions of refinements to the basic model — now universally termed the Dutch cone — and even more so with accounts of its application in soft sediments. Sanglerat (1972) takes the reader through the historical development of the methodology. Other compendia, such as the proceedings of a recent international specialist conference, underline the keen interest which this method claims world-wide. Its application specifically in the marine environment is covered by Semple and Johnston (1980), among many others.

The SPT method was introduced at about the same time. Its mature debut appears to have been about 1927, since when the equipment has undergone little amendment. However, the interpretation of results has come under frequent review, from the early attempts (e.g. Peck, 1953) to derive allowable foundation loads to later summaries of their general interpretation in terms

of settlement by Bazaraa (1967).

Both methods are described in detail by Sanglerat (1972). Both have become established as sensible alternatives to taking samples for later testing in a laboratory wherever shear strength information is required from relatively soft or loose strata. Their advantages over sampling include the avoidance of sample disturbance, acquisition of information from a more continuous profile and a more immediately available comment on site conditions. Their disadvantages crystallize into the single doubt regarding the consistency of their interpretation in terms of soil strength for a range of different lithologies in variable sequence. In addition, the SPT is a dynamic test, which is thought by some to render it forever an empirical tool in the determination of static soil strength properties.

Against this background the use of both methods has grown apace and can be forecast to blossom further. For instance, Sanglerat (1972) commented in relation to the Fugro wireline penetrometer, Wison 90, 'The development . . . is very rapid and improved apparatus can be expected in the near future'. Little wonder, then, that leading international site investigation companies are pressing ahead with the development of a new generation of such devices, concentrating particularly on modifications of the static penetrometer. The methods gain more and more credibility with increasing remoteness of the test site from a base laboratory and with increasing complexity of the mechanical stratification of the soil profile. Both of these conditions are commonly met simultaneously in the rugged investigation of the seabed as well as in the preliminary survey of layered media beneath terrestrial sites.

It is in the light of this industrial history that the present Zetelogic Probe was conceived. Its conceptual design sets out to combine the actions of both the basic static and dynamic devices currently available into one self-contained piece of machinery for insertion into a borehole, thereby eliminating a large part of the support facilities required for the operation of the two separate methods. The realization of this ambition required the shedding of some of the more restrictive parts of the specifications for the standard tests, but this treatment is considered to be more than justified in the interests of producing a more comprehensive and reliably consistent site investigation system which could in any case be calibrated against the individual standard methods if considered appropriate. The outline design of the probe is recounted below, where attention is drawn to the many advanced features which distinguish it from its contemporaries and render it likely to improve considerably the reliability of mechanical data collection when its potential is fully mobilized.

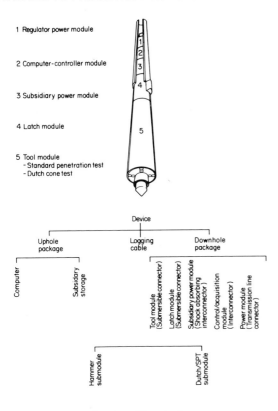

Fig. 1. The device.

The Zetelogic probe

The device (Fig. 1) is a modular system divided into man-size units. It consists of an uphole and downhole package jointed mechanically and electronically by a hauling logging cable. A multiprocessor digital electronic system controls mechanical operations and data acquisition, transmission, storage and display. Communications between the uphole processing/display and storage computer and the downhole tool control and data collecting computer takes place over a paired serial trans-reception line. The electronic package is a general controlling package for a series of tools of which this is the first.

The downhole package

The downhole package. (Fig. 2) comprises five modules joined by five connectors:

(1) *Tool module.* Contains the mechanism of the interactive tools which probe the sediment, sensing the sediment's strength and texture.

(2) *Latch module.* A mechanism for fixing the tool to accommodate any reactive forces generated by the tool. In addition, the latch tests mechanically the sidewall sediments over a broad area.

(3) *Subsidiary power module.* Supplies and controls electrical power either transmitted from the surface or from on-board high-density batteries, to run the interactive tool mechanism.

(4) *Controller/acquisition module.* Contains the downhole controlling and data acquisition electronics which converts multiple channels of analogue data into a single digital channel of transmitted data.

(5) *Power module.* Supplies regulated and conditioned power to the controller/acquisition and signal processing circuits.

Tool module

The tool module incorporates two independent tools — a 5 cm² CPT (Fig. 3a) and a standard penetration test tool (Fig. 3b) — with their individual propulsion mechanism.

The CPT comprises a central supporting axis (containing a wiring conduit); an end cone, which is separated from the axis by a load cell — eight friction sleeves on the initial 152.4 mm of the probe; and a fixed sleeve on the remaining 304.8 mm. This arrangement enables both the measurement of axial load on the probe and also a high-resolution monitoring of the sleeve friction. This reveals the sediment's point strength and the friction resistance envelope under constant velocity penetration. Owing to the spacing of the friction sleeves, a much higher resolution of analysis can be achieved than in present systems.

An electric drive is used to push the cone into the sediment. A clockwise rotation of the motor pushes the cone by means of a screw mechanism, while the motor unit is clamped mechanically to the inside of the container tube by internal latching mechanisms. Optical shaft encoders and a digitally operated motor controller board work in tandem to keep a constant penetration velocity up to 0.02 m/s. With constant velocity, the transducers sense axial cone load/sleeve friction up to a maximum axial load of 10 kN (1 ton) and a maximum penetration of 457.2 mm.

The standard penetration tool is a split-spoon sampler with the addition of a piezoelectric micro-accelerometer. The sampler body consists of an anvil ring, a split cylinder and a shoe ring. The split cylinders assembled together are screwed into the anvil ring in the main tool body and the shoe ring screws the sampler together. An accelerometer is fitted into a cavity in the anvil ring. This senses the instantaneous acceleration and deceleration of the progressing interaction, generating data that quantify dynamic strength parameters and the textural character of the sediment. The SPT is driven over the CPT with a sample depth of 152.4 mm with the CPT in place or 457.2 mm with the CPT removed.

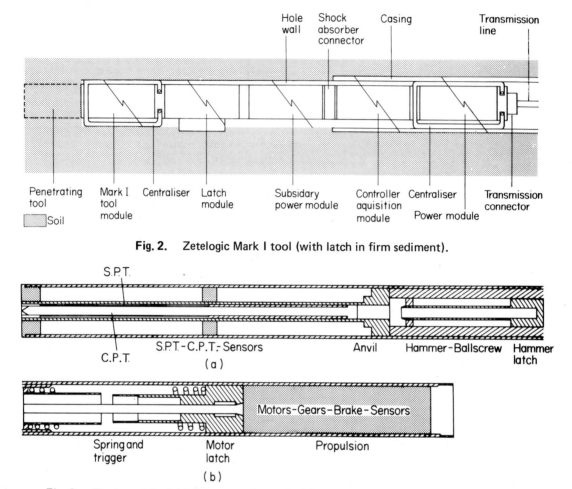

Fig. 2. Zetelogic Mark I tool (with latch in firm sediment).

Fig. 3. Tool module. (a) CPT penetration unit; (b) standard penetration unit (test tool).

The standard condition for the SPT and all critical dimensions are adhered to in this method of propulsion. In the standard test a 63.5 kg mass is dropped 0.76 m accelerated to a terminal velocity of 3.86 m/s by gravity alone. This develops a kinetic energy of 473 N m which is transmitted to the SPT. The Mark I tool augments gravity by storing energy in a spring which is transmitted to a smaller mass moving through a smaller distance but which generates an identical energy. A 26.3 kg mass is propelled 0.25 m accelerated to terminal velocity of 6 m/s by gravity and a spring. This again develops a kinetic energy of 473 N m. These parameters are well within the limits of plastic deformation of the active parts. The sequence of propulsion is:

(1) Hammer, spring, spring compressing motor unit and engaging units are disengaged and allowed to rest on the SPT anvil ring, a ball screw mechanism being used. The optical shaft encoder is used to determine the increment of penetration of the last impact.

(2) Spring compressing motor unit rotating anti-clockwise pulls the hammer 0.25 m back from the SPT, compressing the spring up to 333.5 kP (3270 N). As soon as the motors rotate, the engaging unit latches the compressing motor unit to the inner container wall.

(3) The hammer is released and the SPT driven into the sediment while the sensors are interrogated.

This sequence is repeated for at least 100 cycles or until 0.46 m is achieved.

Multiple signals generated by the sensors are converted into digital form for transmission over one wire (and a common line). This is achieved by use of a multiplexer tree controlled by the downhole computer system. The following control/acquisition functions refer to Fig. 4.

Strain gauge signals. Eight strain gauge signals are selected individually by a multiplexer, the channel selection being controlled by the microcomputer. The signal passes through the processing circuit and through a second multiplexer to the auto ranging circuit.

High-range sensors. Fifteen analogue signals (between

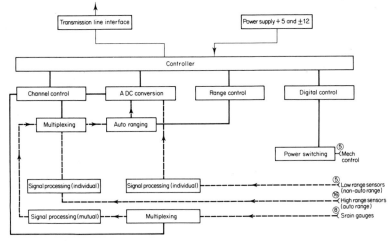

Fig. 4. Downhole package data flow.

− 12 V and + 12 V) are signal processed and selected individually by a dual multiplexer and passed into an auto-ranging circuit.

Auto-ranging circuit. The analogue signals selected by the above multiplexers all have a large range. This circuit analyses the signal in 15 ms and determines whether the signal needs attenuating or amplifying. This results in high efficiency, fast analogue to digital conversion and high sensitivity from the sensor over a large range.

Low-range sensors. Five analogue signals (between − 12 V and + 12 V) are fixed signal processed, resulting in an output to analogue to digital conversion of 0–2.5 V.

Analogue to digital conversion. The microcomputer selects the channel for conversion and using the internal timer converts the analogue voltage into a 16 bit digital word.

Transmission. The 16 bit word, auto-ranging factor and channel number are transmitted to the uphole package in a preformatted phase.

The downhole computer contains all the control and data handling routines on 4K bytes of ROM; of the 128 bytes of RAM, 80 bytes can be used for on-site user-defined program routines which can be incorporated if required. Apart from the transmission line and common line, the communication link has a reception line for the uphole computer to control the downhole computer's program control, a power-up/interrupt control line and a reset line.

Uphole package

The computer system at the surface is a specialized M68000-based package (Fig. 5) used to generate raw data graphical displays or empirically derived strength and texture graphical displays from sensor output. This is achieved with the minimum of operator interaction. All instructions are prompted and answered using single inputs. It is a software driven system.

The data generated and transmitted to the surface are decoded and stored in the buffer RAM in the uphole system. This is downloaded at high rates on to disc for computation on the site or in the laboratory.

Discussion

As has been illustrated, this probe combines several novel features in seeking to introduce a new generation of site investigation facilities. The advantages of its modular construction have already been touched upon. However, in addition, it should be noted that they include easy transport and assembly by one operative as well as simple maintenance and replacement. It is intended to devise additional instrumented working heads for the probe to serve the further requirements of the site investigation industry as it responds to the knowledge that this modern technology can incorporate such a comprehensive range of test facilities into a compact package. Additions or alternatives for other geotechnical and geological applications will be compatible with the basic power, drive, control and data retrieval modules and can be forseen to utilize much of the basic computing hardware merely with the replacement of the software package — also in modular form.

The interactive capability of the system between downhole and uphole microcomputers, with or without a user interface, renders the application of the tool very flexible. It introduces the possibility of adjusting the test programme according to the conditions met during the early stages of penetration if required, either by reference to a predetermined range of optional

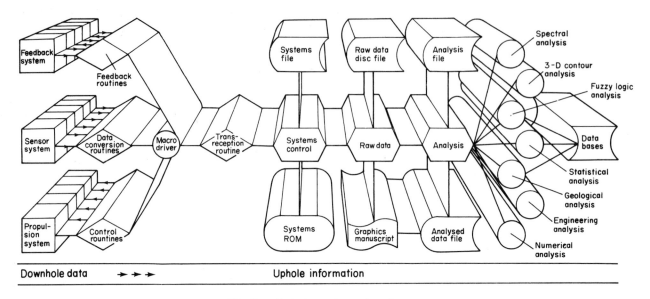

Fig. 5. Uphole package data flow.

responses to a particular event or at the instruction of the operative. A measure of selectivity is thus available in activating either or both of the probe components in sequence. This facility will be extended to further sensing elements as they become available.

The entire package is robust and simple to use. Its software has been designed to be very 'user-friendly'. There are very few external moving parts on the probe. The uphole monitor requires very few instructions and all of these are presented to the operative as options on the LCD screen so that prompts are given as each new operating stage is met. It is to be hoped that such simplicity will be recognized for what it is — an essential gesture to ease operation and optimize the test programme and procedure. By so doing, it brings the operation of the device into the domain of the site engineer or geologist as well as the specialist technician. Consequently it gives a more informed interaction between downhole test site and these responsible individuals than may often by the case when *in situ* test methods of more complex operational procedure and little or no real interactive capability are used.

Returning momentarily to technical matters, it has been shown that the CPT style penetrometer is of 5 cm^2 projected area, the lesser of the two options available from more conventional commercial models. In turn, the shaft consists of eight separate friction recording segments, allowing the compilation of the frictional resistance component acquired by the conventional sleeve and yet also providing for greater sensitivity of interpretation up the shaft, rendering it far more suitable for the recognition of thinly layered strata, a common phenomenon. In future this friction

sleeve will be constructed of multiple segments, using matrix technology, for a higher resolution of the friction envelope. Collectively, then, the features of this static probe can simultaneously repeat the 'standard' test and yet go beyond it to a higher sensitivity of mechanical strata differentiation, albeit with a restricted penetration of 457.2 mm.

The dynamic penetrometer also deserves special comment. It will have been recognized that it is not possible to contain within a machine of the compact size of the Zetelogic Probe a mechanism to deliver the same momentum of the 'standard' test (SPT). Here the downhole device sacrifices the momentum criterion but retains the energy equivalence of blow delivery available in the standard test. As the interpretation of the standard test is itself an empirical matter, there is, of course, nothing absolute about the definition of its design parameters. The derivation of mechanical soil properties from the dynamic probe is consequently merely a matter of different calibration but nevertheless equally valid. In fact, it would be easy to argue that this calibration is far more reliably and easily achieved in uniform soils with the Probe, as it has the facility to perform the static test immediately above the dynamic test in the same hole. Consequently the new dynamic probe can be calibrated to simulate the output from the standard test if required, but, much more importantly, can be more readily related to the soil parameters determined by the static CPT. As does the static probe, this new dynamic penetrometer goes well beyond its standard equivalent in increasing sensitivity of interpretation; this is achieved by allowing the identification of fine layering from the interpretation of the deceleration of the sampler at each separate blow. Whether this interpretation need remain forever empirical or will succumb to similar analysis to that sometimes directed towards the ambitious treatment of dynamic probes (e.g. Young, 1970) remains to be seen. In any case it already contains an extensive package of sensitive information not available from the standard model.

With such an extensive array of data from a single tool it is important to recognize the versatility and comprehensibility of the data handling system. It will record all the raw data from each of the many feeder channels at any time (or penetration interval, for all practical purposes). It will not only store and return this information in digital form, but also repeat the singular raw data in colour-graphic hard copy virtually instantaneously. This provides the assurance that is required by the operative and the client's representative that basic information is being collected. Much of this can be used directly by experienced interpreters in the conventional way. However, this probe can go much further and permit on-line mechanical strata differentiation based on the interpretation of the combined data by a statistical package down-loaded into the working memory of the uphole system. This simultaneously volunteers an interpretation service to the novice, especially by reference to its compound memory, and also encourages the expert to attempt more comprehensive mechanical data resolution.

Conclusion

A new downhole testing device has been introduced. With minimal significant deviation from standard specifications it incorporates the facilities of the conventional static Dutch cone and dynamic standard penetration test. In both cases it can simultaneously mimic the standard test and yet permit much additional information of higher sensitivity to be recorded. Its control and data handling systems are responsive to virtually real-time interaction and interpretation. Raw data retrieval is available and can be presented directly or in reduced mode, either digitally or graphically. Its advanced specification coupled with robust design and simplicity of operation are considered to identify the Zetelogic Probe as a runner in the general drive to upgrade *in situ* geotechnical equipment and procedures.

References

Bazaraa, A. R. S., 1967. Use of the SPT for Estimating Settlements of Shallow Foundations on Sand. Thesis, University of Illinois.

Barentsen, P., 1936. Short description of a field testing method with cone-shaped sounding apparatus. *Proc. 1st Int. Conf. Soil Mechanics and Foundation Engineering, Cambridge*, Vol. 1 (B/3) pp. 7–10.

Peck, R. B., 1953. General report, soil properties — field investigations. *Proc. 2nd. Panamerican Conf. Soil Mechanics and Foundation Engineering, Brazil*, Vol. 2, pp. 449–455.

Sanglerat, G., 1972. *The Penetrometer and Soil Exploration Developments in Geotechnical Engineering*. Vol. 1, Elsevier, Amsterdam.

Semple, R. M. and Johnston, J. W., 1980. Performance of Stingray in soil sampling and in-situ testing. In: D. A. Ardus (ed.), *Offshore Site Investigation*. Graham and Trotman, London, pp. 169–182.

Terzaghi, K., 1953. Static cone penetration tests. *Proc. 3rd Int. Conf. Soil Mechancis and Foundation Engineering, Zurich*, Vol. 3, pp. 229–231.

Young, C. W., 1970. Terradynamic development of a low velocity earth-penetrating projectile. Sandia Laboratory Report SC-DR-70-302, Sandia Corporation, Albuquerque, New Mexico.

8

A New Technique to Identify Sea Sediments in Shallows

Masaharu Fukue, Shigeyasu Okusa and Takaaki Nakamura
Department of Ocean Civil Engineering, Tokai University, Orido, Japan

A simple technique for fast prediction of the grain size characteristics of sea sediments is developed. The device used is a small sled hinged to a detector which is made of steel blade. When the sled is dragged by a ship, the free end of the detector slides on the seafloor. In principle, the detector vibrates at different rates during sliding according to the type of soil encountered. In this study the oscillation characteristics shown by the vibration of the detector were recorded by an oscillator through a bridge amplifier, and were then correlated with the grain size of the identified soils.

Introduction

The identification and classification of seabed sediments has long been one of the main objectives of marine geology, sediment lithology, morphology and marine soil engineering. The conventional technique for the identification of sediments employs sampling and soil testing, while current researches into predicting the physical properties of sediments, such as grain size, bulk density and sediment type, utilize acoustic techniques (Baldwin *et al.,* 1981; Addy *et al.,* 1982). However, the acoustic techniques are based on the qualitative reflection type of echo sounder and difficulty in the quantitative analysis is inevitable.

At present, each type of technique may have both merits and demerits, in regard to simplicity in operation and in analysis, and from the economic point of view and that of precision of data. Therefore a specific technique or combination of two or more techniques should be used for the identification of sediments, and other techniques should be developed for the requirements imposed by sea sediment investigation.

The constraints at present relevant to sea soil investigation can be viewed in terms of: (a) sea environment conditions, e.g. wave, wind, other rigorous sea conditions, etc; (b) utilization of essential large equipment, such as scaffold, platform or ship; (c) development of a quick and simple technique with the minimum

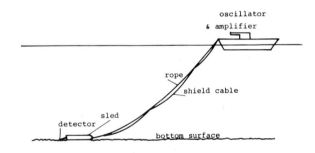

Fig. 1. Illustration of general technique for identification of surface sediments.

expenditure of investigative time, consistent with the constraints identified in (a) and (b).

This study attempts to develop and examine a new technique.

Technique and principle

The general technique developed in this study is illustrated in Fig. 1. In brief, a small sled with a detector is dragged by a ship; the detector, which is made of an elastic thin steel blade, vibrates as a result of the roughness of the grain surface. The oscillation waves produced by the vibration of the detector are recorded by an oscillator through a bridge amplifier by means of a

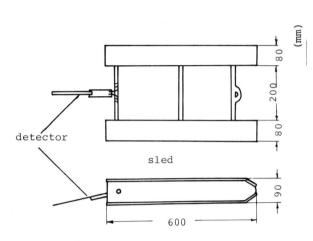

Fig. 2.   Small marine sled used in study.

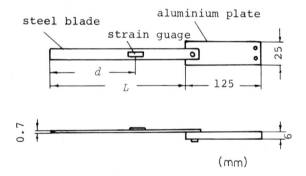

Fig. 3.   Typical example of detector used in study.

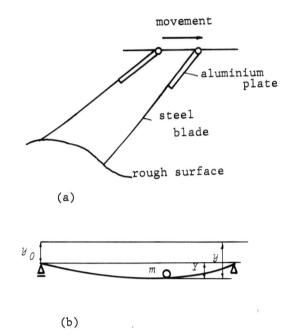

Fig. 4.   Vibration of elastic beam under a forced displacement: (a) moving detector; (b) vibrating simple beam.

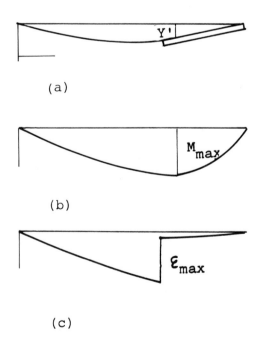

Fig. 5.   Elastic deflection (a), bending moment (b), and axial strain (c) experienced by detector.

strain gauge. Therefore, if the oscillation wave can significantly be correlated with the grain size, it would be possible to identify sediments from the correlation. The sled used is illustrated in Fig. 2. Its mass is 16 kg, i.e. a weight of 156.8 N, in the air. A typical detector is shown in Fig. 3. The steel blade is bolted to an aluminium plate which acts as a weight. The aluminium plate is hinged to the sled as illustrated in Fig. 2.

As far as the detector is concerned, it is clear that for the purpose of this study, it is necessary to examine the vibration or deflection of the detector under the condition illustrated in Fig. 4(a), in which a detector whose upper end is hinged is moving horizontally. The lower end is free and is moving up and down with a certain horizontal speed.

Although the situation may be slightly different, the

problem may be likened to that of vibrations produced in a simple beam under a relative displacement y_0 at one end of the beam (Fig. 4b).

For the simplest case, if we assume that $y_0 = a_0 \cos(pt)$, the differential equation of motion, without viscous damping, is

$$m\ddot{y} = -k(y - y_0) \qquad (1)$$

Therefore,

$$\ddot{y} + q^2 y = q^2 y_0 \qquad (2)$$

where $q = \sqrt{k/m}$, k is the spring constant and m is the mass, as illustrated in Fig. 4(b).

Considering steady state vibration, the solution of equation (2) is

$$y = a \cos(pt) \qquad (3)$$

Differentiating equation (3),

$$\dot{y} = -ap \sin(pt)$$
$$\ddot{y} = -ap^2 \cos(pt) = -p^2 y \qquad (4)$$

From equations (2) and (4) we obtain

$$y = (q^2 y_0)/(q^2 - p^2) = \left\{ q^2 a_0 \cos(pt) \right\}/(q^2 - p^2)$$

Therefore the elastic deflection is

$$Y = y - y_0 = (s-1)y_0 = (s-1)a_0 \cos(pt) \qquad (5)$$

where $s = q^2/(q^2 - p^2)$ and Y is the elastic deflection of the beam. If we assume that an amplitude a_0 is proportional to grain size, D (i.e. $a_0 = cD$), equation (5) may be rewritted as

$$Y = (s-1)cD \cos(pt) \qquad (6)$$

Then the maximum deflection is given by

$$Y_{\max} = \pm(s-1)cD \qquad (7)$$

Thus, in the simplest case, the maximum elastic deflection of the beam depends on a value of s, on parameter c and on grain size, D. Note that the resonance condition occurs when $q = p$.

However, in the actual detector the problem may be complicated by viscosity, jumping of the detector and the discontinuity of flexural rigidity of the detector. Now we consider qualitatively the bending of the detector under a virtual elastic deflection Y', as shown in Fig. 5(a). With a knowledge of applied mechanics, we can obtain the modes of bending moment (Fig. 5b) and the corresponding axial strain (Fig. 5c), under a given elastic deflection. The discontinuity of the axial strain shown in Fig. 5(c) arises from the difference in rigidity between the steel blade and the aluminium plate, as axial strain is proportional to bending moment but is in inverse proportion to flexural rigidity.

Thus, we might expect that from equation (6) and

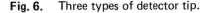

(not to scale)

A-Type   B-Type   (mm)   C-Type

Fig. 6. Three types of detector tip.

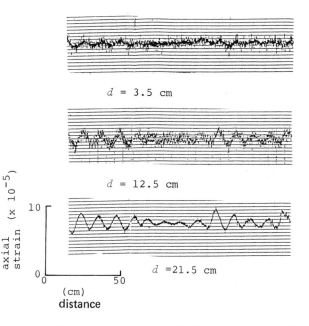

Fig. 7. Typical example of oscillation waves recorded by B-type detector with a length of 22.5 cm (towing speed = $50 \, \text{cm s}^{-1}$; $D_{10} = 2.85$ mm, $D_{60} = 3.98$ mm).

Fig. 5(c) the axial strain produced in the detector during sliding is a function at least of (a) roughness of surface (i.e. grain size), (b) sliding speed, (c) type of detector and (d) position of strain gauge.

Preliminary experiment

In order to examine the vibrational characteristics of steel blades, detectors of many types, with regard to size and tip shape, were used. Six sizes of detector and three types of tip shape were used. The tip shapes used are illustrated in Fig. 6; the detectors using these tip shapes will be referred to as A-type, B-type and C-type, respectively.

In the laboratory study the detectors without the sled

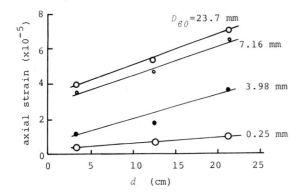

Fig. 8. Relationships for grain size D_{60} between axial strain and distance from free end to strain gauge, obtained by A-type detector with a length of 22.5 cm.

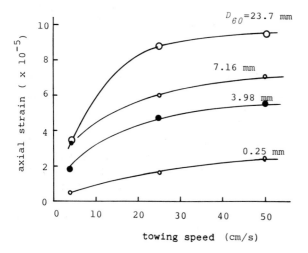

Fig. 9. Relationships between axial strain and towing speed for various types of soil, obtained by B-type detector (L = 15.4 cm; d = 14.5 cm).

Were slid on identified soils, such as fine- and coarse-grained sands and fine- and coarse-grained gravels. The soils sampled were obtained from the Miho coast in Shimizu, Japan.

Figure 7 shows a typical example of oscillation waves, in which d denotes the distance from the free end to the attached position of the strain gauge (see Fig. 3). Figure 7 shows the same frequency pattern but different amplitudes for different ds.

Figure 8 shows the relationships between axial strain and d in an A-type detector, where the axial strain is taken as a frequent maximum amplitude of oscillation waves. It is noted that a little change in water temperature does not affect the response characteristics of the strain gauge. In addition, the calibration can readily be done by the amplifier.

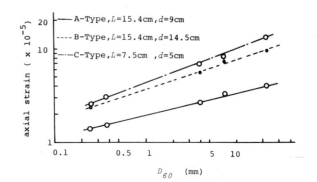

Fig. 10. Linear relationships between axial strain and D_{60}, obtained with various types of detector.

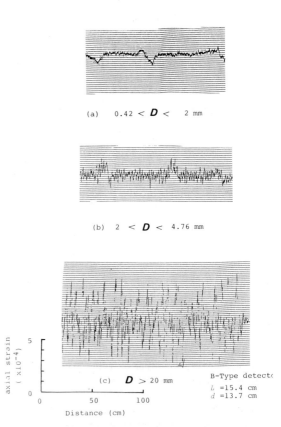

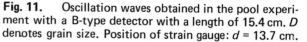

Fig. 11. Oscillation waves obtained in the pool experiment with a B-type detector with a length of 15.4 cm. D denotes grain size. Position of strain gauge: d = 13.7 cm.

In Figs. 7 and 8 it is seen that the greater the d the greater the axial strain, and that axial strain is a strong function of grain size. Figure 8 also shows that the greater the grain size D_{60} (i.e. 60% grain size in the distribution curve) the greater the axial strain for any d. It should also be stated that the same trends were

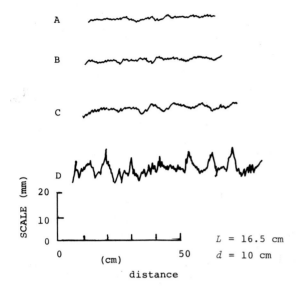

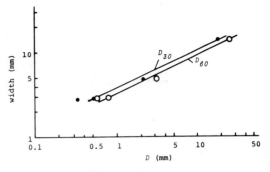

Fig. 14. Relationship between maximum width of envelope of oscillation waves and grain sizes D_{30} and D_{60}.

Thus the axial strain obtained from the oscillation waves is a function of d, D, sliding speed and type of detector. If a speed of more than 40 cm/s is used, the speed effect may be neglected.

Pool experiment

A model experiment using the sled with detectors was performed in an artificial pool. For this experiment five types of soil were used as sediments.

Figure 11 shows typical oscillation waves obtained in the pool experiment, which are a strong function of grain size. This great variation in oscillation waves with grain size characteristics can be significantly used for the identification of sediments. Note that this pool experiment is conducted under basically the same conditions as field investigation.

Field application

The technique described above was studied in Seto Inland Sea, which is located in the western part of Japan and is surrounded by Honsyu, Kyushu and Shikoku Islands. The sled was dragged by a small boat at a speed of 50 cm/s i.e. 1 knot. The detector used was of the A-type; it was 16.5 cm long and the distance d was 10 cm. This A-type detector is not sensitive, in comparison with the B or C types. Therefore, if the B- or C-type detector is used, higher sensitivity can be obtained. If a higher dragging speed is required, some weight should be attached to the aluminium plate of the detector, or the heavier plate should be used. However, the maximum speed is limited by the responses of the detector.

The oscillation waves obtained in the field are shown in Fig. 12, and the grain size distribution curves corresponding to the waves A, B, C and D are presented in Fig. 13.

Figure 14 shows the relationship between the width

Fig. 12. Oscillation waves recorded using an A-type detector in Seto Inland Sea (towing speed = 50 cm s^{-1}).

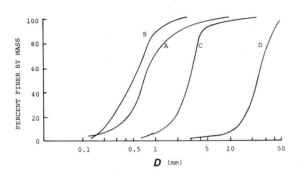

Fig. 13. Grain size distribution curves of samples from Seto Inland Sea.

observed when other detectors and other soil types were studied, and these trends qualitatively agree with the theoretical approach. There are many possibilities for grain size in the relationship between axial strain and grain size. In this case the D_{60} was tentatively chosen.

Figure 9 shows the relationships between the axial strain and the sliding speed of the detector. The axial strain is seen to be a function of speed for the lower range, but it tends to an almost asymptotic value for the higher range.

Figure 10 shows typical relationships between the axial strain and 60% grain size for the various types of detector. It is apparent that the axial strain increases linearly with increase in grain size, in log-log scale. This means that the assumption made in equation (6) may be correct. Note that the same trends were observed when other types of detectors were used.

of the oscillation waves recorded on the recorder chart and grain size. The trend shown in this figure is very similar to the trends described earlier, as the techniques used are the same. It should be noted that little variation of oscillation waves in the smaller-grained soils is due to the disturbance of sediments. This disturbance is caused by the movement of the detector during dragging of the sled. Therefore the more sensitive detector is required for fine-grained soils.

Thus, from the preliminary, pool and field experiments, it is apparent that the oscillation characteristics of the detector reflect the grain size of sediments. As far as granular soils are concerned, the technique developed in this study is applicable and has a future.

Conclusion

A good degree of correlation is found between the grain size of sediments and the axial strain produced in the steel blade detector. The technique is very simple and may satisfy the requirements to a certain degree. In addition, it may be possible that a better analysis may be achieved by some other technique — e.g. spectrum analysis, etc.

Acknowledgements

The authors wish to acknowledge the stimulus afforded by discussion with Professors A. Saito and S. Kosuge, Tokai University.

References

Addy, S. K., Behrens, E. W., Haines, T. R., Shirley, D. J. and Worzel, J. L., 1982. High-frequency subbottom reflection types and lithologic and physical properties of sediments. *Mar. Geotech.*, 5 (1), 27−49.

Baldwin, K. C., Celikkol, B. and Silva, A. J., 1981. Marine sediment acoustic measurement system. *Ocean Engng*, 8 (5), 481−488.

Part II

Sub-seabed Site Investigation and Design

Section 3 — Probes, Corers and Pore Pressure Measurement

Introduction

A. W. Malone (Chairman)

Engineering Development Department, Government Geotechnical Control Office, Hong Kong

Vibrocorers in geotechnical investigations

The first marine use of vibrated corers is attributed by Rosfelder and Marshall (1966) to Kudinov in the 1950s and in recent years a range of bottom standing vibro-coring systems have become commercially available. Perlow and Dette (1983, personal communication) draw attention to the potential of the vibrocorer as both sampler and penetrometer for shallow offshore foundation investigations.

The commercially available vibrocorers are designed to take sand samples 5–6 cm long, essentially for descriptive purposes, sample recovery being improved with the aid of a piston in the liner tube; most can operate successfully in water depths of up to about 100 m. Clay sampling is less successful, though soft clay adaptions are available. Vibrocorer advance relies upon liquefaction and some commercial types incorporate water cutting jets in addition. The resulting structural disturbance may severely modify the mechanical properties of sediments.

Most commercially available vibrocorers are driven under nominally constant energy, and penetration recording devices are commonly attached or available. Hence there is clearly scope for making crude resistance measurements. However, local correlations between vibrocorer penetration rate and the standardised penetration tests, whilst interesting on specific sites, would have limited general application. Unlike the cone penetration test the vibrocorer penetration rate is not found to be especially sensitive to density variations and minor lithological changes as Perlow and Dette's figures show.

The commercially available vibrocorer systems are incapable of penetrating dense coarse granular materials and stiff clays. Penetration capability can be greatly extended by subjecting the corer to repeated impact as well as vibration, as shown by Menard and Gamblin (1965), and impact vibrocorers are now being developed in which impact begins automatically as refusal is approached (A. A. Rodger, personal communication, 1983).

Cone penetration test

In the use of the widely accepted cone penetration test the engineering designer relies on tried and proven empirical correlations. The theories of cone penetration resistance are still rudimentary having apparently progressed only as far as adequately coping with the case of surface indentation, according to Houlsby and Wroth (1983).

Transient and residual pore pressures in seabed deposits

An analysis of the transient flow in a permeable bed which results from pressure wave action on the fluid-bed interface was undertaken by Yamamoto (1977) and Madsen (1978) by adopting Biot's 3-dimensional consolidation equation which takes into account compressibility of the soil skeleton and of the pore fluid.

There appears to be limited experimental verification of the Yamamoto—Madsen approach in the free-field case based on successful prediction of measured transient changes in pore pressures in bed sands under wave action (Finn *et al.*, 1983). The paper by Okusa, Nakamura and Fukue (1983) presents preliminary results of experiments to measure pore pressures in gravel beds, to which the Yamamoto—Madsen approach is inapplicable (Madsen, 1978).

High cyclic stresses due to wave loadings can cause the progressive build-up of pore pressure during loading by a series of waves. Calculation procedures for wave-induced pore pressure build-up using a modified earthquake liquefaction approach have been suggested by Seed and Rahman (1977), and extended by Finn *et al.* (1983) to include progressive soil softening. The method relies on measurements of pore pressure build-up in laboratory specimens under cyclic stresses and this is a serious inherent defect because of the problems of sample disturbance. No experimental field verification exists.

Measurement of ambient excess pore pressures

Ambient excess pore pressures may arise naturally from wave loading, rapid rates of sedimentation and biogenic gas generation. A site investigation for many marine engineering purposes will be incomplete in geological situations conducive to the existence of ambient excess pore pressures unless measurements are made. A number of techniques have been used for seabed pore pressure measurements, as illustrated by the following three representative cases.

Richards *et al.* (1975) report an early successful attempt to measure soil pore water pressures from a survey ship in 274 m deep water in the recent soft clay bed of the Wilkinson Basin, 125 km east of Boston, USA in 1967. An ambient excess pore water pressure measurement of $10 \, \text{kN m}^{-2}$ was made by a differential piezometer probe emplaced by free fall terminating 4 metres below mud line.

In the intervening years, the most notable results are from the USGS SEASWAB (Shallow Experiment to Assess Storm Wave Affects on the Bottom) investigation between 1975 and 1977 in the East Bay of the Mississippi River delta, described by Dunlap *et al.* (1978). Various types of piezometer were pushed under dead weight into a very recent well-laminated, bioturbated, gassy silty clay at various depths below mud line to a maximum depth of about 19 m. Ambient excess pore pressures were generally substantial, giving effective stresses close to zero. High initial pore pressure values were measured in a deformational feature (a depression seen on side-scan sonar records and thought to be a collapse structure) and wave activity during winter storms an summer hurricanes caused significant movement with increased pore pressures.

Ambient excess pore water pressures have also been measured in unlithified deposits at greater depths by a wireline borehole piezometer and one such exercise is described by Hooper and Preslan (1979). The measurements were made at about 72 m below bed in the top of a sand formation during geotechnical investigations at a site south of the South West Pass of the Mississippi River delta where water depth was about 60 m. The sand formation is an extensive medium dense fine siliceous sand, silty at the top and up to 24 m thick; it underlies a sequence of pro-delta and shelf clays and very recent mudslide soils which increase in thickness over the lateral extent of the sand formation away from the coast from about 46 to 76 m. The underlying deposits are shelf clays. The sand contains a high concentration of methane gas. A 200-mm diameter borehole was advanced some 3 m into the sand formation and the wireline piezometer was pushed into the soil below the base. The excess pore pressure stabilised to about $116 \, \text{kN m}^{-2}$ some 5 min. after placement. The cause of the ambient excess pore pressures is postulated as rapid loading from the very recent mudslide deposits, though biogenic gas pressure may explain about one third of the excess.

Pore pressure measurement by the piezo-cone penetrometer

The cone penetration test (CPT) has become a conventional land site investigation test in the last two decades and in recent years there has been a rapidly growing interest in the piezo cone penetrometer (a cone penetrometer which also houses a piezometer), although it is over 15 years since the first trials took place. Conventional use involves the recording of the dynamic pore pressures generated during constant penetration and measurements of the dissipation of pore pressures when the piezo-cone is temporarily halted, the latter enabling calculation of coefficient of horizontal consolidation as described by Tortensson (1975). Ambient pore pressures may be measured when excess pore pressure generated by penetration have dissipated, as illustrated, for example, by Battaglio *et al.* (1981) who recorded ambient pore pressures in excess of hydrostatic values in a 24 m thick consolidating bed of soft organic clay in Italy, and by Campanella and Robertson (1981), Jones *et al.* (1981) and Sugarawa and Chikaraichi (1982) who all illustrate the use of the piezo-cone for measuring non-hydrostatic flow regimes in tailings deposits in Canada, South Africa and Japan respectively. The following remarks concentrate on this application.

Standards exist now for the CPT but lack of standardisation of the piezo cone, especially as regards the cone angle and location of the porous element, makes it difficult to compare dynamic measures taken by different instruments. It seems probable that a shaft-mounted porous element will be most suitable for ambient pore pressure measurement.

In the use of the piezo-cone on land it has been found to be highly sensitive to lithological variations, hence its unusual merit in enabling *in situ* pore pressure measurements to be made in chosen 'lithological targets', for example in very thin sand lenses.

One of the first major applications offshore of piezo-cone testing was the site investigation for Hong Kong's replacement airport. During the winter of 1981/82 about 90 piezo-cone tests were made in a sequence of Late Pleistocene and Holocene deposits (consisting of silty clay, clayey sand and silty sand with gravel and cobbles) at a site in water depths of 4–5 m located just offshore of Lantao Island, which defines the eastern margin of the Pearl River estuary. The tests successfully measured the coefficients of horizontal consolidation of these sediments but conclusions cannot be drawn regarding ambient excess pore pressures in the clays (geologically unlikely) because dissipation tests were conventionally terminated before equilibrium pore water pressure was reached. Wave-induced transient pore pressures are evident on the dissipation curves for test locations to a depth of several metres in the soft bed clays.

It should be noted that in order to make such dynamic measurements with accuracy, high frequency response must be ensured and careful field experimental procedures must be followed. Problems of unwanted gas intrusion into the measuring system are especially acute in gassy bed soils, as illustrated by the SEASWAB case histories. Sills and Nageswaran (1984) suggest that it is not enough to employ a high entry porous element and ensure a stiff measuring system, because gas bridging across narrow passages between porous element and transducer can cause continuous gas to exist from soil to transducer with the result that soil pore gas pressure is recorded. Careful design of the piezometer element can however ensure reliable measurements of gas and water pressures.

The CPT has now become a normal part of offshore investigation in the North Sea and most of the other major offshore oil and gas fields of the world. It is clear that the piezo cone has great potential in offshore investigations.

References

Battaglio, M., Jamiolkowski, M., Lancellotta, R. and Maniscalco, R. 1981. Piezometer Probe test in cohesive deposits. In: G. M. Norris and R. D. Holtz (Eds.), *Proc. Session on Cone Penetration Testing and Experience*, ASCE National Convention, St. Louis, Missouri, pp. 264–302.

Campanella, R. G. and Robertson, P. K. 1981. Applied cone research. In: G. M. Norris and R. D. Holtz, *Proc. Session Cone Penetration Testing and Experience*, ASCE National Convention, St. Louis, Missouri. pp. 343–362.

Dunlap, W. A., Bryant, W. R., Bennett, R. H. and Richards, A. F., 1978. Pore pressure measurements in underconsolidated sediments. Pap. No. 3168, *Proc. Offshore Technology Conf.*, Houston.

Finn, W. D. L., Siddharthan, R. and Martin, G. R. 1983. Response of seafloor to ocean waves. *J. Geotech. Eng. Div. ASCE* **109** (4) 556–572.

Jones, G. A., van Zyl, D. and Rust, E. 1981. Mine tailings characterisation by piezometer cone. In: G. M. Morris and R. D. Holtz, *Proc. Session Cone Penetration Testing and Experience*, ASCE National Convention, St. Louis, Missouri, pp. 303–324.

Hooper, J. R. and Preslan, W. L. 1979. Pressurized shallow sands in the Mississippi Delta region. Pap. No. 3483, *Proc. Offshore Technology Conf.*, Houston.

Houlsby, G. T. and Wroth, C. P. 1984. Calculation of stresses on shallow penetrometers and footings. This volume, pp. 107–112.

Madsen, O. S. 1978. Wave-induced pore pressures and effective stresses in a porous bed. *Geotechnique* **28** (4) 377–393.

Menard, L. and Gambin, M. 1965. Application du vibro-foncage hydraulique aux sondages sous-marins. *Sols Soils* **14**, 24–35.

Okusa, S., Nakamura, T. and Fukue, M. 1984. Measurements of wave-induced pore pressure and coefficents of permeability of submarine sediments during reversing flow. This volume, pp. 113–122.

Richards, A. F., Øien, K., Keller, G. H. and Lai, J. Y. 1975. Differential piezometer probe for an *in situ* measurement of sea-floor pore-pressure. *Geotechnique* **25** (2) 229–238.

Rosfelder, A. M. and Marshall, N. F. 1967. Obtaining large, undisturbed, and orientated samples in deep water. In: A. F. Richards (ed.), *Marine Geotechniques* University of Illinois Press, pp. 243–263.

Seed, H. B. and Rahman, M. S. 1977. *Analysis for wave-induced liquefaction in relation to ocean floor stability*. Rep. No. UCB/TE-77/02 Univ. California, Berkeley, CA.

Sills, G. C. and Nageswaran, S. 1984. The preparation and characteristics of reconstituted soil containing undissolved gas bubbles. (in press).

Sugawara, N. and Chikaraishi, M. 1982. On estimation of ϕ' for normally consolidated mine tailings by using the pore pressure cone penetrometer. *Proc. 2nd Europ. Symp. Penetration Testing*, Amsterdam, pp. 883–888.

Tortensson, B. A. 1975. Pore pressure sounding instrument. *Proc. Conf. In situ Measurement of Soil Properties*, ASCE, Raleigh, North Carolina, Vol. 2 pp. 48–54.

Yamamoto, T. 1977. Wave-induced instability in seabed. *Proc. ASCE Spec. Conf. Coastal Sediments*, pp. 898–913.

9

Calculation of Stresses on Shallow Penetrometers and Footings

G. T. Houlsby and C. P. Wroth
Department of Engineering Science, Oxford University, UK

A short review is made of the types of analyses available for cone penetration tests. The assumptions involved in analyses using the method of characteristics (slip line methods) are discussed, and some analyses of problems of shallow cone penetration and bearing capacity are presented. No details of the mathematics of the technique are given, but reference is made to sources where these details are available. The cone penetration calculations have application both to soil testing and to the analysis of spudcan foundations. Graphical results are presented for the variation of the cone factor, N_c, with the angle and roughness of a shallowly penetrating cone. The second example of the application of the method is to problems of bearing capacity on soils with strength increasing with depth, and is relevant to the design of very large concrete gravity oil-production platforms. Numerical values of bearing capacity factors for smooth and rough circular footings on non-homogeneous soils are presented and compared with the equivalent values for strip footings.

Introduction

Penetrometer devices are being used increasingly in the investigation of the engineering properties of soils in both the laboratory and the field. The fall cone test is now used in Britain as the standard laboratory test to determine the liquid limit, and has been used for some time as a simple strength test in Scandinavia. More importantly, the cone penetration test (CPT) is now a well-established site investigation technique, and is particularly well suited to seabed investigation because of its simplicity, robustness and capability for remote logging.

All penetration tests are recognized as being essentially strength tests, but the back analysis of the tests to provide soil strength values is not straightforward. From this point of view, the penetration test may be contrasted with, for instance, the vane test, for which only a very simple analysis is required, and the pressure meter test, for which certain reasonable assumptions allow a complete analysis in terms of soil strength and stiffness values. However, at present penetration tests are almost universally interpreted by using empirical correlation factors, with only moderate degrees of success. For instance, even in the correlation of cone pressures with the undrained shear strength, a variation of the N_k factor from below 10 to over 30 is reported by many authors (e.g. Jamiolkowski *et al.*, 1982; Nash and Duffin, 1982; O'Riordan *et al.*, 1982), with no definite indication of how N_k correlates with other soil properties.

The analysis of cone penetration problems is also of interest in several other areas of geotechnics. Analysis of deep penetration is relevant to problems of piling and also to possible methods for the sub-seabed disposal of high-level nuclear waste using penetrators. Shallow penetration of the seabed by conical footings is relevant to the design of spudcan foundations. The term 'shallow' means here that only the conical part of a penetrator is in contact with soil (Fig. 1a), as opposed to deep penetration (Fig. 1b).

Various attempts have been made to analyse cone penetration problems, and four main categories of analysis are currently available.

(1) Cavity expansion analysis, in which the advancing cone is modelled approximately by the expansion of a cylindrical or spherical cavity. This approximate solution leads to cone factors depending on the ratio between strength and stiffness of the soil.

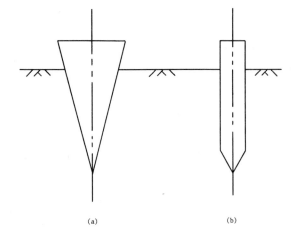

Fig. 1. (a) Shallow penetration and (b) deep penetration of cone penetrometer.

(2) Slip line methods, in which the equations of the plastically deforming soil are integrated along characteristic directions (slip lines) to obtain cone loads. The method is successful for shallow cones but involves difficulties for deep penetration problems.

(3) Finite element analysis has many advantages, but the combination of large deformations during plastic flow under conditions of axial symmetry poses a very severe test for the accuracy of the technique.

(4) Strain path methods, in which a deformation field is assumed and the constitutive equations integrated along streamlines to give cone loads. The method has the disadvantage that it implies a set of stresses which are not necessarily in equilibrium.

In this paper some analyses of type (2) are presented. At present it has not been possible to achieve a complete analysis of a deeply penetrating cone by this method, since it has not yet proved possible to obtain a kinematically admissible deformation field for this problem as well as a stress field which is in equilibrium. Some approximate solutions have, however, been obtained (Houlsby and Wroth, 1982a). It is anticipated that the analysis of such problems may best be achieved by a combination of slip line analysis of the region adjacent to the cone and cavity expansion analysis of the surrounding region of soil. An approximate analysis of this sort has been made by Last (1982) for plane strain wedge indentation, but no results are yet available for a cone.

For shallow penetration problems (i.e. when only the conical part of a penetrometer is buried, as in the fall cone test rather than the CPT), the method of characteristics (or slip line analysis) can give an exact solution for the loads on a penetrometer. The application of the method will therefore be illustrated by

this simpler problem. Note that the analysis of very blunt shallow cones becomes the same as that of the bearing capacity problem, and many of the results presented here are more applicable to the capacity of spudcan foundations than to cone penetration tests.

A second area of interest in problems of seabed geotechnics where the method of characteristics can usefully be applied is the analysis of footings on non-homogeneous materials. The undrained strength of seabed soils can often be fitted as a linear increase of strength with depth, and this increase of strength may have an important significance for the design of footings on the seabed. Some results of calculations for the bearing capacity of plane strain and circular footings on non-homogeneous materials are presented.

Method of analysis

The method of integration of the equations of a plastically deforming material along characteristics (or slip lines) is well developed for plane strain problems in metal plasticity, where the material is usually idealized as rigid-plastic with strength given by the Tresca condition (which is identical with the Mohr–Coulomb condition with zero friction angle). The method has also been adapted for frictional materials in plane strain, and, for instance, Sokolovskii (1965) gives many applications of this method of calculation.

Problems of axial symmetry may also be studied by this method, although they must invariably be solved by numerical methods rather than by closed-form solution (which is often possible in plane strain). Cox *et al.* (1961), for instance, presented some analyses for both frictionless and frictional materials.

Soils, however, present a special problem in that their properties may be non-homogeneous, a common case being that of an increase of undrained strength with depth, and Davis and Booker (1974) presented analyses of plane strain footings on such materials. Similar methods may be used for axially symmetric problems, and a brief study of this type of problem is presented later.

The mathematics of these problems was presented by Houlsby and Wroth (1982b) and it would be inappropriate to repeat it here, but the assumptions underlying the theory may be given. The analysis depends first on combining two sets of equations: (1) the equations of equilibrium for an element of soil; (2) the equation which defines the condition that each element of soil is in a state of plastic yield. Clearly the first set of equations is always valid. The second represents two assumptions: first, about the form of the yield criterion (the Mohr–Coulomb condition is usually used, but other relations are also possible); and second, that *every* element of soil is at yield (in practice, zones of soil in which yield does not occur may be embedded within the plastically deforming material). Combination of the above equation leads to a set of partial differential equations which are 'hyperbolic' and can be integrated from boundaries, on which the tractions are known, along the 'characteristics' (which are identical with the slip lines) to give stresses throughout the plastic region and tractions on other boundaries of interest.

Such analyses are, in fact, familiar to geotechnical engineers in the form of slip line analyses of footings and retained walls, where the starting point for the analysis is always the free surface on which the tractions are known and the final result is, for instance, the active thrust on a retaining wall.

In the case of axially symmetric rather than plane strain problems an additional assumption must be made about the value of the circumferential stress. The Haar–von Karmann assumption that the circumferential stress is equal to either the major or the minor principal stress is usually made, and this is sometimes regarded as an assumption of special significance. It represents, in fact, no more than an application of the Mohr–Coulomb yield criterion to specify the flow rule as well as the yield surface.

For materials which obey the 'associated flow rule' (which is reasonable only for the undrained behaviour of clays) this method yields a 'lower bound' or safe solution for collapse loads. It is widely believed that even for 'non-associated' materials the method still gives reasonably accurate estimates of true collapse loads.

After determination of the stresses it is possible to carry out calculations for the displacements and strains within the soil, provided that certain displacement boundary conditions are known. This calculation is straightforward for cohesive materials but no completely satisfactory analysis of the displacements in frictional materials is yet possible. No further discussion of the displacement problem is given here.

Example calculations

Bearing Capacity Factors for Cones

The following calculations have two principal applications.

The first is at the very small scale (cones of a few tens of millimetres diameter) where the calculations relate to cone penetration tests. Since only shallow penetration is considered here, the analysis is directly relevant to the laboratory fall cone test, but only indirectly applicable to the CPT. The relevance to the CPT is principally in giving some indication of a possible deformation mechanism, and may also indicate the importance of cone angle and cone roughness if similar

trends of behaviour occur for both shallow and deep cones.

The second major application is to the penetration of spudcan footings into the seabed. Many spudcans can be treated as rather blunt shallow cones of a diameter of several metres. The N_c factors presented here may be used to calculate the penetration of such a footing, given the leg load, the cone angle and an assumption about the cone roughness. No indication is given of the reduction in bearing capacity factor due to the presence of horizontal load or overturning moment.

Houlsby (1982) presented some results of calculations of the loads on a cone of $30°$ included angle penetrating a cohesive soil, as part of a study of the fall cone liquid limit test. The force on a cone is analysed in the same way as a bearing capacity problem, and may be calculated as $P = N_c c_u \pi r^2$, where N_c is a non-dimensional bearing capacity factor, c_u is the undrained strength of the clay and r is the radius of the cone in contact with the soil surface. In this preliminary study it was found that the factor N_c depends on the cone roughness α (defined as a_u/c_u, where a_u is the undrained adhesion of the clay to the cone surface) and the semi-angle of the cone β. More complete results of a study of this problem are given in Fig. 2.

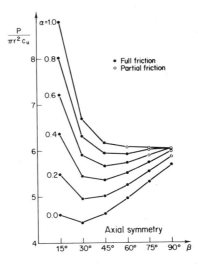

Fig. 2.   Results of calculations of loads on cones: ●, full friction; ○, partial friction.

Consider first the fully smooth cone. At $\beta = 90°$ (i.e. a flat, circular punch) the bearing capacity factor is 5.69 (a result previously obtained by Cox *et al.*, 1961). As the cone angle is decreased, the bearing capacity factor drops, owing to the presence of the inclined surface on which the soil can slip. As the cone becomes very sharp, the depth of penetration becomes large for a given cross-section at the surface, so that the deformed region increases in volume. The result is that at small cone angles the factor rises again. It is not possible (owing to numerical problems) to make calculations for very thin cones, but it appears that the factor will approach inifinity for a very slender cone. The result is that there is an optimum angle of cone (semi-angle about 27°) for which the bearing capacity factor has a minimum value of about 4.44.

Although the maximum shear stress on the surface of the cone is specified as a_u, in some cases (for rather blunt cones) the slip line field is such that the full a_u value is not mobilized on the surface of the cone. In Fig. 2 distinction is made between calculated points for which the full a_u value is mobilized and those for which only partial friction is mobilized on the cone surface (i.e. the contact shear stress is less than a_u).

Considering now the fully rough cone, the bearing capacity factor for a rough circular punch ($\beta = 90°$) is 6.05, and as β is reduced the factor rises steadily, owing to the presence of a fixed plane on which the full friction must be mobilized. At intermediate roughness intermediate curves of bearing capacity factor against cone angle are obtained.

The results presented in Fig. 2 may be compared with the equivalent calculation for a plane strain wedge penetration. For this case the bearing capacity factor may be calculated explicitly as:

$$N_c = 1 + 2\beta + (1 - \alpha^2)^{1/2} + \arcsin(\alpha) + \alpha \cot(\beta)$$

where full friction is mobilized at the wedge surface and

$$N_c = 2 + \pi$$

for cases when full friction is not mobilized. This latter case applies for $2\beta > \pi - \arcsin(\alpha)$. The results of this calculation are given in Fig. 3 (note that for the plane strain case P is defined as the force per unit length on the footing). It can be seen that approximately the same pattern emerges as for the axially symmetric case. Two notable exceptions are that the bearing capacity factor of $2 + \pi$ applies for $\beta = \pi/2$ independent of the roughness, and that $N_c = 2$ for the case of $\alpha = 0, \beta = 0$.

Bearing Capacity on Soils with Strength Increasing with Depth

The second example considered here relates to the problem of the calculation of a soil in which the cohesion increases linearly with depth in the form $c_u = c_0 + \rho z$, where c_0 is the undrained strength at the soil surface, ρ is a constant and z is the depth below the soil surface. The application of this type of calculation is to very large foundations – for example, for concrete gravity platforms for oil production. The importance of the effect of soil non-homogeneity is measured by a non-dimensional factor $\rho B/c_0$, where B is the width of

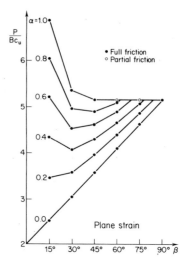

Fig. 3.   Results of calculations for plane strain wedge penetration: ●, full friction; ○, partial friction.

a plane strain footing or diameter of a circular footing. This factor becomes more important as B becomes larger, and for typical values of ρ and c_0 the effect of non-homogeneity is only important for footings in excess of about 50 m diameter. Clearly, very little information from case histories is available for such large footings, so that it has not yet proved possible to confirm the validity of the calculations. However, the analyses presented here are believed to represent reasonably accurately the behaviour of very large circular foundations on non-homogeneous soils. The important effects of horizontal loading and overturning moment are again, unfortunately, excluded from the analysis.

The problem of the bearing capacity on a non-homogeneous soil was studied by Davis and Booker (1974) for the case of smooth and rough footings in plane strain. In Tables 1 and 2 results are given for both plane strain and axially symmetric footings on the non-homogeneous material. The bearing capacity is given by a non-dimensional bearing capacity factor N_c (defined as P/Ac_0, where P is the load on the footing and A is the area of the footing) which is a function of $\rho B/c_0$. The factor N_{cps} refers to a plane strain footing and N_{cas} to a circular footing. The values for plain strain footings agree closely with the results obtained by Davis and Booker.

The final column in Tables 1 and 2 gives the ratio N_{cas}/N_{cps}, which is usually referred to as the shape factor, and is conventionally taken as 1.2 when Terzaghi's bearing capacity expressions are used. It is clear from the tables that the value of 1.2 is unrealistically high for the non-homogeneous material. Even for the homogeneous material ($\rho = 0$) the factor given by these

TABLE 1 Bearing capacity factors for smooth footings

$\rho B/c_0$	N_{cps}	N_{cas}	N_{cas}/N_{cps}
0.0	5.14	5.69	1.107
2.0	6.66	6.74	1.102
4.0	7.82	7.58	0.970
6.0	8.84	8.33	0.943
8.0	9.78	9.03	0.923
10.0	10.67	9.67	0.906

TABLE 2 Bearing capacity factors for rough footings

$\rho B/c_0$	N_{cps}	N_{cas}	N_{cas}/N_{cps}
0.0	5.14	6.05	1.176
2.0	7.57	7.61	1.005
4.0	9.08	8.71	0.960
6.0	10.37	9.67	0.932
8.0	11.52	10.54	0.915
10.0	12.67	11.33	0.894

calculations is only in the range 1.107–1.176, but for both smooth and rough footings the shape factor drops significantly as $\rho B/c_0$ becomes larger. Where non-homogeneity is very significant (usually for very large footings), the bearing capacity factor for a circular footing is actually smaller than that for a strip footing.

Conclusions

The two types of problem considered above are simply examples of the type of calculation which can be carried out by the method of characteristics to calculate loads on shallowly penetrating cones or footings. The method is fast and cheap, so that the sort of parametric study presented above is easy to carry out. It is, of course, possible to perform more complex calculations, and, for instance, it would be simple to carry out a study of cones of different angles and roughnesses in a non-homogeneous soil. Some more complex geometries may be studied (e.g. a two-stage cone), although it is not always possible to formulate a problem of a given geometry in such a way that it can be analysed by this method. Non-homogeneity can be extended to include layered soils, with, for instance, a piecewise linear representation of the soil strength. In principle it is also possible to extend the method to account for anisotropic strength characteristics. Finally, it must be noted that all the analyses in this paper have been undrained total stress analyses using zero friction angle. It is also possible to carry out effective stress analyses using frictional properties.

The examples presented here relate entirely to the relatively simple problem of shallow cone penetration, which can be solved exactly and is relevant to the fall cone test (at the very small scale) and spudcan and gravity platform foundations (at the very large scale).

The more complex problem of deep cone penetration is also being studied by this method, and approximate solutions which are relevant to the CPT and to high-level radioactive waste disposal have been obtained (Houlsby and Wroth, 1982a).

It is clear from all the above possibilities that there is a multiplicity of parameters that can be varied, and parametric studies of only a few selected areas of interest may be carried out. A second application of the technique is therefore in carrying out one-off analyses of specific structures with realistic soil property values. Because of the rapidity of the method, it is again possible to study a certain range of parameter values.

The method of characteristics is severely limited in the class of problems that it can be used to study, but for those areas where a solution is possible it is thought to yield realistic results. The intention of this paper has been to outline some of the possible uses for this method of analysis, which is an extension to the rigorous calculation of conventional bearing capacity factors. In particular, the extension to axial symmetry and to non-homogeneous materials is important. The method is particularly well suited to parametric studies.

Acknowledgement

This work has been commissioned by the Department of the Environment, UK as part of the radioactive waste management research programme.

References

Cox, A. D., Eason, G. and Hopkins, H. G., 1961. Axially symmetric plastic deformation of soils. *Phil. Trans. Roy. Soc., Ser. A,* **254**, 1–45.

Davis, E. H. and Booker, J. R., 1974. The effect of increasing strength with depth on the bearing capacity of clays. *Geotechnique,* **23**, 551–563.

Houlsby, G. T., 1982. Theoretical analysis of the fall cone test. *Geotechnique,* **32**, 111–118.

Houlsby, G. T. and Wroth, C. P., 1982a. Determination of undrained strengths by cone penetration tests. *Proc. 2nd European Symp. Penetration Testing,* Amsterdam, Vol. 2, pp. 585–590.

Houlsby, G. T. and Wroth, C. P., 1982b. Direct solution of plasticity problems in soils by the method of characteristics. *Proc. 4th Int. Conf. Numerical Methods in Geomechanics,* Edmonton, Vol. 3, pp. 1059–1071.

Jamiolkowski, M., Lancellotta, R., Tordella, L. and Battaglio, M., 1982. Undrained strength from the CPT. *Proc. 2nd European Symp. Penetration Testing,* Amsterdam, Vol. 2, pp. 599–606.

Last, N. C., 1982. The Cone Penetration Test in Granular Soils. Ph.D. Thesis, University of London.

Nash, D. F. T. and Duffin, M. J., 1982. Site investigation of glacial soils using cone penetration tests. *Proc. 2nd European Symp. Penetration Testing*, Amsterdam, Vol. 2, pp. 733–738.

O'Riordan, N. J., Davies, J. A. and Dauncey, P. C., 1982. The interpretation of static cone penetration tests in soft clays of low plasticity. *Proc. 2nd European Symp. Penetration Testing*, Amsterdam, Vol. 2, pp. 755–760.

Sokolovskii, V. V., 1965. *Statistics of Granular Media*. Pergamon Press, Oxford.

10

Measurements of Wave-induced Pore Pressure and Coefficients of Permeability of Submarine Sediments during Reversing Flow

Shigeyasu Okusa, Takaaki Nakamura and Masaharu Fukue
Faculty of Marine Science and Technology, Tokai University, Orido, Japan

Wave-induced pore pressure was measured at a location on the Pacific coast of central Japan with two types of probe. Analysis revealed that the damping of wave-induced pore pressure with respect to the wave pressure at the seafloor was greater than that predicted by published theories, even at shallow depths proximate to the mudline.

Coefficients of permeability for porous media during reversing flow were measured by using water oscillations in a U-tube. The coefficients increased with increasing period of oscillations for coarser media and approached a constant value for long-period oscillations. The permeable response of a seabed to wave action would vary according to the wave period.

Introduction

Wave-induced pore pressure and effective stress in submarine sediments are important to the stability of seabeds. Recently, wave-induced pore pressure in submarine sediments has been measured in relation to wave-induced seafloor instability. Measurements have been carried out in normally consolidated silty clays in the Gulf of Mexico (Richards *et al.*, 1975) and in underconsolidated Mississippi Delta clayey sediments (Bennett *et al.*, 1976; Bennett, 1978; Dunlap *et al.*, 1978; Bennett and Fairs, 1979; Hulbert and Bennett, 1982). Several theoretical considerations have been presented (Madsen, 1978; Yamamoto, 1981), but are yet to be proved, owing to a paucity of verification data.

The authors have collected observational data on wave-induced pore pressure at the innermost part of Suruga Bay, located on the Pacific coast of central Japan. With initial results being reported elsewhere (Okusa and Uchida, 1980), the present paper records the results of observations carried out at a shallow water site. Damping and time lag of wave-induced pore pressure with respect to the wave pressure at the seafloor were examined.

In the second part of the present paper the results of laboratory tests on the permeable response of coarse media to reversing flow are discussed. One of the causes of the damping and time lag may be related to resistance of the soil skeleton to water flow. The water flow through submarine sediments due to waves is a type of reversing flow. Experiments were conducted on changes in the soil skeleton resistance in respect of this reversing flow through a specimen consisting of a porous medium submerged in a U-tube. The water surface in one side of the tube was subjected to oscillation for periods ranging from several seconds to several tens of seconds, while the water surface oscillation of the other side of the tube was measured under steady state oscillation. The resistance of the soil skeleton to the reversing flow was examined from the damping of the output oscillations. Glass balls, gravel and sand were tested. The coefficients of permeability obtained through this method are relatively low for short-period oscillations of several seconds when compared with the steady values for long-period oscillations.

Measurements of wave-induced pore pressure

Site

The observations were carried out at the head of a sand spit located on the Pacific coast of central Japan. The

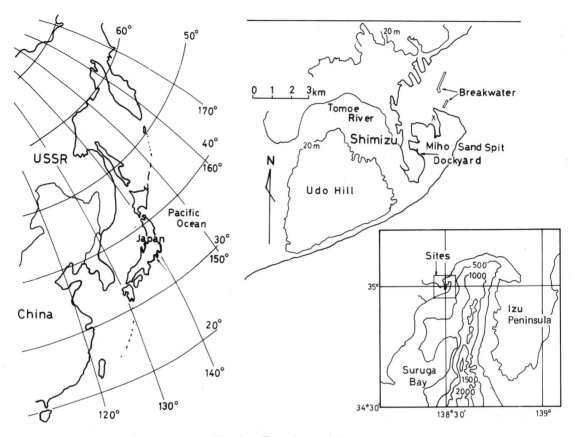

Fig. 1. Experimental site.

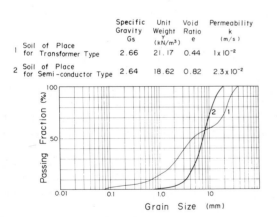

	Specific Gravity G_s	Unit Weight γ (kN/m³)	Void Ratio e	Permeability k (m/s)
1 Soil of Place for Transformer Type	2.66	21.17	0.44	1×10^{-2}
2 Soil of Place for Semi-conductor Type	2.64	18.62	0.82	2.3×10^{-2}

Fig. 2. Physical and permeable properties of sediment.

site was located beneath a pier for sightseeing vessels (35°01′03″N, 138°31′34″E), and was protected from rough waves by two offshore breakwaters, as is shown in Fig. 1. Water depth measured about 1 m at high water, while the bottom sediment consisted of sand and gravel, as can be seen in Fig. 2. Small wind waves and swells were observed at the site, but larger waves occurred

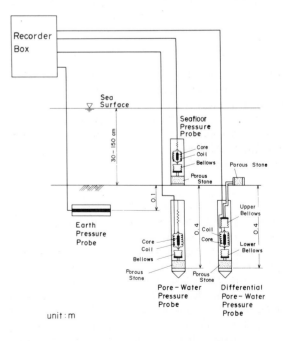

unit : m

Fig. 3. Arrangement of measurement system.

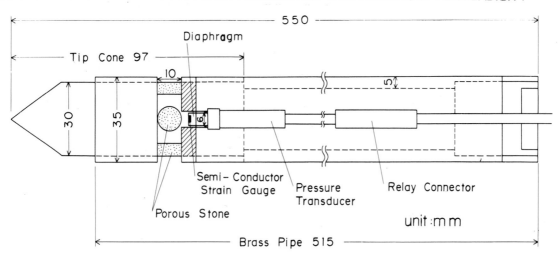

Fig. 4.   Semiconductor-type piezometer.

during the passage of typhoons and atmospheric depressions.

Probes

Two types of probe were used in this study. One was a transformer type capable of monitoring vibrations in absolute pore pressure and differential pore pressure, and earth pressure as induced by waves. Figure 3 shows the schematic diagram of the measurement arrangement. The sediment pore pressure, u_1, is expressed by,

$$u_1 = u_0 + u_w(t) \qquad (1)$$

where u_0 is static pressure from mean sea level at the lower bellows and $u_w(t)$ is wave-induced variation of pore pressure with time, t. Of these, $u_w(t)$ is the only component fluctuating with time. The upper bellows in Fig. 3 senses the variation in wave pressure, u_2, through the filter installed at the seafloor. The relation is represented by

$$u_2 = u' + u(t)$$

where u' is static pressure from mean sea level at the upper bellows and $u(t)$ is wave-induced fluctuation at the seafloor with time. Of these, $u(t)$ is the only component fluctuating with time. Therefore, the fluctuation component of the difference $(u_1 - u_2)$, namely $\Delta u = u_w(t) - u(t)$, acts to displace the core in the probe over time, thereby generating a current. The absolute pressure fluctuation at the lower bellows is expressed by equation (1).

The semiconductor pressure transducer shown in Fig. 4 was specially designed for small wind waves and

swells. The pressure transducer used in this piezometer had a design pressure range of 49 kPa, with a repeatability of 0.098 kPa.

Results and discussions

Observation of rough waves was conducted from 8 to 9 October 1982, when Typhoon-21-82 passed in close proximity to central Japan, as can be seen in Fig. 5. For the initial trial, the disc-like earth pressure probe measuring 0.3 m in diameter was implanted at a depth of approximately 0.1 m from the mudline. The pore pressure and differential pressure probes were installed at a depth of about 0.4 m from the mudline (Fig. 3). Figure 6 shows an example of the records collected.

Observations using the semiconductor-type probe were carried out in 1980, when sea conditions were calm. The implantation depth of the pore pressure probe was about 0.4 m below the mudline. The pore pressure probe and seafloor pressure probe were aligned in an almost vertical position. Figure 7, which presents an example of the collected records, clearly shows the filter effect of the sediment on the seafloor pressure due to short-period waves.

Figure 8 shows pressure spectra for the recording illustrated in Fig. 6. It will be seen that the spectra of seafloor pressure, pore pressure, and earth pressure in the sediment are nearly equal except for a short period, where the relation is complex. If the spectra of the seafloor pressure and pore pressure in the sediment are exactly the same, the earth pressure indicating the total pressure in the sediment would also show the same value, while the differential pressure between the two would not show any fluctuation. Therefore a

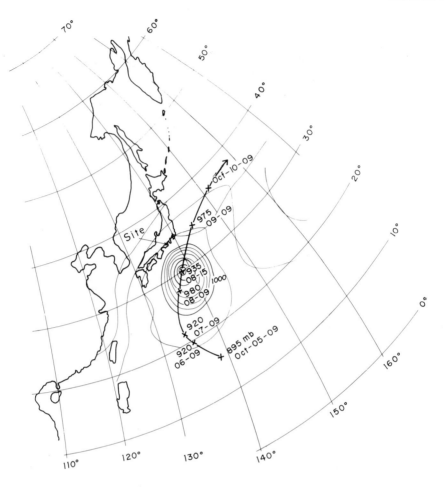

Fig. 5. Path of Typhoon-21-82.

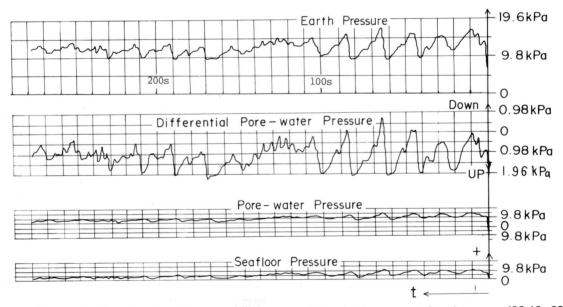

Fig. 6. Wave comparison for seafloor pressure, pore pressure, differential pressure and earth pressure (23.42—23.47, 8 October, 1982)

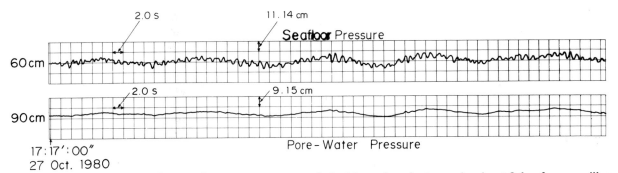

Fig. 7. Wave-induced seafloor and pore pressures recorded with semiconductor probe about 0.4 m from mudline.

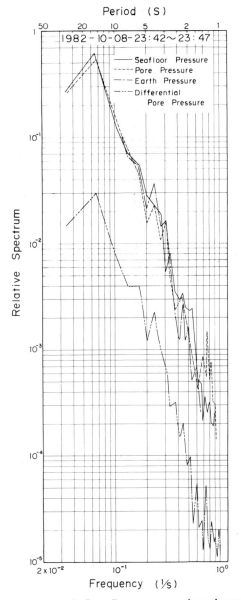

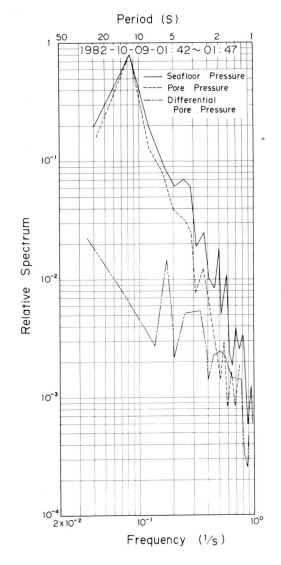

Fig. 8. Spectra of seafloor, pore and earth pressures: ——, seafloor pressures; ———, pore pressure; —·—, earth pressure ———-, differential pore pressure.

Fig. 9. Spectra of seafloor, pore and differential pore pressures: ——, seafloor pressure; ———, pore pressure; ———-, differential pore pressure.

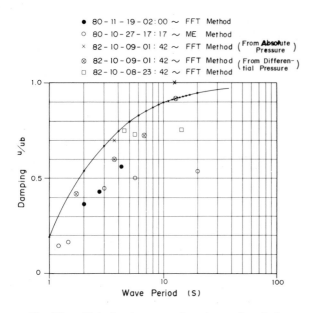

• 80 – 11 – 19 – 02 : 00 ∼ FFT Method
○ 80 – 10 – 27 – 17 : 17 ∼ ME Method
× 82 – 10 – 09 – 01 : 42 ∼ FFT Method (From Absolute Pressure)
⊗ 82 – 10 – 09 – 01 : 42 ∼ FFT Method (From Differential Pressure)
□ 82 – 10 – 08 – 23 : 42 ∼ FFT Method

Fig. 10.   Relation between damping and period.

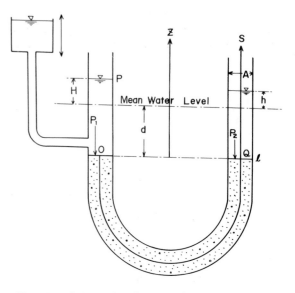

Fig. 11.   Schematic of U-tube for measuring permeability

considerable amount of fluctuation in the differential pore pressure spectrum implies that the seafloor pressure would be transferred to the pore water with damping and/or time lag.

Figure 9 shows power spectra for the data from 01.42 to 01.47 on 9 October 1982. The damping cannot be observed for the 12.5 s period waves, but the spectrum for the differential pore pressure indicates that a slight damping occurred even for long-period waves. For short-period waves the damping is remarkable. The power spectra for the seafloor and pore pressures shown in Fig. 7 display a remarkable damping. The discrepancy between the damping information derived from the differential pressure and received directly from the absolute pressures may be attributable to variation in the accuracy of the recordings. Figure 10 shows the relation between the damping and the wave period. The solid line in Fig. 10 represents a theoretical prediction (Madsen, 1978) for a water depth of 0.5 m and when $z = 0.4$ m. Measured pore pressures in the submarine sediment decrease faster than predicted pore pressures.

A slight time lag is observed from the cross-correlation between the seafloor pressure and pore pressure for the data shown in Fig. 6. However, the time lag information obtained from the cross-correlation and the other data is rather confusing. This again may be due to the reading accuracy of the data. Therefore future readings will be performed with equal accuracy by use of a magnetic tape, after which the taped data will be directly converted into digital values.

Coefficients of permeability for porous media during reversing flow

Theory and Methods

Consider the U-tube shown in Fig. 11, in which area A is filled with a porous medium having a length l. Take the axial system as shown in Fig. 11. The relation between the discharge velocity, v', and the seepage velocity, v, of water through a porous medium with porosity n is $vn = v'$. Consider a small element, $A \times ds$, along the U-tube. If we assume that the water flow through a porous medium experiences resistance 2α from the porous skeleton in proportion to the floor velocity, v, per unit mass of water, the equation of motion of the water in the small element is expressed by

$$\frac{Dv}{Dt} An\rho\, ds = -\frac{\partial p}{\partial s} An\, ds - g\frac{\partial z}{\partial s} An\rho\, ds - 2\alpha v An\rho\, ds \quad (2)$$

where p is water pressure at s, ρ is density of water, g is acceleration of gravity, $-\partial z/\partial s$ is the gradient of the tube relative to the z-axis and $Dv/Dt = \{v(\partial v/\partial s + \partial v/\partial t)\}$. If we assume that the flow through a U-tube is uniform along the s-axis, then $\partial v/\partial s = 0$ and equation (2) becomes

$$\frac{\partial v}{\partial t} = -g\frac{\partial}{\partial s}\left(\frac{p}{\rho g} + z\right) - 2\alpha v = -g\frac{\partial \zeta}{\partial s} - 2\alpha v \quad (3)$$

and $$\zeta = \frac{p}{\rho g} + z$$

where ζ is the piezometric head in the tube. In the case of stationary flow, i.e. $\partial v/\partial t = 0$, equation (3) gives Darcy's law and the coefficient of permeability is

$$k = \frac{ng}{2\alpha} \qquad (4)$$

The governing equation, equation (3), can be represented in terms of k by

$$\frac{\partial v}{\partial t} + \frac{ng}{k}v = -g\frac{\partial}{\partial s}\left(\frac{p}{\rho g} + z\right) \qquad (5)$$

When the water tank shown in Fig. 11 is periodically moved up and down with amplitude H_0 and period T, the boundary conditions are as follows:

$$H = H_0 \sin\frac{2\pi}{T}t = H_0 \sin \omega t \qquad (6)$$

$$\left.\begin{array}{l} p_1 = p_0 + \rho g(d + H) \quad \text{at point 0 in Fig. 11} \\[4pt] p_2 = p_0 + \rho g(d + h) \quad \text{at point Q in Fig. 11} \end{array}\right\} \qquad (7)$$

where p_0 is atmospheric pressure, d is mean water depth to the surface of a porous medium and h is amplitude of the output oscillation in the right-hand side of the tube. Integrating equation (5) along the s-axis from 0 to Q under the boundary conditions of equations (6) and (7) and using $nAv = Av' = A\,\mathrm{d}h/\mathrm{d}t$, we get, from equation (5),

$$(D^2 + 2D + \omega_0^2)h = R \exp(i\omega t) \qquad (8)$$

$$\omega_0^2 = \frac{ng}{l}, \quad R = \frac{ngH_0}{l} = \omega_0^2 H_0, \quad \omega = \frac{2\pi}{T}, \quad D = \frac{d}{dt} \qquad (9)$$

and the imaginary part of equation (8) expresses the motion, h. The general solution of equation (8) is expressed by (Jaeger, 1963)

$$h = C_1 \exp(\beta_1 t) + C_2 \exp(\beta_2 t)$$

$$+ \frac{R}{\left\{(\omega_0^2 - \omega^2)^2 + (2\alpha\omega)^2\right\}^{1/2}} \sin(\omega t - \theta) \qquad (10)$$

where C_1 and C_2 are integral constants, respectively, and

$$\beta_1 = -\alpha + (\alpha^2 - \omega_0^2)^{1/2}$$

$$\beta_2 = +\alpha + (\alpha^2 - \omega_0^2)^{1/2}$$

$$\tan\theta = \frac{2\alpha\omega}{(\omega_0^2 - \omega^2)}$$

The first two terms of equation (10) tend towards zero in an oscillating or steady manner according to $\alpha = ng/2k \gtrless \omega_0$ for a sufficiently long time, t. At steady state the water surface motion of the right-

hand side of the U-tube becomes a forced harmonic oscillation represented by the third term,

$$h = \frac{R}{\left\{(\omega_0^2 - \omega^2)^2 + (2\alpha\omega)^2\right\}^{1/2}} \sin(\omega t - \theta) \qquad (11)$$

By measuring the amplitude, $C = R / \left\{(\omega_0^2 - \omega^2)^2 + (2\alpha\omega)^2\right\}^{1/2}$, or the time lag, θ, at steady state in the U-tube, we can obtain the coefficient of permeability of a porous medium in the tube for each value of the oscillation period as

$$k = \frac{ng\omega}{\left\{(R/C)^2 - (\omega_0^2 - \omega^2)^2\right\}^{1/2}}$$

$$= \frac{2\pi l}{T[(H_0/C)^2 - \left\{1 - l(2\pi/T)^2/ng\right\}^2]^{1/2}} \qquad (12)$$

$$k = \frac{ng\omega}{(\omega_0^2 - \omega^2)\tan\theta} = \frac{2\pi l}{T[1 - l(2\pi/T)^2/ng]\tan\theta} \qquad (13)$$

Experiments and Results

Three types of U-tube were used for the experiment. The dimensions of the tubes are shown in Table 1.

TABLE 1   Dimensions of U-tubes

	Height (m)	Width (m)	Diameter of tube (m)	Length of porous medium (m)	Volume of porous medium (m³)
Type 1	1.5	2.0	0.3	1.03	0.092 7
Type 2	0.54	0.38	0.09	0.382 4	0.002 433
Type 3	1.4	0.9	0.2	0.30	0.009 425

The oscillation of the water surface in the Type 1 tube was generated by the up and down movement of a concrete cylinder measuring 16.5 cm in diameter and 27 cm in length. For Type 2 and Type 3 U-tubes, a bucket connected with the U-tube with a vinyl plastic hose was moved up and down to generate harmonic oscillations (Fig. 11). The period of the waves generated in the above manner ranged from 1 s to about 30 s, and the wave height was 10–100 mm. Porous media were introduced manually into the U-tubes so that each medium was in a homogeneous state. Characteristics of these media are given in Table 2. After several waves, oscillations of the input and output waves became steady. While Fig. 12 provides a record for damping output to input waves, Fig. 13 shows plottings of the amplitude ratio H_0/C against period. Since the reading of the phase lag, θ, was relatively uncertain compared with that of the amplitude ratio, equation (12) was used for calculation of the permeability.

Figure 14 shows the relationship between the coef-

TABLE 2   Characteristics of porous media

Type of U tube	Porous medium	Specific gravity of particles	Range of particle size (mm)	10% diameter (mm)	60% diameter (mm)	Mean diameter (mm)	Uniformity coefficient	Porosity (%)
Type 1	Gravel	2.62	19.10–15.90	16.20	17.80	17.40	1.1	34.0
		2.63	9.52–4.76	5.05	7.20	6.70	1.1	34.9
	Sand and gravel	2.63	2.38–1.19	1.26	1.79	1.66	1.4	36.3
	Glass ball	2.49	20.50	20.50	20.50	20.50	1.0	36.1
		2.48	13.10	13.10	13.10	13.10	1.0	38.8
	Mixture of glass balls	2.49	20.50 (70%) 13.10 (30%)			18.86		36.6
		2.48	20.50 (50%) 13.10 (50%)			17.58		35.5
		2.48	20.50 (30%) 13.10 (70%)			16.08		35.7
Type 2	Gravel	2.65	9.52–4.76	5.10	7.15	6.70	1.4	43.3
		2.68	4.76–2.00	2.18	3.17	3.03	1.5	41.8
	Sand	2.57	2.00–0.105	0.845	1.610	1.510	1.9	38.6
	Glass ball	2.58	13.60–11.70	11.80	12.80	12.60	1.1	44.4
Type 3	Gravel	2.67	19.52–9.52	10.20	14.70	13.60	1.4	37.3
		2.69	9.52–4.76	5.03	7.20	6.70	1.4	35.3
		2.69	4.76–2.00	2.18	3.30	3.03	1.5	34.0
	Glass ball	2.54	21.50–19.20	19.30	20.70	20.30	1.1	40.2

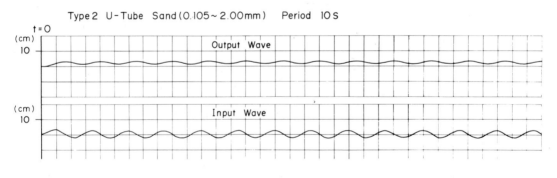

Type 2   U-Tube   Sand (0.105~2.00mm)   Period 10 S

Fig. 12.   Damping of output wave to input wave.

ficients of permeability calculated from the amplitude ratio and the time periods. It is clear that for coarser media the coefficients of permeability increase with increasing period of oscillations, and approach a constant value for long-period oscillations. The period over which the coefficients remain steady is greater, being approximately 10 s for large-diameter media and around 5 s for small-diameter media. For coarser media the coefficient of permeability at a steady range is ten or more times greater than that at around 2 s in period. For finer gravel and sand it seems that the coefficient of permeability is independent of the oscillation period.

If the coefficient of permeability of coarser media depends on the oscillation period of reversing flow in a short period range, the permeable response of the seabed to wave action would vary according to the wave period. However, experiments continue, and many problems such as the relation between the coefficient and absolute seepage velocity remain to be solved.

Conclusions

Wave-induced pore pressure was measured in sand and gravel sediment at a depth of about 0.4 m from the mudline in about 1 m of water. Wave-induced seafloor pressure was transferred in the sediment with damping, which was a little greater than that predicted by current theories (Madsen, 1978; Yamamoto, 1981), with a probable time lag. Permeability and deformation properties of submarine sediments would play an important role in the damping transfer of wave pressure at the

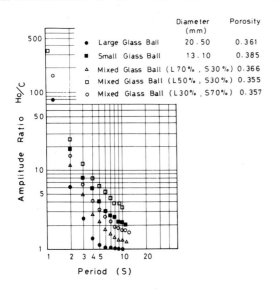

Fig. 13.   Relation between amplitude ratio and period.

seafloor to the pore pressure with a time lag. A more elaborate theory and accurate measurements are necessary for the future study of this problem.

The soil skeleton resistance to reversing flow was measured by using the forced oscillation of water in a U-tube filled with a porous medium. The coefficient of permeability, which is inversely proportional to the resistance, increased with increasing oscillation period from around 2 s and became steady for long-period oscillations for gravel materials. The varying permeable response of coarser sediments to ocean waves of different periods would be an important consideration in the behaviour of seabeds.

Acknowledgements

The authors express their thanks to Dr Sanae Unoki, the director of Physical Oceanography Laboratory, The Institute of Physics and Chemistry, who first suggested the fundamental idea of the use of a U-tube for

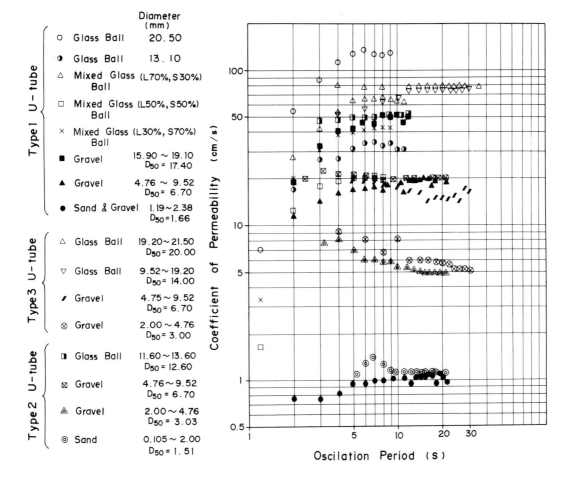

Fig. 14.   Relation between coefficients of permeability and oscillation period.

determining the porous medium resistance against reversing flow and derived the equations concerned. However, the conclusions are the authors' responsibility.

References

Bennett, R. H., Bryant, W. R., Dunlap, W. A. and Keller, G. H., 1976. Initial results and progress of the Mississippi Delta sediment pore-water pressure experiment. *Mar. Geotech.,* **1** (4), 327–335.

Bennett, R. H., 1978. Pore-water pressure measurement: Mississippi Delta submarine sediments. *Mar. Geotech.* **2** (2), 177–189.

Bennett, R. H. and Fairs, J. R., 1979. Ambient and dynamic pore pressure in fine-grained submarine sediments: Mississippi Delta. *Appl. Ocean Res.* **1** (3), 115–123.

Dunlap, W. A., Bryant, W. R., Bennett, R. H. and Richards, A. F., 1978. Pore pressure measurements in unconsolidated sediments. Pap. No. 3168. *Off-shore Technology Conf.* Houston, Vol. 2, pp. 1049–1058.

Hulbert, M. H. and Bennett, R. H., 1982. Anomalous pore pressure in Mississippi Delta sediments: Gas and electrochemical effects. *Mar. Geotech.* **5** (1), 51–62.

Jaeger, J. C., 1963. *An Introduction to Applied Mathematics.* Clarendon Press, Oxford, pp. 76–88.

Madsen, O. S., 1978. Wave-induced pore pressure and effective stresses in a porous bed. *Geotechnique,* **28** (4), 377–393.

Okusa, S. and Uchida, A., 1980. Pore-water pressure change in submarine sediments due to waves. *Mar. Geotech.* **4** (2), 145–161.

Richards, A. F., Øien, K., Keller, G. H. and Lai, J. Y., 1975. Differential piezometer probe for an *in situ* measurement of sea-floor pore-pressure. *Geotechnique,* **25** (2), 229–238.

Yamamoto, T., 1981. Wave-induced pore pressure and effective stresses in inhomogeneous seabed foundations. *Ocean Engng,* **8** (1), 1–8.

11

Non-linear Finite Strain Consolidation of Soft Marine Sediments

Robert L. Schiffman and Vincenzo Pane

Department of Civil Engineering, University of Colorado, Boulder, Colorado, USA

This paper presents an analytical procedure for hindcasting the genesis of a soft marine sediment which is consolidating under its own weight. The procedure is based upon the use of non-linear finite strain consolidation. This theory is reviewed. A formulation is developed which permits the computation of the progress of consolidation during the accumulation of a sediment column. A numerical procedure for solving the governing equation by the method of lines is outlined. The procedure is applied to the development of the geotechnical stratigraphy of Mississippi delta sediments. It is shown that an accurate calculation can be made of the present state of effective stresses and pore-water pressures. It is further shown that an accurate assessment of the present void ratio profile can be obtained.

Introduction

The theory of non-linear finite strain consolidation (Gibson *et al.,* 1967) is unrestricted with respect to the magnitude of strain and the nature of the variation of compressibility and permeability with respect to the void ratio of the sediment. These factors allow the theory to be a powerful tool in the development of the genesis of marine deposits and in the prediction of the behaviour of these deposits when subjected to continuing sediment accumulation.

This paper presents a short review of the theory of nonlinear finite strain consolidation and its numerical implementation to problems of sediment accumulation. A procedure for performing a geotechnical stratigraphic analysis is described. This procedure is applied to hindcasting the formation of the prodelta sediments of the Mississippi delta.

Physical assumptions

The fundamental assumptions of the one-dimensional theory of non-linear finite strain consolidation are:

(1) The saturated soil system consists of three components: soil particles, soil skeleton and pore fluid.

(2) The soil particles and pore fluid are incompressible.

(3) The soil skeleton deforms in either a linear or nonlinear manner with no restriction on the magnitude of strain.

(4) The pore fluid is Newtonian.

(5) The flow of fluid through the porous skeleton is governed by the Darcy–Gersevanov law.

(6) The fluid flow velocities are small.

The governing equation

The governing equation for this theory is based upon the balance and constitutive equations for each phase of the mixture. Two fundamental balance equations are required for both the fluid and the soil skeleton. These are the conservation of linear momentum and conservation of mass (continuity). This theory is quasi-static (no inertial forces). Thus the conservation of linear momentum reflects equilibrium.

In addition to the balance laws, the governing equation reflects a set of constitutive relationships. First, the flow relationship must be specified. This is accomplished by the Darcy–Gersevanov law. Second, since the soil is assumed to be a two-phase mixture (soil skeleton and water), a relationship between the phases must be specified. This is provided by the effective stress principle. In addition, the relationship between the void ratio and effective stress must be specified as a single-valued function independent of position and time. A similar relationship governs the expression relating the void ratio and the coefficient of permeability. This implies that the consolidation is monotonic, the soil is normally consolidated, the soil deposit is homogeneous and there are no secondary consolidation effects. It has been shown that, in spite of these restrictions, the theory of non-linear finite strain consolidation is an excellent predictor of consolidation processes in soft soils (Croce, 1982; Scully, 1983).

Based upon the above, a general theory of one-dimensional non-linear finite strain consolidation has been developed by Gibson et al. (1967). The governing equation[*] for this theory is:

$$\frac{\partial}{\partial z}\left[g(e)\frac{\partial e}{\partial z}\right] \pm f(e)\frac{\partial e}{\partial z} = \frac{\partial e}{\partial t} \qquad (1a)$$

where

$$g(e) = -\frac{k(e)}{\gamma_w(1+e)}\frac{\mathrm{d}\sigma'}{\mathrm{d}e} \qquad (1b)$$

$$f(e) = -\left(\frac{\gamma_s}{\gamma_w}-1\right)\frac{\mathrm{d}}{\mathrm{d}e}\left[\frac{k(e)}{1+e}\right] \qquad (1c)$$

in which e is the void ratio, γ_s and γ_w are the solid and fluid weights per unit of their own volume, respectively, k is the coefficient of permeability, σ' is the effective stress and z is a reduced coordinate. This coordinate is defined as the volume per unit area of solids lying between the datum plane and the Lagrangian (initial) coordinate point (Terzaghi, 1927; McNabb, 1960). The Lagrangian coordinate, a, is related to the reduced coordinate, z, by

$$z(a) = \int_0^a \frac{\mathrm{d}a'}{1+e(a',0)} \qquad (2)$$

Equations (1) with appropriate boundary and initial conditions and constitutive properties provide the governing relationships from which a solution can be developed.

The governing equation for one-dimensional non-linear finite strain consolidation differs from conventional theory in two important aspects. First, the theory as given by equations (1) is not bound by any specified void ratio–effective stress and void ratio–permeability relationships. These relationships may be linear or nonlinear as determined by laboratory tests. It is restricted, however, to the monotonic behavior of homogeneous soils without intrinsic time effects

(creep) and chemical interactions between constituents. Second, the magnitude of the strain is unrestricted.

The formulation in terms of void ratio explicitly includes the effect of self-weight in the governing equation. Other formulations, such as when the excess pore-water pressure is the dependent variable, require external specification of the self-weight effect.

The governing equations (1) are mildly nonlinear equations of the diffusion–convection type. When one deals with problems of sediment accumulation, the moving boundary value problem can be transformed to a problem with a fixed geometry by defining a new independent variable, y, as

$$y = z(t)/h_z(t) \qquad (3)$$

where h_z is the total height of solids in the sediment layer at any given time. The governing equation is then

$$\frac{1}{h_z^2(t)}\frac{\partial}{\partial y}\left[g(e)\frac{\partial e}{\partial y}\right] \pm \frac{f(e)}{h_z(t)}\frac{\partial e}{\partial y} + \frac{\partial e}{\partial t} = 0 \qquad (4)$$

This equation can by solved numerically.

Numerical analysis

Boundary value problems concerned with the simultaneous accumulation and consolidation of sediments can be solved numerically by a variety of methods. One convenient means is the method of lines (Ames, 1977). This is a classical method which reduces a partial differential equation to a system of ordinary differential equations by discretizing all but one of the independent variables. Given a partial differential equation of the form of equation (4) and using a centered difference approximation for the spatial y coordinate, one obtains a system of ordinary differential equations of the form

$$\frac{\partial e_i}{\partial t} = f(t, e_1, e_2, \ldots, e_N) \quad i = 1, 2, \ldots, N \quad (5)$$

where there are N spatial mesh points i.

There are many numerical methods for solving this system of ordinary differential equations, and quality-assured software packages are readily available.

The geotechnical stratigraphy of Mississippi delta sediments

We now consider the application of the theory presented above to the hindcasting of the formation of the pro-delta sediments of the Mississippi delta. We consider the data collected and tested by Shephard et al. (1978, 1979) for the upper 37 m of the South Pass Outer Continental Shelf lease area, Block 47. These sediments are predominantly massive bedded clays and silty clays. The age of this deposit is approximately 13 000 years,

---

[*]The upper/lower sign is taken if the coordinate direction is measured against/with gravity.

TABLE 1 Consolidation characteristics of sediments from boring 1 (Shephard *et al.,* 1978).

Depth below mudline (m)	Effective overburden stress, σ_0' (kPa)	Effective preconsolidation stress, σ_c' (kPa)	Void ratio e_0	Void ratio e_c
1.52	6.5	4.5	2.85	2.90
3.05	12.9	21.0	2.70	2.55
6.10	26.3	13.0	2.44	2.75
6.71	29.1	11.0	2.30	2.55
9.14	40.7	19.8	2.07	2.25
13.72	63.4	13.5	1.40	1.75
20.73	104.8	58.5	1.20	1.30
30.48	168.9	120.0	1.03	1.30
36.58	208.7	85.0	1.65	1.95

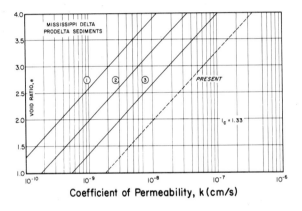

Fig. 1. Void ratio—permeability relationships.

with the upper 8 m deposited in the last 500 years (Parker, 1977; Shephard *et al.,* 1978, 1979).

The available soil property data were obtained from consolidation tests performed on 'undisturbed' cores for Boring 1 (Shephard *et al.,* 1978, 1979). The pertinent information is presented in Table 1 and Fig. 1.

The void ratios e_0 and e_c were taken from the consolidation test data (Shephard *et al.,* 1979) at σ_0' and σ_c', respectively. The compression indices, C_c, for these tests varied from 1.19 at the mudline to 0.5 at the deeper portions of the sediment column.

The relationship between the void ratio and the coefficient of permeability, k, as determined by laboratory tests for present conditions, is shown in Fig. 1. The dotted line represents the lower bound of the reported data (Shephard *et al.,* 1979), which has been formulated as

$$e = 4.0 + 1.33 \log (k/3.5 \times 10^{-7}) \qquad (6)$$

where k has the units of cm/s.

The geotechnical stratigraphic analysis is based upon the solution of equation (4). This analysis requires data

on the void ratio—permeability and void ratio—effective stress characteristics of the sediment at the time of deposition , and on the history of deposition; that is, the height—time relationship for the sediment. This may be either the total height of solids, h_z, or the Lagrangian height, h_a. The Lagrangian (initial) height is the height of the sediment that would occur if no consolidation took place.

History of Depostion

In order to estimate the history of deposition as input to the calculation process, it is first necessary to determine the total volume of solids in the 37 m profile being analyzed. This can be deduced from the relationship between the effective overburden stress, σ_0', and the height of solids, z. This relationship is

$$\sigma_0' = (\gamma_s - \gamma_w)z + q_0' \qquad (8)$$

where q_0' is the effective stress at the mudline, which, in this case is small. From the data available and presented in Table 1, the value of σ_0' at the 37 m level is 208.7 kPa. Then, from equation (8), the maximum height of solids, h_z, is 12.3 m (at 13 000 years after the start of deposition). The height of solids, h_z, can be converted to a Lagrangian height, h_a, by

$$h_a = h_z(1 + e_{00}) \qquad (9)$$

where e_{00} is the void ratio at arrival and the persistent boundary condition at the top of the accumulating sediment. It was assumed that this void ratio was 4.0. This value is an estimate based upon studies of the nature of seabed sediments in the Mississippi delta (Parker, 1977).

The geologic information (Shephard *et al.,* 1978) indicates that the full 37 m was deposited over the last 13 000 years and that the upper 7–9 m was deposited over the last 500 years. A deposition history was then obtained by proportionality, a uniform distribution of solids through the sediment column being assumed. The calculated deposition history in terms of Lagrangian height is presented in Table 2.

TABLE 2 Calculated depositional history

Time BP (years)	Lagrangian height, h_a (m)
13 000	0.5
500	48.6
0	61.9

It is noted that an initial value of 0.5 m is set. This is for the purpose of starting the numerical analysis. The low void ratio, for bedded silty clays below 37 m, was assumed to provide an impervious boundary.

Consolidation Properties

The only available data concerning the consolidation properties are those taken at present (Shephard *et al.,* 1978, 1979). These data reflect the aging of the deposit over the past 13 000 years and do not truly represent the consolidation properties at the time of deposition. For purposes of this analysis, the void ratio—effective stress relationship was assumed to have the form

$$e = e_{00} - C_c \log (\sigma'/\sigma'_{00}) \qquad (10)$$

where e_{00} is the void ratio at arrival and σ'_{00} is the effective stress at arrival. Also, the void ratio—permeability relationship was assumed in the form

$$e = e_{00} + I_c \log (k/k_{00}) \qquad (11)$$

where k_{00} is the coefficient of permeability at e_{00}. The permeability index, I_c was chosen to be 1.33, the same as the indicated laboratory data. The void ratio—permeability properties used in the analysis are shown in Fig. 1 as solid lines and are tabulated in Table 3.

TABLE 3   Void ratio—permeability data

Curve number	Void ratio e_{00}	Permeability index, I_c	Coefficient of permeability k_{00} (cm/s)
1	4.0	1.33	1.17×10^{-8}
2	4.0	1.33	3.50×10^{-8}
3	4.0	1.33	1.05×10^{-7}
Present	4.0	1.33	3.50×10^{-7}

The void ratio—effective stress properties used in the analysis are shown in Fig. 2 and are tabulated in Table 4.

TABLE 4   Void ratio—effective stress data

Curve number	Void ratio e_{00}	Compression index, C_c	Reference effective stress, σ'_{00} (kPa)
A	4.0	1.1	1.0
B	4.0	1.3	1.0
C	4.0	1.5	1.0

The chosen permeabilities are between 3 and 30 times less than the lower bound value reported for present conditions. In addition, the chosen compression indices are somewhat greater than the values reported for present conditions. These chosen values, which

should be representative of the soil at arrival, are estimates which, owing to lack of information, cannot be verified in the laboratory. The chosen values were based upon consolidation/permeability tests performed on remoulded Gulf of Mexico sediments by the authors.

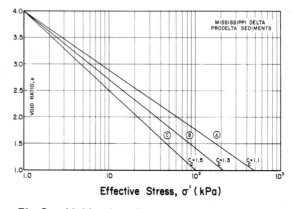

Fig. 2.   Void ratio—effective stress relationships.

Geotechnical Stratigraphic Analysis

A series of geotechnical stratigraphies were developed for the nine combinations of consolidation/permeability properties given above. These analyses were obtained by solving equation (4), the deposition history previously described being used. Figure 3 presents a linearized plot of the present excess pore-water pressure profiles obtained from the analysis. These are shown as solid lines. The curves are referenced by their endpoints, the designations given in Tables 3 and 4 and Figs. 1 and 2 being used. Also shown, as dotted lines, are the bounds measured by *in situ* pore water pressure probes. The lower bound (Bennett, 1977; Shephard *et al.,* 1978) is formulated as

$$u = 2.94\xi - 8.35 \qquad (12)$$

where u is the excess pore-water pressure (in kPa) and ξ is the convective coordinate (in meters) originating at the mudline. The upper bound (Hirst and Richards, 1977; Dunlap *et al.,* 1978; Shephard *et al.,* 1978) is formulated as

$$u = 4.24\xi - 12.57 \qquad (13)$$

Table 5 presents a summary of the results of the analysis. Also shown in this table are the values measured by laboratory consolidation tests. It is noted that the excess pore-water pressure, u, is the difference between the overburden pressure, σ'_0, and the present effective stress, σ'_c. These positive values indicate that at present the sediment is underconsolidated.

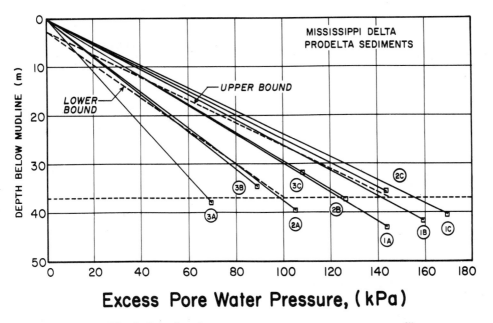

Fig. 3. Calculated present excess pore-water pressure profiles.

TABLE 5 Summary of geotechnical stratigraphic analysis

Permeability relationship	Compressibility relationship	$h_{present}$ (m)	Stress at bottom (kPa)		
			σ_0'	σ_c'	u
1	A	42.98	211.39	67.50	143.89
2	A	39.78	211.39	106.44	104.95
3	A	37.95	211.39	142.16	69.23
1	B	41.52	211.39	52.43	158.96
2	B	37.39	211.39	85.39	126.00
3	B	34.64	211.39	122.23	89.17
1	C	40.41	211.39	41.70	169.69
2	C	35.48	211.39	68.45	142.94
3	C	31.83	211.39	103.38	108.01
	Measured	37.00	208.70	85.00	123.70

The first refinement of the analysis is to choose those property combinations which fall within the measured excess pore-water pressure bounds shown in Fig. 3. These are 1A, 2B and 3C. It is noted that combination 2B matches the present height of 37 m quite satisfactorily. Combination 1A is approximately 6 m too high, while combination 3C is approximately 5 m too low. The calculated void ratio profiles, at present, are shown for the three combinations in Fig. 4. Also plotted are the estimated current void ratios, e_c, as reported in Table 1. It appears that, given the accuracy of the measured void ratios, the analysis can be further refined to include only combinations 2B and 3C.

Figure 5 presents the calculated and measured effective stress profiles, σ_0' and σ_c', for combinations 2B and 3C. Figure 6 presents the calculated and measured

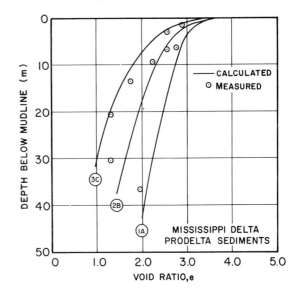

Fig. 4. Present void ratio profile: (——) calculated; ⊙, measured.

profile of present excess pore-water pressure. In addition to the upper and lower bounds previously discussed, a line representing the mean of the measured values (Shephard *et al.*, 1978, 1979) is also shown. This line is formulated as

$$u = 3.77\xi - 9.42 \qquad (14)$$

Finally, Fig. 7 presents the height–time relationships

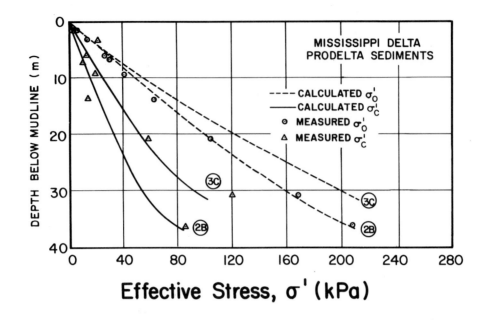

Fig. 5. Present effective stress profile: (———) calculated σ_0'; (———) calculated σ_c'; (⊙) measured σ_0'; (△) measured σ_c'.

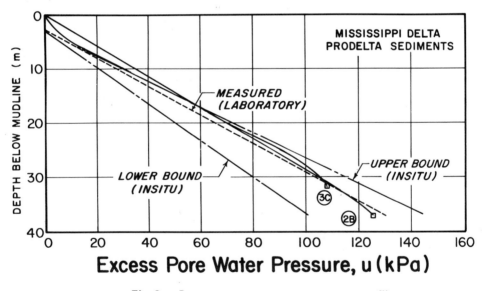

Fig. 6. Present excess pore-water pressure profiles.

for combinations 2B and 3C. Also shown on this figure is the deposition history in terms of the Lagrangian height, h_a.

A review of Figs. 5 and 7 shows that while combination 3C does not match the calculated effective overburden stress, σ_0', it appears to match the present effective stress σ_c' better than combination 2B. When one considers the approximations and uncertainties in calculating σ_c' from laboratory consolidation curves, we tend to place less credence on σ_c' data than on the other controls such as matching the height and the measured effective overburden stress, σ_0'. We therefore conclude that combination 2B, or a small variation thereof, is the appropriate one to use for a geotechnical stratigraphic analysis of the sediments in question.

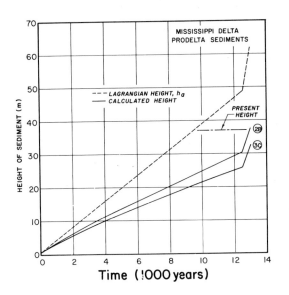

Fig. 7. History of deposition: (– – –) Lagrangian height, h_a; (——) calculated height.

Conclusions

In this paper we have presented a method of hindcasting the genesis of a sediment column which is consolidating under its own weight. An example has been given, based upon previously published data. It has been shown that non-linear void ratio consolidation theory can be properly applied to these problems with optimistic prospects of success. It should be noted, however, that the success of the techniques presented relies on appropriate consolidation/permeability properties of the sediment in its state of arrival, an accurate assessment of the deposition history of the sediment column and a present void ratio profile of the sediment column. The last item forms a benchmark with which one can match the analysis to *in situ* conditions.

The agreement shown in the presented example is remarkable, in spite of the lack of information on consolidation/permeability properties of the sediment on arrival. Thus information can be achieved by testing remoulded samples taken close to the mudline. Also, it is noted that the theory can be used to qualitatively estimate the range of consolidation/permeability properties by matching dependent variables (excess pore-water pressure, void ratio and height history) to measured *in situ* values.

In spite of this, it is noted that there is a need to refine the theory to account for several important effects. First, consideration should be made of the process of sedimentation where the effective stress principle is not fully applicable (Michaels and Bolger, 1962; Been and Sills, 1981). Second, the effect of gas in the sediment should be considered. Third, changes due to aging of the consolidation/permeability properties should be investigated. While the effect of aging on the void ratio—effective stress characteristics has been studied for terrestrial soils (Bjerrum, 1967), there is a gap in our knowledge of this effect with respect to marine sediments. No information exists, to our knowledge, on the effect of aging on the void ratio—permeability relationship. Fourth, the material deposited may come from several sources during the period of deposition. This will lead to nonhomogeneous material deposits. The theory should be augmented to account for this phenomenon. Fifth, no account has been taken of skeleton creep. Sixth, the formation of marine deposits is hardly one of monotonic deposition. Account should be taken of erosional processes where sediment is removed. All these are areas of further research.

Acknowledgments

The first-named author is pleased to acknowledge the assistance of R. E. Gibson (Kings College, London), who first suggested the use of the concept of reduced co-ordinates in solving problems of sediment accumulation. We are further pleased to acknowledge the stimulating and helpful comments by W. R. Bryant (Texas A & M University), H. Y. Ko (University of Colorado), I. Noorany (San Diego State University), H. W. Olsen (US Geological Survey) and D. Znidarcic (University of Zagreb). The financial support was provided by the National Science Foundation.

References

Ames, W. F., 1977. *Numerical Methods for Partial Differential Equations*, 2nd edn. Academic Press, New York.

Been, K. and Sills, G. C., 1981. Self-weight consolidation of soft soils: an experimental and theoretical study. *Geotechnique*, **31**, 519–535.

Bennett, R. H., 1977. Pore-water pressure measurements: Mississippi Delta submarine sediments. *Mar. Geotech.* **2**, 177–189.

Bjerrum, L., 1967. Engineering geology of Norwegian normally-consolidated marine clays as related to settlements of buildings. *Geotechnique*, **17**, 81–118.

Croce, P., 1982. Evaluation of Consolidation Theories by Centrifugal Model Tests. M.S. Thesis, Department of Civil Engineering, University of Colorado.

Dunlap, W. A., Bryant, W. R., Bennett, R. H. and Richards, A. F., 1978. Pore pressure measurements in underconsolidated sediments. Pap. No. 3168. *Proc. 10th Offshore Technology Conf.* Houston, Vol. 2, pp. 1049–1058.

Gibson, R. E., England, G. L. and Hussey, M. J. L., 1967. The theory of one-dimensional consolidation of saturated clays, I. Finite non-linear consolidation of thin homogeneous layers. *Geotechnique*, **17**, 261–273.

Hirst, T. J. and Richards, A. F., 1977. *In situ* pore-pressure measurement in Mississippi Delta front sediments. *Mar. Geotech.,* **2**, 191–204.

McNabb, A., 1960. A mathematical treatment of one-dimensional soil consolidation. *Q. Appl. Math.,* **17**, 337–347.

Michaels, A. A. and Bolger, J. C., 1962. Settling rates and sediment volumes of flocculated kaolin suspensions. *Ind. Engng Chem.,* **1**, (1), 24–33.

Parker, R. A., 1977. Radiocarbon Dating of Marine Sediments. Ph.D. Dissertation, Department of Oceanography, Texas A & M University, College Station, Texas.

Scully, R. W., 1983. Determination of the Consolidation Properties of a Phosphatic Clay at Very High Void Ratios. M.S. Thesis, Department of Civil Engineering, University of Colorado.

Shephard, L. E., Bryant, W. R. and Dunlap, W. A., 1978. Consolidation characteristics and excess pore water pressures of Mississippi Delta sediments. Pap. No. 3167 *Proc. 10th Annual Offshore Technology Conf.,* Houston, Vol. 2, pp. 1037–1048.

Shephard, L. E., Bryant, W. R. and Dunlap, W. A., 1979. Geotechnical properties and their relation to geologic processes in South Pass Outer Continental Shelf lease area blocks 28, 47 and 48, offshore Louisiana. Rec. Rep. 79-5-T, Department of Oceanography, Tex A & M University, College Station, Texas.

Terzaghi, C., 1927. Principles of final soil classification. *Publ. Roads,* **8**, (3), 41–53.

Section 4 — Cyclic Loading and Sub-seabed Stability

Introduction
Liquefaction of seabed sediments: triaxial test simulations

R. K. Taylor (Chairman)

Science Laboratories (Engineering Geology), University of Durham, UK

A distinction is drawn between (critical state) 'actual liquefaction' of loose materials and 'cyclic liquefaction' of denser sediments. The first is triggered by shock loading, the second resulting from sustained cyclic stresses such as earthquake loading or storm loading. Load control and completely reversing cyclic triaxial tests, respectively, are considered to provide appropriate simulations.

The cyclic loading response of argillaceous sediments and of colliery tailings under a simulated 'British earthquake' is discussed in relation to the stability of sediments of similar plasticity accumulating on continental slopes. In particular, reliquefactions of mica fines, and the apparently similar flow rate behaviour of granular and plastic mine tailings during cyclic liquefaction, are demonstrated.

The liquefaction response of North Sea sediments is also commented upon.

Introduction

Research into liquefaction and related processes has not been restricted to seismically active land areas or to seabed sediments from off-shore oilfields. In October 1966 a liquefaction failure of coarse colliery discard in a tip at Aberfan, South Wales, led to a flowslide which killed 144 persons, of whom 116 were young children. A flowslide resulting from a complex piping-type failure of a colliery tailings dam at Buffalo Creek, West Virginia, in 1972 killed 118 people and made 4000 homeless. The huge Montaro flowslide in the Peruvian Andes completely destroyed the village of Mayumarca in 1974. Between July 1964 and June 1974 a total of 305 bulk carriers were lost at sea with no indication of cause. Liquefaction was suspected in a number of cases (see Green and Hughes, 1977; Green and Kirby, 1981). It is the concern over events such as these that has broadened the field of liquefaction investigations.

Subaerial liquefaction, in which the fluid phase is mainly gas, has been termed 'fluidization' by Casagrande (1971). Indeed, the fastest flowing slides, at speeds of between 160 and 320 km/h, would appear to be of this type (see, for example, Bishop, 1973). Recent analysis of mine discard flows treat the materials as a Bingham plastic rheological model (Jeyapalan *et al.,* 1983).

This section centres around four important areas of seabed sediment behaviour under wave loading. They are: cyclic triaxial shear test simulation of (storm) wave forces leading to liquefaction; the use of shear waves in the laboratory for assessing *in situ* liquefaction susceptibility; submarine slope stability evaluation; leg penetration hazards of jack-up rigs. The behaviour of the latter type of rig as a temporary seabed structure was aptly described as a 'grey area' in the Workshop Session, reported on in the Appendix to this section.

It is not intended in this introductory review to comment on all these aspects of seabed mechanics, but rather to concentrate on the most common laboratory test equipment (the triaxial rig) used for evaluating liquefaction susceptibility. More particularly, tests on argillaceous (waste) fines produced by the mining and quarrying industries will be discussed in relation to the stability of cohesive sediments on the continental shelf.

Liquefaction: historical

Quick(sand) conditions most frequently result from the seepage pressure of water percolating upwards through a sediment. The sediment loses its bearing capacity entirely when the hydraulic gradient is unity or more — e.g.

$$\frac{h}{L} > \frac{G_s - 1}{1 + e} \; \hat{=} \; 1$$

where: h = head; L = flow path length; G_s = specific gravity of grains ($\hat{=} 2.65$); e = void ratio ($\hat{=} 0.65$).

Piping may be regarded as a more extreme variant of the quick(sand) condition. However, when the sudden decrease in shearing resistance of the sediment to almost zero occurs without the aid of seepage pressure, the phenomenon is referred to as 'liquefaction' (originally termed 'spontaneous liquefaction'). Terzaghi and Peck (1948) describe the collapse of a metastable grain skeleton when subject to a mild shock. A decrease in volume occurs at an unaltered value of total stress, σ, which is preceded by a temporary increase in pore pressure, u, to a value almost equal to the total stress. Consequently, the effective stress ($\sigma' = \sigma - u$; and the second term of the Coulomb equation) is reduced to negligible proportions, and the material will flow like a viscous liquid. This sort of mechanism invokes a sufficiently loose grain packing for volume change to occur so that a liquefaction failure may develop.

It is here that two different mechanics have been recognized. The first is the 'critical void ratio' (CVR) concept established in the pioneering work of Castro (1969) and Casagrande (1971, 1976). The second concerns the response of denser (dilative) sands to cyclic loading (Seed and Lee, 1966; Seed, 1976). Both approaches may be conceived as particular trigger mechanisms required to promote two different types of liquefaction response:

(1) CVR concept (load control triaxial tests on loose materials): to simulate shock movements such as a sudden rotational (toe) failure in a bank of loose material, sudden differential settlements in loose granular soils, mining subsidence ground movements, short-term local earth tremors and blasting effects.

(2) Cyclic loading (by means of triaxial, simple shear torsion tests, shaking tables): to simulate cyclic stresses experienced during more general earthquake conditions, wave loading of seabed sediments, or sustained vibrations from heavy earth-moving equipment and rail traffic.

The centrifuge is also used for testing models (Schofield, 1980), as well as sand samples by the 'bumpy road' technique. However, as pointed out by Schofield, flights may be expensive.

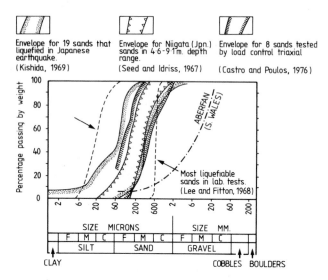

Fig. 1.   Particle size distributions of silts/sands and coarse-grained colliery discard which liquefy (data after Castro and Poulos, 1976, Taylor *et al.*, 1978a; De Herrera et al., 1980).

Before we consider the CVR approach it should be recognized that, like the quick(sand) condition, liquefaction phenomena are not necessarily restricted to fine-grained sediments (see Aberfan grading, Fig. 1). Nevertheless, it has been shown that the most susceptible sediments are in the coarse/medium sand to medium silt sizes, with fine to medium sands being particularly susceptible (Fig. 1). This is in broad agreement with earlier observations of Terzaghi and Peck (1948), who cite an effective grain size of less than 0.1 mm and a uniformity coefficient of less than 5 as susceptibility indicators.

Monotonic load control triaxial tests

Castro (1969) conducted special back-saturated, consolidated-undrained triaxial tests mainly on 'banding sand'. Following consolidation, the axial load was increased with small dead-load increments applied at about 1 min intervals. A liquefaction response to the increasing load is depicted by stress path A in Fig. 2(b). The counterclockwise 'strain softening' path indicates a rise in pore-water pressure during loading. The drop in deviator stress (q) beyond maximum is customarily marked by rapid straining (possibly by as much as 30 per cent in a fraction of a second).

The clockwise (strain hardening) stress path C shows the response of a dense sample which dilates so that the

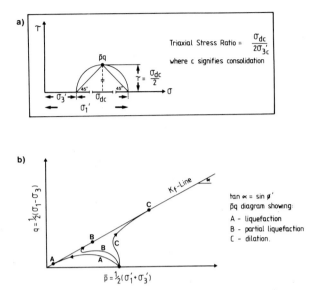

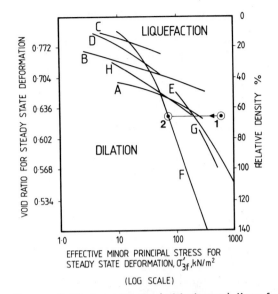

Fig. 2. (a) Triaxial stress ratio in terms of Mohr circle; (b) stress paths for three different responses to load control triaxial test.

Fig. 3. Critical void ratio (critical state) lines for eight different sands (after Castro and Poulos, 1976). Initial and final effective minor principal stress points shown for (diagrammatic) test example.

pore water is in tension and the sample gains strength, moving up the shear strength envelope (K_f line) at failure.

An intermediate, partial liquefaction stress path (path B in Fig. 2b) has also proved to be a common response of fabricated and remoulded specimens of both fine- and coarse-grained colliery discards (Taylor *et al.*, 1978a; Taylor and Morrell, 1979). With partial liquefactions the pore pressure behaves rather erratically after passing q maximum and the specimen shows a tendency to liquefy with increased shear strain. Ultimately, the pore-water pressure decreases, however, as the specimen dilates such that the shear strength is increasing again at failure (strain hardening).

By plotting the ($\log_{10}$) stress difference across the sample (σ'_3) against void ratio at the start and finish of the test, Castro (1969) found that no matter what the value of σ'_3 was initially, all specimens that liquefied or partially liquefied ended up along a reasonably well-defined line. This was defined as the critical void ratio or F line, for which liquefaction with a flow structure develops.

Figure 3, taken from Castro and Poulos (1976), illustrates the overall wide CVR line spectrum of the different sands which they tested. It will also be seen that there is considerable variation in F line gradients for the individual sand types. Diagrammatically on this figure is shown an initial σ'_3 value (point 1) and the failure value (point 2), for a specimen with liquefaction tendencies (CVR line, sand F).

Importantly, sands which dilate and do not liquefy lie below the respective CVR line, while sands which liquefy lie above the line.

The critical state was defined by Hvorslev (1937) as "a condition . . . under which continued flow does not cause further changes in the shearing resistance and void ratio". It has important implications in modern soil mechanics (see, for example, Schofield and Wroth 1968). However, it is currently not possible to recognize any useful correspondence between the liquefaction behaviour of a sand under unidirectional monotonic loading ('actual liquefaction') and under cyclic loading ('cyclic liquefaction'). For example, the void ratios of the Leighton Buzzard sand specimen shown in Fig. 6 were lower than the 'critical void ratio' at the same initial confining pressure of 115 kN/m² ($e = 0.69 -$ CVR value).

Completely reversed cyclic triaxial tests

Although the basic test has a number of limitations (Yoshimi, 1977), it is easy to maintain a constant volume and the surface of the specimen is largely free from shear stresses. Moreover, conventional triaxial equipment is readily modified for cyclic loading (Taylor, *et al.*, 1978b). Because only isotropic consolidation can be simulated, the cyclic *triaxial* stress ratio, TSR (defined in Fig. 2a), has to be modified to take into account K_0 and the limitations of the apparatus. To conform with field (earthquake) estimations of cyclic stress

Fig. 4. A, Sand boil occurring in bed of small river during Nihonkai Chibu earthquake, May 1983 (taken by Mr Naofumi Matsumori, Koho-shitsu (Information Department, Minehama-mura Village Office, Akita Prefecture, Japan. Courtesy of Mr Matsumori). B, Sand boils (volcanoes) photographed a few days later.

ratio, a correction factor, c_r, is required. For the tests reported in this paper, $c_r = 0.56$. It is based on comparisons with multidirectional shaking tables (De Alba *et al.*, 1976).

Seed and Idriss (1971) introduced a simplified procedure so that an appropriate cyclic stress ratio can be used to determine the cyclic strength of the sediment under (level ground) earthquake conditions. The cyclic triaxial stress ratio (TSR = 0.15) used in the tests illustrated in Figs. 5 and 6, for example, is considered to represent a fair simulation of an irregular earth tremor with a peak acceleration of 0.08 g. It is believed that these test conditions model a 200 year return period UK event of about Intensity VI on the Modified Mercali Scale.

Under earthquake loading, pore pressures build-up and the reduction in effective stresses leading to liquefaction is due primarily to shear waves propagating upwards from bedrock. Sand boils (or volcanoes) are a

well-known surface manifestation of liquefaction during earthquakes[*]. The dramatic photographs of the upwards seepage pore pressures in Fig. 4 were taken during the main shock of the Nihonkai Chibu earthquake (middle of Sea of Japan earthquake) on 26 May 1983. The sand volcanoes were photographed a few days later.

Liquefaction of seabed sediments

The same general principles apply in large measure to liquefaction induced by either earthquake loading or wave loading. With the latter, however, there are differences: (a) storm durations are considerably longer than seismic events; (b) wave periods are longer than earthquake cycle periods; (c) wave loading is at the mudline, whereas shear waves induced by earthquakes propagate upwards from a lower level in the ground.

Suitable modification of earthquake-induced liquefaction analysis can be made (for example, Nataraja *et al.*, 1980), and in this section of the Conference Proceedings a 'step by step' analysis is given by Ishihara and Yamazaki (1984). The importance of this latter paper is that a cyclic triaxial torsion shear test is used to accommodate the continuous rotation of principal stresses revealed by stress analysis of seabed conditions. It is pertinent to note that this results in a 20% reduction in the cyclic TSR and a 35% reduction in the conventionally defined cyclic stress ratio.

An objection to cyclic triaxial shear tests has been the contention that stress gradients are unavoidably created near the end platens, such that premature liquefaction is artificially induced just beneath the top platen. Casagrande (1976) refers to the 'pumping action' which the test induces. It is interesting to record that this is precisely the phrase used by Kee and Ims (1984) in the discussion on the difficulties in differentiating between scour and liquefaction in respect of the punching through of legs of jack-up rigs. There was also discussion as to whether scour development at the corners and around the edges of seabed structures might be produced by liquefaction. In this writer's opinion the completely reversed cyclic triaxial test is a singularly appropriate simulation of these storm-loading effects.

Although the North Sea suffers intense winter storms, it would appear that most sands on which platforms have been placed are very dense. A number of contributors believed that the Dutch sector has the more unstable bed, but it was suggested that the burial of a pipeline in the sector may have been caused by scour rather than by liquefaction.

[*]Sometimes seen temporarily in shallow, coastal waters. Observed by Dr R. C. Chaney that sand boils in 3–4.5 m of water off the Californian coast, disappeared after 2 weeks.

Trench wall failure and failures of pipelines by flotation have been examined by Machemehl (1978); two such incidents in the North Sea are reported by Strating (1981).

Liquefaction failures with coastal flowslides occur in the Dutch province of Zeeland (Koppejan *et al.*, 1948), and flowslides in Norwegian fjords are described by Terzaghi (1957) and Bjerrum (1971). A number of these were initiated by the dumping of dredgings and other fills.

Metastable fine sands and coarse silts, subject to collapse and downslope flow, are likely to accumulate in various continental shelf localities (Almagor *et al.*, 1983). In another contribution to this volume (Booth *et al.*, 1984) the bottleneck morphology and lobate nature of the distal end of some debris flows and slides on the continental slope of the USA were strikingly similar to subaerial flowslides. Significantly, (cohesive) sediments with plasticity indices (PI) in the same range as those of colliery tailings and slurries (PI = 4–25) whose cyclic liquefaction response is known accumulate in the shelf environment. The response of argillaceous fines to cyclic loading is undoubtedly pertinent to the stability of these latter *in situ* sediments.

Cyclic loading of argillaceous sediments

The devastating Tangshan (China) earthquake of 1976 demonstrated that soils containing large amounts of plastic fines can experience liquefaction (Chung and Wong, 1982). However, the latter authors concur with earlier observations based on behaviour during earthquakes that a clay content greater than 20% might well signify a liquefaction-resistant soil. Their tests also showed that soils with a PI greater than 15 were likely to be resistant. In the latter context it was concluded by Taylor *et al.* (1978b) that fine colliery discards with a PI > 10 are unlikely to liquefy within the 15 cycles of simulated ground shaking prescribed for the smaller, MMS Intensity VI 'British earthquake'.

It is well known that specimens taken from undisturbed samples are more resistant to liquefaction than those fabricated (or remoulded) in the laboratory[*] (for example, Seed *et al.*, 1977 – sands; Taylor and Morrell, 1979 – fine-grained discards). It is possible that this could in part be due to a sediment's history of preshearing, as suggested by Yoshimi (1977). Samples subjected to moderate levels of cyclic loading (preshearing) exhibit a considerably higher resistance to liquefaction. In marked contrast, Finn *et al.* (1970)

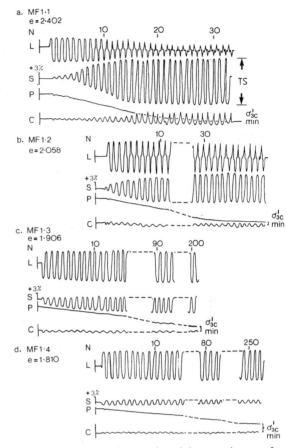

Fig. 5. Condensed from ultraviolet read-out of reliquefaction tests on mica fines. Operating conditions and notation as given in Fig. 6.

and subsequent authors have shown that soil specimens which have undergone liquefaction and are then reconsolidated will display an increased susceptibility to liquefaction. Youd (1977) summarized this phenomenon in terms of two possible explanations: (a) the creation of an unstable fabric – 'potential collapse nodes' – as a function of dilationary tendencies during the first liquefaction, or (b) redistribution of pore-water pressures during prestraining, so producing a loosened zone within the specimen.

The implications of reliquefaction are considerable. Deposits previously affected by liquefaction should be treated as being even more suspect than before the event. It is not unreasonable to suggest that both preshearing and reliquefaction may obtain in the shelf environment. That reliquefaction of argillaceous sediments is possible is illustrated by Fig. 5 (a–d), which is taken from ultraviolet records of cyclic loading sequence conducted by G. R. Morrell at Durham University.

The fabricated test specimen was of silt-sized muscovite (mica fines) with 7% kaolinite impurity: Md = 0.01 mm,

---

[*]Most tests reported in these Conference Proceedings are on fabricated specimens. Results will consequently be on the conservative side.

PI = 10%. It can be seen in Fig. 5(a) that cyclic liquefaction has occurred after about 722 cycles when the
residual effective stress across the specimen ($\sigma'_{3C,min}$)
is approximately zero. The series of reconsolidations
and reliquefactions performed on this specimen (Fig.
5b–d) illustrates that the progressive reduction in void
ratio, which is an expression of increasing density, is
matched by a decrease in the propensity for structural collapse.

Importantly, the rise in pore-water pressure parallels
the relief of intergranular stress as the platelets reorgan-
ize under cyclic shear. The total increase in pore-water
pressure may be expected to be relatively small as
the fabric becomes denser and, hence, the minimum
effective stress reached during each reliquefaction
increases. If the deviator stress is sufficient, cyclic
liquefaction is possible (albeit with more limited shear
strain). Eventually, however, the deviator stress becomes
too small to cause liquefaction at load maxima (Fig. 5d).

A feature which is apparent in this test suite is the
characteristically slow increase in pore-water pressure.
Effective stresses may not be uniform throughout
argillaceous or clay-rich specimens and external measurement of pore-water pressure is probably impeded by
low permeability.

With reliquefied Leighton Buzzard sand illustrated in
Fig. 6, the pore-water pressure build-up is very fast;
much more than for the original specimen, which
was fabricated in a relatively dense state. Nevertheless,
the void ratio of the reliquefied sand is lower than that
of the original. A third reconsolidation followed by
cyclic liquefaction is not shown in Fig. 6. However, in
this final stage a further decrease in void ratio to 0.591
was matched by a decrease in shear strain, but with a
corresponding increase of approximately 88% in residual
effective stress, $\sigma'_{3C,min}$. Here again, although the sand
becomes mobile more rapidly, it only remains so during
parts of the cycle because it is flowing under a greater
residual effective stress.

Effective stress is believed by Taylor and Morrell
(1979) to be an important control on the velocity of
flow of sediments during cyclic liquefaction. Using the
subaqueous mass transport terminology of Carter
(1975), the types of flow initiated by liquefaction in
fine-grained colliery discards are *inertia* flows. These
are divided into *grain flows* and *slurry flows,* with the
latter containing clay and flowing with enhanced viscosity. However, at zero effective stress (cyclic liquefaction conditions) the total shear strain amplitudes
(defined as TS in Fig. 5) and the maximum shear strains,
as a percentage of initial specimen length, were the
same in both plastic and non-plastic discards. Since
the particular plastic discards had clay size contents
of up to 15 per cent by weight and PI of up to 16,
they were presumably flowing as slurries (*sensu* Carter).

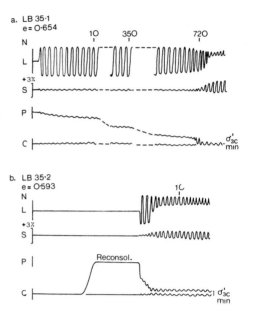

Fig. 6.   Condensed ultraviolet charts for reliquefaction
tests on Leighton Buzzard sand. Notation: (N) = No. of
load cycles, (L) = load trace; (S) = strain trace, (P) =
pore-water pressure; (C) confining pressure; ($\sigma'_{C,min}$) =
residual effective confining pressure; initial effective
confining pressure = 115 kN m^{-3}; triaxial stress ratio =
0.15; (TS) = total strain.

Under the same effective stress conditions there is thus
no tangible evidence to suggest that non-plastic (granular)
sediments should move faster as *grain flows* than the
more argillaceous sediments with up to 15 per cent
plastic fines.

Summary and conclusions

In addition to observed earthquake-induced liquefaction
phenomena on land (sand boils and large settlements of
structures), flowslides have caused much concern in
many parts of the world. Liquefaction of bulk cargoes
is also of great consequence in marine engineering.

A distinction can be drawn between 'actual liquefaction', pertaining to essentially loose granular materials,
and 'cyclic liquefaction' of denser sediments, including
argillaceous types. Earlier pioneering work concerned
with loose sands adopted a 'critical state approach'.
Liquefaction occurs in response to the appropriate
trigger mechanism: the sudden shock in loose materials
compared with sustained, small cyclic (completely
reversing) stresses in denser materials.

The load control triaxial test can be used to simulate
shock loading of loose materials and the completely
reversing cyclic triaxial shear test is a fair simulation of
earthquake and wave(storm) loadings. Objections to the

'pumping action' of the cyclic triaxial test are unlikely to be valid in storm loading conditions at the mudline. Ishihara and Yamazaki (1983) advocate a cyclic triaxial torsion shear test to accommodate the continuous rotation of principal stresses which stress analysis of seabed conditions reveals.

Drawing a distinction between liquefaction and scour effects was identified in discussion (largely on the North Sea) as an unresolved area of seabed mechanics. Appropriate *in situ* instrumentation was agreed to be lacking.

Observations following earthquake loading have shown that soils containing large amounts of plastic fines can experience liquefaction. There is some evidence to suggest that a clay size content greater than 20% may signify liquefaction resistance.

Under a 100 year return period Intensity VI (MMS scale) 'British earthquake', it is unlikely that liquefaction of colliery tailings with a PI index greater than 10 would occur. Cyclic tests of Chung and Wong (1982) tentatively suggest a PI greater than 15 for soils under more severe conditions.

With sustained cycling (more in conformity with storm loading), cyclic liquefaction can be induced in fine-grained colliery discards in the PI range 4–25. Tests on both mica fines and Leighton Buzzard sand are reported in which subsequent reliquefactions are induced. With each successive cyclic liquefaction the decreasing void ratio is matched by a decreasing shear strain. The total pore-water pressure build-up is progressively smaller as the soil fabric becomes denser, and consequently the minimum (residual) effective stress increases with each reliquefaction. There is also reason to believe (Taylor and Morrell, 1979) that during cyclic liquefaction argillaceous sediments containing up to 15% plastic fines (PI up to 16) will flow at a similar velocity to that of non-plastic, granular sediments.

The significance of the above tests is directed to those sediments of similar plasticity which accumulate on the continental slope. There is presumably a balance between processes such as consolidation and preshearing history, which will increase strength, as opposed to earth tremors and sustained storm loadings, which may cause liquefaction and possibly reliquefaction.

Acknowledgements

Warm thanks are due to Professor Shigeyasu Okusa (Tokai University), who forwarded the photographs of the 1983 Nihankai Chibu earthquake. The writer also acknowledges laboratory assistance of Mr G. R. Morrell, formerly Senior Research Assistant, University of Durham.

References

Almagor, G., Bennett, R. H., Lambert, D. N. and Forde, L. S., 1984. Analysis of slope stability, Wilmington and Lindenkohl Canyons, US Atlantic margin. This volume, pp. 77–86.

Bishop, A. W., 1973, The stability of tips and spoil heaps. *Quart. J. Engng Geol.*, **6**, 335–376.

Bjerrum, L., 1971. Subaqueous slope failures in Norwegian fjords. Norwegian Geotechnical Institute Publication No. 88.

Booth, J. S., Silva, A. J. and Jordan, S. A., 1984. Slope-stability analysis and creep susceptibility of Quaternary sediments on the Northeastern United States continental slope. This volume, pp. 65–76.

Casagrande, A., 1971. On liquefaction phenomena (see Green and Ferguson, 1971).

Casagrande, A., 1976. Liquefaction and cyclic deformation of sands; a critical review, Harvard Soil Mechanics Series No. 88, Harvard University.

Carter, R. M., 1975. A discussion and classification of subaqueous mass-transport with particular application to grain-flow, slurry flow, and fluxoturbidites. *Earth Sci. Rev.*, **11**, 145–177.

Castro, G., 1969. Liquefaction of Sands. Ph.D. thesis, Harvard University.

Castro, G. and Poulos, S. J., 1976. Factors affecting liquefaction and cyclic mobility. In: *Liquefaction Problems in Geotechnical Engineering – ASCE National Convention*, Philadelphia, pp. 105–131.

Chung, Kin Y. C. and Wong, I. H., 1982. Liquefaction potential of soil with plastic fines. *Proc. Soil Dynamics and Earthquake Engineering Conf.* Southampton, Vol. 2, pp. 887–897.

De Alba, P., Seed, H. B. and Chan, C. K., 1976. Sand liquefaction in large scale simple shear tests. *J. Soil Mech. Foundns Div., ASCE*, **102** (GT9), 909–927.

De Herrera, M. A., Zsutty, T. C. and Abom, C. A., 1980. The analysis of liquefaction potential based on probabilistic ground motions. *Proc. Int. Symp. Soils under Cyclic and Transient Loading*, Swansea, Vol. 1, pp. 517–521.

Finn, W. D. L., Bransby, P. L. and Pickering, D. J., 1970. Effect of strain history on liquefaction of sand. *J. Soil Mech. Foundns Div., ASCE*, **96** (SM6), 1917–1934.

Green, P. A. and Ferguson, P. A. S., 1971. 'On liquefaction phenomena' by Professor A. Casagrande: Report of lecture. *Géotechnique*, **21**, 197–202.

Green, P. B. and Hughes, T. H., 1977. Stability of bulk mineral cargoes. *Trans. Inst. Min. Metall.*, **86**, A150–A158.

Green, P. B. and Kirby, J. M., 1981. Behaviour of damp fine-grained bulk mineral cargoes. *Trans. Inst. Mar. Engrs*, **94**, Paper 19.

Hvorslev, M. J., 1937. Uber di fesigkeitseigenschaften gestörter bindiger böden. Ingidensk. Skr. A, No. 45 [English translation No. 69–3. Waterways Experiment Station, Vicksburg, Miss., 1969].

Ishihara, K. and Yamazaki, A., 1984. Wave-induced liquefaction in seabed deposits of sand. This volume pp. 139–148.

Kee, R. and Ims, B. W., 1984. Geotechnical hazards associated with leg penetration of jack-up rigs. This volume, pp. 169–174.

Jeyapalan, J. K., Duncan, J. M. and Seed, H. B., 1983. Investigation of flow failures of tailings dams. *J. Geotech. Engng Div., ASCE,* **109** (2), 172–201.

Koppejan, A. W., Wamelen, B. M. and Weinberg, L. J., 1948. Coastal flowslides in the Dutch province of Zeeland. *Proc. 2nd Int. Conf. Soil Mechanics and Foundation Engineering,* Rotterdam, Vol. 5, pp. 89–96.

Machemehl, J. L., 1978. Pipelines in the coastal ocean. In: *Pipelines in Adverse Environments — A State of the Art.* ASCE Special Conference Pipelines Div., T.P., pp. 187–203.

Nataraja, M. S., Singh, H. and Maloney, D., 1980. Ocean wave-induced liquefaction analysis: A simplified procedure. *Proc. Int. Symp. Soils under Cyclic and Transient Loading,* Swansea, Vol. 1, pp. 509–516.

Schofield, A. N., 1980. Cambridge geotechnical centrifuge operations. *Geotechnique,* **30**, 227–268.

Schofield, A. N. and Wroth, C. P., 1968. *Critical State Soil Mechanics.* McGraw-Hill, New York.

Seed, H. B., 1976. Evaluation of soil liquefaction effects on level ground during earthquakes. In: *Liquefaction Problems in Geotechnical Engineering.* ASCE National Convention, Philadelphia, pp. 1–104.

Seed, H. B. and Idriss, I. M., 1971. Simplified procedure for evaluating soil liquefaction potential. *J. Soil Mech. Foundns Div., ASCE,* **97** (SM9), 1249–1273.

Seed, H. B. and Lee, K. L., 1966. Liquefaction of saturated sands during cyclic loadings. *J. Soil Mech.*

Foundns Div., ASCE, **92** (SM6), 105–134.

Seed, H. B., Mori, K. and Chan, C. K., 1977. Influence of seismic history on liquefaction of sands. *J. Geotech. Engng Div., ASCE,* 103 (GT4), 257–270.

Strating, J., 1981. A survey of pipelines in the North Sea: incidents during installation, testing and operation. *Proc. Offshore Technology Conf.,* Pap. No. 4069, pp. 25–32.

Taylor, R. K., Kennedy, G. W., and MacMillan, G. L., 1978a. Susceptibility of coarse-grained coal-mine discard to liquefaction. *Proc. III Int. Congr., International Association of Engineering Geology,* Madrid. Spec. Sess. Vol. 3, pp. 91–100.

Taylor, R. K., Morrell, G. R. and Macmillan, G. L., 1978b. Liquefaction response of coal-mine tailings to earthquakes. *Proc. III Int. Congr., International Association of Engineering Geology,* Madrid. Spec. Sess. Vol. 3, pp. 79–90.

Taylor, R. K. and Morrell, G. R., 1979. Fine-grained colliery discard and its susceptibility to liquefaction and flow under cyclic stress. *Engng Geol.,* **14**, 219–229.

Terzaghi, K., 1957. Varieties of submarine slope failures. Publ. No. 25, Norwegian Geotechnical Institute.

Terzaghi, K. and Peck, R. B., 1948. *Soil Mechanics in Engineering Practice.* Wiley, New York, 566 pp.

Yoshimi, Y., 1977. Liquefaction and cyclic deformation of soils under undrained conditions. *Proc. 9th Int. Conf. Soil Mechanics and Foundation Engineering,* Tokyo, Vol. 2, pp. 613–623.

Youd, T. L., 1977. Packing changes and liquefaction susceptibility. *J. Geotech. Engng Div., ASCE,* **102** (GT8), 918–922.

12

Wave-induced Liquefaction in Seabed Deposits of Sand

Kenji Ishihara and Akira Yamazaki
Department of Civil Engineering, University of Tokyo, Japan

To explore the nature of cyclic stress alteration in seabed deposits of sand due to travelling waves, two-dimensional stress analysis was carried out on a homogeneous elastic half-space subjected to a series of harmonic loads moving on its surface. The analysis indicated that changes in shear stress occur in such a way that, while its amplitude is maintained constant, the directions of the principal stresses rotate continuously. With a view to simulating such stress changes in the laboratory test, a series of cyclic triaxial torsion shear tests was conducted on loose specimens of sand. The test results indicated that the conventionally defined cyclic stress ratio is reduced by about 30 per cent if the rotation of the principal stress directions is involved in the cyclic loading. It is known that the magnitude of wave-induced pressure at the seabed changes as ocean waves move in towards the shore. In unison with the changes in the pressure wave, the cyclic stress ratio induced in the seabed also changes. However, owing to the wave breaking, there exists an upper limit in the magnitude of the wave pressure and, hence, in the induced cyclic stress ratio. With these characteristics in mind, a set of charts was provided to facilitate computation of the induced cyclic stress ratio in the seabed deposit of water of any depth, on the assumption that the seabed consists of a homogeneous isotropic elastic half-space. The charts are organized so that the induced cyclic stress ratio can be obtained for known wave parameters that are specified by design storms at an offshore location. The cyclic stress ratio thus obtained was compared with the corresponding cyclic stress ratio causing failure in the sand, which was determined in the laboratory with a cyclic triaxial torsion shear test equipment in which continuous rotation of the principal stress direction was executed in simulation of the wave-induced stress conditions occurring in the seabed deposit.

Introduction

The phenomenon of ocean wave-induced liquefaction has attracted increasing attention in recent years in relation to the integrity of near-shore and offshore installations such as pipelines, anchors and platform structures. Differential loading on the seafloor by the pressure wave induces a sequence of cyclic shear stress in the underlying soil, and if the induced shear stress exceeds the strength, significant deformation or liquefaction failure may occur, thereby exerting damaging influences on nearby engineering installations. This issue was put in proper perspective by Henkel (1970, 1982), and several investigators have elaborated analytical frameworks for liquefaction analysis under wave loadings. Seed and Rahman (1978) developed a methodology in which generation and contemporaneous dissipation of excess pore-water pressures during strong

wave loadings can be evaluated. Finn et ·al. (1983) extended this approach to stability analyses of the seabed deposit on the basis of the effective stress principle. The more practical aspect of the wave-induced liquefaction was dealt with by Nataraja and Gill (1983), who tried to explain several cases of seabed failure on the basis of the simplified procedure of liquefaction. Liquefaction potential of the seafloor deposits at particular sites was studied by Lee and Focht (1975) and Clukey et al. (1980) on the basis of the simplified procedure.

Although the study described in this paper is an extension of the works mentioned above, two novel features appear to be important in establishing a framework for the study of wave-induced liquefaction. The first aims to clarify the nature of cyclic alteration of the wave-induced shear stress, which is characterized by rotation of the principal stress directions; the second is to propose a methodology for evaluating the magnitude of cyclic stress on the basis of design storm parameters specified at an offshore location in deep water. Considering these aspects, a new methodology for analysis of wave-induced liquefaction is described.

Characteristics of wave-induced cyclic stress

Water waves propagating on the ocean may be considered to consist of an infinite number of wave trains having a constant amplitude and wavelength. Passage of such an array of waves on the seafloor creates harmonic pressure waves on the sea-floor, increasing the pressure under the crest and reducing it under the trough. The stresses induced on the seabed are, therefore, analysed by applying a sinusoidally changing load on the horizontal surface from minus to plus infinity, as illustrated in Fig. 1. If the seabed deposit is assumed to consist of a homogeneous elastic material extending to an infinite depth, the stresses can be readily determined by using Boussinesq's classical solution for the two-dimensional plane strain problem. Assume that a harmonic load

$$p(x) = p_0 \cos\left(\frac{2\pi}{L}x - \frac{2\pi}{T}t\right) \tag{1}$$

is distributed on the surface of an elastic half-space, where p_0 is the amplitude of the load, L is the wavelength and T is the period of waves. The vertical normal stress, σ_v, horizontal normal stress, σ_h, and shear stress, τ_{vh}, induced in the half-space by this load are determined (Yamamoto, 1978; Madsen, 1978) as

$$\left.\begin{array}{l} \sigma_v = p_0 \left(1 + \frac{2\pi z}{L}\right) \exp\left(-\frac{2\pi z}{L}\right) \cdot \cos\left(\frac{2\pi x}{L} - \frac{2\pi t}{T}\right) \\[2mm] \sigma_h = p_0 \left(1 - \frac{2\pi z}{L}\right) \exp\left(-\frac{2\pi z}{L}\right) \cdot \cos\left(\frac{2\pi x}{L} - \frac{2\pi t}{T}\right) \\[2mm] \tau_{vh} = p_0 \frac{2\pi z}{L} \cdot \exp\left(-\frac{2\pi z}{L}\right) \cdot \sin\left(\frac{2\pi x}{L} - \frac{2\pi t}{T}\right) \end{array}\right\} \tag{2}$$

where x and z indicate the spatial coordinates in the horizontal and vertical directions, respectively, as illustrated in Fig. 1. It is well known that the major components of stress associated with the shearing deformation of a soil body are the stress, τ_{vh}, and the stress difference, defined as $(\sigma_v - \sigma_h)/2$. For the present problem, the stress difference is calculated from equation (2) as follows:

$$\frac{\sigma_v - \sigma_h}{2} = p_0 \frac{2\pi z}{L} \exp\left(-\frac{2\pi z}{L}\right) \cdot \cos\left(\frac{2\pi x}{L} - \frac{2\pi t}{T}\right) \tag{3}$$

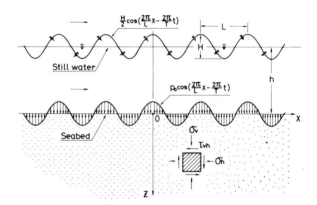

Fig. 1.   Definition of notation.

Comparison of the expressions for τ_{vh} and $(\sigma_v - \sigma_h)/2$ indicates that the amplitude of these two components of stress is identical and they differ only in the time phase of cyclic load application. In order to obtain an insight into the nature of the cyclic loading, let the variable z be eliminated between τ_{vh} and $(\sigma_v - \sigma_h)/2$. Then one obtains

$$\tan\left(\frac{2\pi x}{L} - \frac{2\pi t}{T}\right) = \frac{2\tau_{vh}}{\sigma_v - \sigma_h} \tag{4}$$

According to elementary stress analysis, the right-hand side of equation (4) is equal to the tangent of twice the angle of the major principal stress direction to the vertical axis, β, as illustrated in Fig. 2(a). Hence, one obtains

$$\beta = \frac{\pi}{L}x - \frac{\pi}{T}t = \tfrac{1}{2}\tan^{-1}\left(\frac{2\tau_{vh}}{\sigma_v - \sigma_h}\right) \tag{5}$$

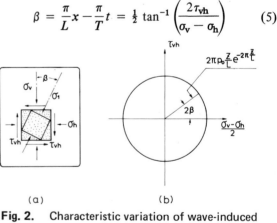

(a)   (b)

Fig. 2.   Characteristic variation of wave-induced shear stress.

This equation implies that, when attention is drawn to stress changes in a soil element at a fixed point x, the direction of the principal stress is rotating continuously through $180°$ during one period, T, of cyclic load application. Similar rotation of the principal stress axes is taking place simultaneously throughout the depth of the half-space. If the variable $(2\pi x/L - 2\pi t/T)$ is eliminated between the shear stress, τ_{vh}, and $(\sigma_v - \sigma_h)/2$, one obtains

$$\left(\frac{\sigma_v - \sigma_h}{2}\right)^2 + \tau_{vh}^2 = p_0^2 \left(\frac{2\pi z}{L}\right)^2 \cdot \exp\left(-\frac{4\pi z}{L}\right) \quad (6)$$

Elementary stress analysis shows that the left-hand side of equation (6) is equal to the square of the deviator stress, which is defined as the difference between the maximum and minimum principal stresses, σ_1 and σ_3, respectively, divided by 2. Hence one obtains

$$\frac{\sigma_1 - \sigma_3}{2} = \left\{\left(\frac{\sigma_v - \sigma_h}{2}\right)^2 + \tau_{vh}^2\right\}^{1/2} = p_0 \frac{2\pi z}{L} \exp\left(-\frac{2\pi z}{L}\right) \quad (7)$$

This equation indicates that, when attention is drawn to stress changes in a soil element at a fixed depth, the deviator stress is unchanged at all times and at all points on the horizontal plane during the cyclic load application. The same nature of cyclic alteration of stress is taking place at each depth of the half-space. As can easily be understood from equation (7), the two components of stresses, τ_{vh} and $(\sigma_v - \sigma_h)/2$, are increasing or decreasing alternately so as to keep the deviator stress at a constant value during the entire period of cyclic load application. When the two components of stresses are represented in a rectangular coordinate system, equation (6) is an equation of a circle with its radius equal to the deviator stress, $(\sigma_1 - \sigma_3)/2$. Such a plot is presented in Fig. 2(b). In this type of plot, the angle which a stress vector makes to the horizontal coordinate axis represents twice the angle of the maximum principal stress axis to the vertical, 2β. It can be mentioned in summary that the cyclic change of shear stress induced in an elastic half-space by a harmonic load moving on its surface is characterized by a continuous rotation of the principal stress direction, the deviator stress being maintained constant. Therefore, if the load on the sea bottom is represented by a propagating harmonic load and if the seabed deposit is assumed to be an elastic half-space, the cyclic stress induced in the deposit has characteristics such that the cyclic alteration of shear stress is executed with a continuous rotation of the principal stress direction.

Next consider the distribution of the amplitude of the cyclic deviator stress versus depth. Since the deviator stress amplitude is equal to either the amplitude of shear stress, τ_{vh}, or the stress difference, $(\sigma_v - \sigma_h)/2$, it will suffice to consider only the amplitude of shear stress, τ_{vh}, which is given by equation (2). Numerical values of the amplitude of τ_{vh} are plotted against depth in Fig. 3. It can be seen that the shear stress takes a maximum value of $0.368p_0$ at a depth equal to 15.9 per cent of the wavelength. A more

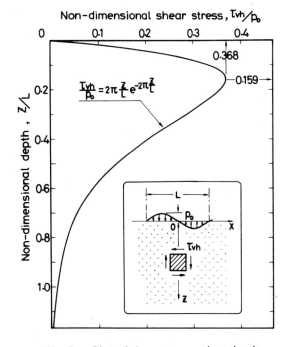

Fig. 3. Plot of shear stress against depth.

significant variable associated with the deformability or strength of soils under cyclic loading conditions is the cyclic stress ratio, which is defined as the ratio between the amplitude of shear stress and that of the effective confining stress. In the horizontal deposition of soils the vertical effective overburden pressure, σ_v', is generally used as a parameter to represent the confinement of soils *in situ*. Since the magnitude of σ_v' is expressed by $\sigma_v' = \rho'gz$, the cyclic stress ratio is given by

$$\frac{\tau_{vh}}{\sigma_v'} = \frac{2\pi}{\rho'g} p_0 \frac{1}{L} \exp\left(-\frac{2\pi z}{L}\right) \quad (8)$$

where z is the depth of the deposit and ρ' is the submerged unit mass of soils in the seabed. The value of the cyclic stress ratio at the mudline is derived immediately by putting $z = 0$ in equation (8):

$$\left(\frac{\tau_{vh}}{\sigma_v'}\right)_0 = \frac{2\pi p_0}{\rho'g L} \quad (9)$$

Therefore the cyclic stress ratio at any depth normalized to the value at the mudline is obtained as

$$\left(\frac{\tau_{vh}}{\sigma_v'}\right)\Bigg/\left(\frac{\tau_{vh}}{\sigma_v'}\right)_0 = \exp\left(-\frac{2\pi z}{L}\right) \quad (10)$$

This relationship is presented numerically in Fig. 4, from which it can be seen that the cyclic stress ratio decreases sharply with depth near the surface of the seabed.

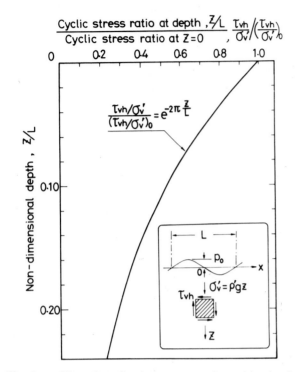

Fig. 4.   Plot of cyclic shear stress ratio against depth.

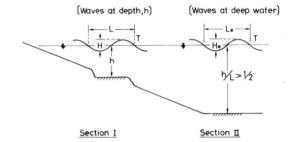

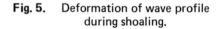

Fig. 5.   Deformation of wave profile during shoaling.

Wave-induced pressure at sea bottom

The height of waves in the ocean is generally very small compared with the length of the waves. The basic equation of water waves solved on this assumption leads to the well-known theory of small-amplitude waves (Horikawa, 1978). According to this theory, the amplitude of the pressure fluctuation exerted on the sea bottom, p_0, due to the travelling wave is given by

$$p_0 = \rho_w g \frac{H}{2} \frac{1}{\cosh\left(2\pi \frac{h}{L}\right)} \qquad (11)$$

where H denotes the height and L is the length of the wave at the place where the water depth is h, as illustrated in Fig. 1. The term ρ_w is the unit mass of the water and g is gravitational acceleration. The values of H and L in equation (11) can not be arbitrarily chosen, but are derived, as follows, from the basic solution of the small-amplitude wave theory,

$$L = \frac{gT^2}{2\pi} \tanh\left(\frac{2\pi h}{L}\right) \qquad (12)$$

and from the energy continuity equation,

$$H = H_0 \left[\left(1 + \frac{\left(4\pi \frac{h}{L}\right)}{\sinh\left(4\pi \frac{h}{L}\right)}\right) \tanh\left(2\pi \frac{h}{L}\right)\right]^{-1/2} \qquad (13)$$

where H_0 denotes the height of waves at deep water condition where the depth of water is large compared with half of the wavelength. Equation (13) is derived on the assumption that no energy loss is involved during wave propagation over the sloping bottom towards the shore whereby the water depth decreases gradually. The wave profiles at two cross-sections over the sloping bottom are shown in Fig. 5. Section I is assumed to be located in the nearshore region, off the surf zone, where the waves are still stable. Section II is assumed to be sufficiently offshore so that the length and height of the waves are determined by fully developed wind waves irrespective of the depth of water. The wave profile at a location such as section II is referred to as deep water condition. The energy continuity equation of equation (13) is established by equating the energy being transmitted by the wave train at the section I to the energy being transmitted by the wave system in deep water. In deriving equation (13) it is assumed that the wave period remains unaltered in water of any depth, whereas the wavelength, velocity and height vary as the wave proceeds towards the shore. For waves travelling in deep water, equation (12) is reduced to

$$L_0 = \frac{gT^2}{2\pi} \qquad (14)$$

where L_0 denotes the wavelength in deep water. Combining equations (12) and (14), one obtains

$$L = L_0 \tanh\left(\frac{2\pi h}{L}\right) \qquad (15)$$

This equation indicates how the wavelength changes as the wave propagates in water of decreasing depth towards the shore.

A most commonly used procedure to evaluate the pressure, p_0, exerted on the sea bottom at a location of given water depth, h, is to envisage an overall profile of waves propagating in water of decreasing depth from the deep water towards the shore, as illustrated in Fig. 5, and to determine the wave height, H, and length, L from equations (13) and (15) on the basis of the key wave parameters, T, L_0 and H_0, specified at a location in deep water. The values of L and H thus determined in water depth h are then introduced in equation (11) to determine the amplitude of pressure p_0.

Cyclic stress ratio at mudline

With the amplitude of pressure p_0 known by the procedure as described above, it is now possible to evaluate the cyclic stress ratio at mudline. By introducing equation (11) into equation (9), a more explicit expression can be obtained, as follows:

$$\left(\frac{\tau_{vh}}{\sigma'_v}\right)_0 = \pi \frac{\rho_w H}{\rho' L} \frac{1}{\cosh\left(2\pi\frac{h}{L}\right)} \qquad (16)$$

Introducing the expressions for H and L from equations (13) and (15), one can rewrite equation (16) as

$$\left(\frac{\tau_{vh}}{\sigma'_v}\right)_0 = \frac{\pi \dfrac{\rho_w H_0}{\rho' L_0}}{\sinh\left(\dfrac{2\pi h}{L}\right)\left\{\tanh\left(\dfrac{2\pi h}{L}\right)\left(1 + \dfrac{4\pi h/L}{\sinh\left(\dfrac{4\pi h}{L}\right)}\right)\right\}^{1/2}}$$

$$(17)$$

Equation (16) is numerically presented in Fig. 6 for different values of H_0/L_0 ranging between 0.008 and 0.13. As mentioned in the foregoing section, values of L_0 and H_0 in deep water are first specified. Then, using equation (15), a value of wavelength, L, at the location of given depth, h, is determined. By inserting all these values into equation (17), the cyclic stress ratio at the mudline can be determined. To facilitate the computation, equation (15) is expressed in the form of a chart in Fig. 7 by plotting L against h. In some cases the wave conditions in deep water are more conveniently specified in terms of the period, T, and the wave steepness, H_0/L_0. In such cases it is expedient first to compute the wavelength, L_0 through equation (14) and then proceed in the same way as mentioned above.

Equation (17) indicates that the cyclic stress ratio at the mudline can increase infinitely when the water depth becomes small relative to wavelength. In such a situation the wave height is large compared with wavelength, as is easily seen from equation (16). However, it

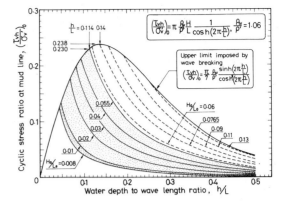

Fig. 6.   Cyclic stress ratio at the mudline.

has been shown that the wave steepness defined by H/L can not increase beyond a certain critical value at which the wave breaks. According to some studies in the field of coastal engineering (Horikawa, 1978), the critical condition for progressive waves to break is given by

$$\frac{H}{L} \leq \frac{1}{7} \tanh\left(\frac{2\pi h}{L}\right) \qquad (18)$$

Equation (18), setting an upper limit for the wave steepness, also sets an upper boundary for the cyclic stress ratio. By merging the condition of equation (18) with equation (16) the limiting condition for the cyclic stress ratio can be obtained as

$$\left(\frac{\tau_{vh}}{\sigma'_v}\right)_0 \leq \frac{\pi}{7} \frac{\rho_w}{\rho'} \frac{\sinh\left(\dfrac{2\pi h}{L}\right)}{\cosh^2\left(\dfrac{2\pi h}{L}\right)} \qquad (19)$$

This equation is numerically presented in Fig. 6, setting an upper boundary for the cyclic stress ratio. It is to be remembered that the cyclic stress ratio lying above this boundary curve in Fig. 6 cannot exist in reality in the seabed deposit because of the breaking occurring at the crest of propagating waves. It is also to be noted that cyclic stress ratio at the wave breaking is a function of only h/L and can be determined without knowing a value of the wave steepness in deep water condition. However, if the wave steepness, H_0/L_0, corresponding to a given value of h/L needs to be known, it can easily be determined through the following formula, which has been obtained by equating equations (17) and (19):

$$\frac{H_0}{L_0} = \frac{1}{7}\left[\tanh\left(\frac{2\pi h}{L}\right)\right]^{5/2}\left\{1 + \frac{4\pi h/L}{\sinh(4\pi h/L)}\right\}^{1/2} \qquad (20)$$

It is of interest to note that the limiting condition of equation (19) specifies an absolute maximum for the cyclic stress ratio that can exist under the conditions

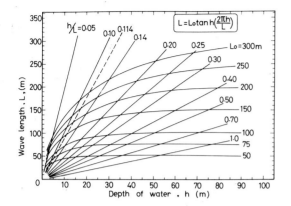

Fig. 7.   Plot of wavelength against water depth.

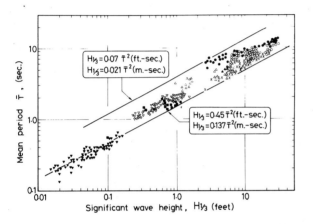

Fig. 8.   Plot from observed data of mean period against significant wave height.

of wave breaking. Simple analytical treatment of equation (19) shows that the maximum value of the cyclic stress ratio is equal to

$$\left(\frac{\tau_{vh}}{\sigma'_v}\right)_0 = 0.2379 \qquad (21)$$

and this occurs when $h/L = 0.140$. The value of the wave steepness, H_0/L_0, in deep water condition corresponds to the above maximum cyclic stress ratio can be determined by introducing $h/L = 0.140$ into equation (20). The value thus obtained was $H_0/L_0 = 0.0765$ and is indicated accordingly in Fig. 6.

Wave profile in deep water

As pointed out in the foregoing, a knowledge of the wave length, L_0, and wave height, H_0, in deep water is necessary for evaluating the amplitude of wave pressure, p_0, at any water depth on a shoaling shore.

It has also been shown that the maximum cyclic stress ratio calculated from the pressure p_0 occurs when waves break and can be determined if the value of wave steepness, H_0/L_0, is known for deep water condition. Therefore, when considering a state of wave breaking, there is no need to know the individual value of H_0 and L_0.

The wave steepness in deep water condition may be inferred from various empirical formulae established on the basis of laboratory and field observations. One such formula that can be used for the present purpose is that derived from data plotted as shown in Fig. 8 (Wiegel, 1964). On the basis of a straight line in Fig. 8, Wiegel proposed an empirical formula,

$$H_{\frac{1}{3}} = 0.137\bar{T}^2 \qquad (22)$$

where $H_{\frac{1}{3}}$ is the significant wave height, defined as the average height of the one-third highest waves of a given wave group, and $\bar{T}$ denotes the mean period in the given wave group. In equation (22) $H_{\frac{1}{3}}$ is expressed in meters and $\bar{T}$ in seconds. It is also known (Horikawa, 1978) that the significant wave height is related to the average wave height, $\bar{H}$, through

$$H_{\frac{1}{3}} = 1.60\bar{H} \qquad (23)$$

Merging equations (23) and (22), one obtains

$$\bar{H} = 0.0856\bar{T}^2 \qquad (24)$$

This empirical formula may be taken as being approximately valid also for correlating the wave height, H_0, and wave period, T, for the case of harmonic waves. Then we obtain

$$H_0 = 0.0856T^2 \qquad (25)$$

In deep water condition the wave length, L_0, is related to the wave period through equation (14). Therefore, by eliminating T between equations (14) and (25), the wave steepness in deep water can be obtained as

$$\frac{H_0}{L_0} = 0.055 \qquad (26)$$

The above consequence indicates that the wave steepness in deep water takes a unique value irrespective of many other factors which seem to influence the formation of wave profile in deep water. However, data points in the chart by Wiegel show some degree of scatter, and it appears desirable to consider a variation within certain limits for the choice of the constant appearing in equation (22). A reasonable range of variation of this constant may be read off from Fig. 8 to be between 0.021 and 0.137 m s^{-2}. Using this range of variation together with equation (14), the range of variation in the wave steepness in deep water may be written as

$$0.008 \leq \frac{H_0}{L_0} \leq 0.055 \qquad (27)$$

The value of wave steepness computed from Wiegel's empirical relationship of equation (22) gives the highest possible value at $H_0/L_0 = 0.055$.

Upper limit in the cyclic stress ratio

It has been shown that the cyclic stress ratio induced by waves at the mudline is determined by equation (17) as a function of h/L, if the wave steepness in deep water, H_0/L_0, is known. It has been pointed out that the cyclic stress ratio thus determined does not become infinitely large, but is bounded by the condition of wave breaking as represented by equation (18). Since the variation of the wave steepness, H_0/L_0, is practically limited within a certain range, as indicated above, the value of the cyclic stress ratio is accordingly limited to a range of variation indicated in Fig. 6 by a dotted area. It is important to notice that in practice there exists a maximum value in the cyclic stress ratio which is equal to 0.230, and this maximum occurs with $H_0/L_0 = 0.055$ when $h/L = 0.114$.

Cyclic triaxial torsional shear test

In order to study the resistance of sands under the cyclic loading conditions simulating the characteristic variation of stresses induced in the seabed by waves, a series of laboratory tests was conducted using a triaxial torsional shear test apparatus in which the horizontal shear stress component, τ_{vh}, and the component of the stress difference, $(\sigma_v - \sigma_h)/2$, can be varied independently. In this apparatus a hollow cylindrical test specimen 10 cm in outer diameter, 6 cm in inner diameter and 10.4 cm in height was encased in rubber membranes and positioned in the triaxial chamber.

The sand used was a type of Japanese standard sand known as Toyoura sand. Its mean particle size, D_{50}, is 0.17 mm and the uniformity coefficient is 2.0. Subrounded to subangular in grain shape, the sand has maximum and minimum void ratios of 0.98 and 0.60, respectively. The specific gravity is 2.65. In the present study, test specimens were prepared by the method of pluviation to a loose state of packing having a relative density of approximately 40–50%. The specimens were consolidated under a confining stress of 294 kN/m² and then subjected to cyclic loads under undrained conditions. Three types of cyclic loading tests were conducted by changing the two components of shear stress either singly or in combination. The loading schemes of these tests are illustrated in the stress space shown in Fig. 9. In the cyclic test with circular rotation of principal stresses, two components of shear stresses were increased or decreased alternately so that the

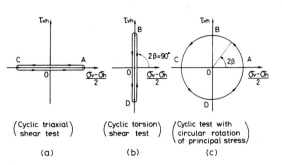

Fig. 9. Stress paths in cyclic loading tests.

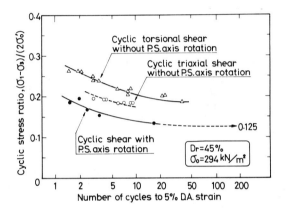

Fig. 10. Plot of cyclic stress ratio against number of cycles for three types of test, with or without principal stress axis rotation.

deviator stress, $(\sigma_1 - \sigma_3)/2$, defined by equation (7) was maintained at a constant value. The other test used were the conventional cyclic triaxial test and the cyclic torsion shear test, as illustrated in Fig. 9(a, b). The test scheme and test results are described in more detail elsewhere (Ishihara and Towhata, 1983).

The results of the cyclic loading tests are presented in Fig. 10, in which cyclic stress ratio is plotted against the number of load cycles required to cause 5 per cent double-amplitude strain. To represent the results of the three types of test in a consistent manner, the cyclic stress ratio is defined as the deviator stress divided by the initial confining stress $(\sigma_1 - \sigma_3)/2\sigma_0'$. Therefore, when the test is the conventional cyclic triaxial test, the corresponding cyclic stress ratio is reduced to $(\sigma_v - \sigma_h)/2\sigma_0'$, and in the case of the cyclic torsional shear test the cyclic stress ratio is τ_{vh}/σ_0', as readily derived from equation (7). Inducement of 5 per cent double-amplitude strain in the sample is used to define failure in each mode of deformation.

Among the results of the three types of test, it appears appropriate to compare the result of the cyclic torsional test with that of the cyclic test with rotation of principal stress axes, in order to single out the effects of the principal stress rotation. It can be seen in Fig. 10 that the cyclic stress ratio causing a failure strain in a given number of cycles — say 20 — is about 30% smaller in the cyclic test without the rotation, compared with the cyclic test involving the rotation of the principal stress direction.

Liquefaction analysis

The likelihood of liquefaction-type failure occurring in the seabed deposit can be assessed by comparing the induced cyclic stress ratio with the magnitude of the cyclic stress ratio required to cause liquefaction in the sand deposit. Since the wave-induced cyclic stresses involve the rotation of principal stress directions, as discussed above, this effect ought to be taken into account in evaluating the cyclic strength used in the analysis. While various factors have been known to exert influence on the liquefaction resistance, the cyclic stress ratio causing liquefaction in a given number of load cycles has been shown to increase in proportion to the relative density, D_r, of the sand. Therefore, with reference to the test result shown in Fig. 10, the cyclic stress ratio causing liquefaction in 50–100 cycles in the case of the rotating principal stress direction can be expressed as

$$\left(\frac{\tau_{vh,1}}{\sigma'_v}\right) = 0.002\,78 \cdot D_r \qquad (28)$$

where $\tau_{vh,1}$ is the amplitude of shear stress to cause liquefaction. The number of load cycles would probably represent the significant number of waves occurring in the ocean during the storm period. The choice of 50–100 load cycles is also warranted when it is considered that no significant variation occurs in the cyclic stress ratio for cycles exceeding 50 in number, as indicated in Fig. 10. Another important factor that must be considered is the effect of anisotropic consolidation which is evaluated in terms of the earth pressure coefficient at rest, K_0. Using the K_0 value, the cyclic strength formula of equation (28) is further modified as

$$\left(\frac{\tau_{vh,1}}{\sigma'_v}\right) = 0.002\,78 \cdot D_r \frac{1 + 2K_0}{3} \qquad (29)$$

In making the analysis of liquefaction, it will be assumed that a train of harmonic waves is travelling over the sloping seafloor in water of decreasing depth from the deep water location towards the shore, as illustrated in Fig. 5. It will also be assumed that a sand is uniformly deposited in the seabed throughout the depth and through the distance of wave travel. The analysis of liquefaction involves the determination of water depth at a location where liquefaction occurs

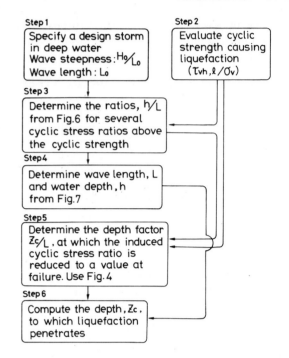

Fig. 11. Flow chart for wave-induced liquefaction analysis of seabed deposit.

and the determination of the depth in the deposit to which liquefaction advances below the sea bottom. The analysis will be performed by following the steps illustrated in Fig. 11.

Step 1

It is necessary first of all to specify conditions of a design storm. The design storm conditions need to be specified at a location far offshore, where the assumption of deep water is satisfied. It would be convenient to specify the wave steepness, H_0/L_0 and the wave length, L_0. Where the wave period, T, is specified, the value of L_0 can readily be determined by equation (14).

Step 2

It is necessary to evaluate the cyclic strength of the sand composing the seabed deposit. The cyclic strength is expressed in terms of the cyclic stress ratio causing liquefaction or 5% double-amplitude shear strain in the sand after a sufficient number of load cycles have been applied. For all practical purposes, 50 load cycles is considered appropriate. In what follows, the cyclic strength thus defined will be referred to as cyclic stress ratio at failure. In evaluating the cyclic strength, the effects of the K_0 condition and the rotation of the principal stress direction must be taken into account.

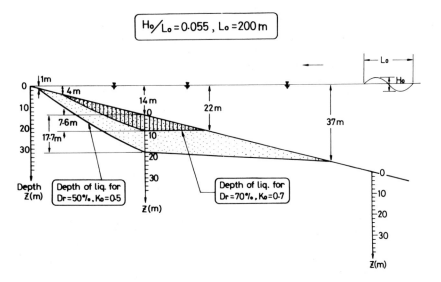

Fig. 12. Examples of wave-induced liquefaction analysis of shoaling seabed deposit.

Step 3

If the cyclic stress ratio at failure is greater than 0.230, nowhere in the seabed deposit is liquefaction-type failure expected to occur. Otherwise, liquefaction may take place down to some depth of the seabed deposit over some distance, with water depth varying in the path of wave shoaling. For given values of the wave steepness, H_0/L_0, and the cyclic stress ratio at failure estimated from equation (29), a value of h/L can be read off from the chart in Fig. 6. It is also convenient to read off values of h/L corresponding to several arbitrarily chosen values of the cyclic stress ratios which are in excess of the cyclic stress ratio at failure.

Step 4

For each value of h/L determined in Step 3, the water depth, h, and the wavelength, L, can be determined using the chart of Fig. 7 and referring to the prescribed value of the wavelength, L_0.

Step 5

The value of the water depth determined from the cyclic stress ratio at failure indicates the water depth at which liquefaction occurs at the fluid–bed interface. In other words, this water depth was determined in such a way that the wave-induced cyclic stress ratio becomes equal, at the mudline, to the cyclic stress ratio of sand at failure. In areas of shallower depth the cyclic stress ratio induced at the mudline becomes greater than the cyclic stress ratio at failure. In such cases it may well be assumed that the induced cyclic stress ratio at the mudline decays with increasing depth, in accordance with the relationship of equation (10). Then, using the

chart in Fig. 4, it is possible to determine a depth factor, Z_c/L, at which the depthwise decreasing cyclic stress ratio is reduced to a value at failure, where Z_c denotes the depth to which failure due to liquefaction penetrates.

Step 6

Using the wavelength, L, determined in Step 4 and the depth factor, Z_c/L, calculated in Step 5, it becomes possible to assess the depth, Z_c, to which liquefaction-type failure penetrates.

Examples of analysis

Using the analysis procedure as described above, example cases will be considered in this section. Assume the most severe storm conditions in deep water, $H_0/L_0 = 0.055$ and $L_0 = 200\,\text{m}$. The liquefaction resistance of the seabed sand sediment is assumed to be given by a representative formula in equation (29). For the sake of comparison, two cases will be considered: a loose sand sediment with a relative density of $D_r = 50\%$ and $K_0 = 0.5$, and a medium-dense sediment with $D_r = 70\%$ and $K_0 = 0.7$. The results of the analysis for the case of the medium-dense sediment are presented in Fig. 12. It may be seen that liquefaction begins to occur when the waves arrive at a location with a water depth of 22 m. The liquefaction penetrates deeper in the seabed deposit, until the waves reach 14 m water depth, where they break. The sediment is liquefied to a depth of 7.6 m. The waves propagating further are continuously subjected to breaking and consequently, with decreasing magnitude of induced cyclic stress ratio, the depth of liquefaction penetration becomes again equal to zero at 4 m water depth. For the case of loose sediment, liquefaction is shown to develop deeper in the deposit and wider in the lateral extent.

It is to be noted that the location of the deepest liquefaction penetration is the same for both cases considered above. This is due to the fact that the location of the maximum induced cyclic stress is determined by the condition of wave breaking, which, in turn, is related only to the wave profile factors.

Conclusions

By assuming the seabed deposit to be a homogeneous elastic half-space and also by assuming the wave-induced load on the seabed to be a train of harmonic waves, two-dimensional stress analysis was carried out to examine the nature of cyclic stresses induced in the seabed deposit. The analysis showed that the wave-induced stresses are characterized by the continuous rotation of the principal stress directions. In an effort to simulate such conditions of cyclic loading, a series of cyclic triaxial torsion shear tests was conducted on sand specimens prepared in the laboratory. On the basis of the results of these tests and also with reference to the results of conventional cyclic triaxial tests, an empirical formula was proposed to evaluate the cyclic strength, allowing for the effect of rotation of the principal stress directions.

In order to evaluate the magnitude of cyclic stresses induced by waves in the seabed deposit, the linear theory of wave propagation was used and the changing feature of wave profile as the wave propagates shorewards on the sloping seabed was clarified. It was shown that the wave steepness (i.e. the height-to-length ratio) increases as the waves travel over the sloping seabed with decreasing water depth and that the wave starts breaking when the wave steepness reaches a certain limit which is specified by the water depth and wavelength. It was further shown that the condition imposed by the wave breaking leads to a maximum value of the cyclic stress ratio of 0.230 that can be induced in the seabed deposit. A methodology was proposed to evaluate the wave-induced cyclic stress ratio based on the wave conditions given offshore in deep water.

The methodology to evaluate susceptibility of the seabed sediment to liquefaction-type failure was then established by incorporating the above-mentioned empirical formula for cyclic strength into the methodology to evaluate the cyclic stress ratio induced by waves. The above methodology for liquefaction analysis was applied to some typical cases in which the most

severe storm conditions are considered. The results of analysis showed that, for a medium-dense deposit of sand with a 70% relative density, liquefaction could extend down to a depth of 17.7 m at a location of 14 m water depth where the wave breaks for the first time in the course of the wave propagation.

Acknowledgements

In establishing the framework of this paper, valuable comments by Professor Kiyoshi Horikawa, Department of Civil Engineering, University of Tokyo, were of great help. The authors wish to express their sincere thanks to him.

References

Clukey, E., Cacchione, D. A. and Nelson, C. H., 1980. Liquefaction potential of the Yukon Prodelta, Bering Sea. *Proc. 12th Annual Offshore Technology Conf.*, Houston, Vol. 2, pp. 315–326.

Finn, W. D. L., Siddharthan, R. and Martin, G. R., 1983. Response of seafloor to ocean waves. *J. Geotech. Eng. Div. ASCE*, **109** (GT4), 556–572.

Henkel, D. J., 1970. The role of waves in causing submarine landslides. *Geotechnique*, **20** (1), 75–80.

Henkel, D. J., 1982. Geology, geomorphology and geotechnics. *Geotechnique*, **32** (3), 175–194.

Horikawa, K., 1978. *Coastal Engineering*. University of Tokyo Press.

Ishihara, K. and Towhata, I., 1983. Sand response to cyclic rotation of principal stress directions as induced by wave loads. *Soils Foundns.*, **23** (4), 11–26.

Lee, K. L. and Focht, J. A., 1975. Liquefaction potential of Ekofisk tank in North Sea. *J. Geotech. Eng. Div. ASCE*, **101** (GT1), 1–18.

Madsen, O. S., 1978. Wave-induced pore pressures and effective stress in a porous bed. *Geotechnique*, **28**(4), 377–393.

Nataraja, M. S. and Gill, G. S., 1983. Ocean wave-induced liquefaction analysis. *J. Geotech. Eng. Div. ASCE*, **109** (GT4), 575–590.

Nataraja, M. S., Singh, H. and Maloney, D., 1980. Ocean wave-induced liquefaction analysis: a simplified procedure. *Proc. Int. Symp. Soils under Cyclic and Transient Loading*, Swansea, Vol. 2, pp. 509–516.

Seed, H. B. and Rahman, M. S., 1978. Wave-induced pore pressure in relation to ocean floor stability of cohensionless soils. *Mar. Geotech.*, **3** (2), 123–150.

Wiegel, R. L., 1964. *Oceanographical Engineering*. Prentice-Hall, Englewood Cliffs, N.J., p. 205.

Yamamoto, O. S., 1978. Wave-induced pore pressures and effective stresses in a porous bed. *Geotechnique*, **28**(4), 377–393.

13

Liquefaction Prediction in the Marine Environment

P. Strachan

Ocean Engineering Research Group, University of Newcastle-upon-Tyne, UK

Cyclic shear stress reversals are induced in seabed sediments either by pressure fluctuations caused by the passage of a wave train or by transmission through an offshore structure subjected to wave forces. In both cases a consequent build-up of excess pore-water pressure has been observed with the use of piezometer probes. The resulting loss of effective stress can potentially lead to a state of liquefaction in marine sands and silts.

The present methods for assessing the liquefaction resistance of sand deposits have been criticized. Attempts have been made, based on *in situ* measurements, to improve such assessments. The most promising technique appears to be the use of acoustic, particularly shear wave, methods. This paper reviews the advantages and disadvantages of the various procedures for assessing liquefaction resistance, with particular emphasis on the use of shear waves. It also introduces the results of some shear wave experiments recently conducted at the University of Newcastle-upon-Tyne.

Introduction

The importance of sand liquefaction was first studied in a comprehensive manner in the 1960s, following extensive damage during major earthquakes in Japan and Alaska. However, the effects of liquefaction were observed much earlier, as evidenced by a contemporary account of an earthquake in 1865 on the coast near Barrow-in-Furness in England, which is not generally recognized as a seismically active area. One eyewitness made the following statement:

'John Thompson and I were coming from Fowla Island, one mile from Rampside, and when we had got nearly half way we saw at a distance from us, a great mass of sand, water, and stone, thrown up into the air higher than a man's head. It was nearly in a straight line between us and Rampside, and when we got to the place there were two or three holes in the sand, large enough to bury a horse and cart, and in several places near them, the sand was so soft and puddly that they would have mired any one if he had gone on to them. We thought this very strange, but we supposed it was owing to the frost, for we did not feel the least shock, or know anything of an earthquake

until we got to Rampside, and saw that everybody was in terror, and the houses sadly shattered. We then went to Conkle, and found a crack in the ground at the foot of the railway embankment, about thirty yards in length, and the water was boiling up in a great many places.'

Liquefaction and associated phenomena result from the dynamic loading of fully saturated soils. Under undrained conditions dynamic loading will give rise to increases in pore-water pressure in both sands and clays. This results in a loss of strength of the soil (reduction of bearing capacity) and a consequent reduction in the safety factor. The reason for the increase in pore-water pressure is that reversing shear stresses in a soil tend to cause a reduction in volume of the soil framework. For a saturated material this must result in an outflow of the pore water. If such an outflow is not possible, as in undrained conditions, then excess pore-water pressures result. This in turn causes a reduction in the effective stress between the individual grains, and in the case of sands, a reduction in frictional resistance. Such an effect can also occur due to static shear stresses in the case of loose sands.

If the pore-water pressure increases until the effective

stress becomes zero, then liquefaction can occur in non-cohesive soils. Liquefaction is the condition where the material has lost all its structural strength. It behaves as a heavy liquid with no rigidity, and can therefore flow. Under conditions of high soil density, only a small volume change (dilation) is necessary for the soil to regain its strength. Displacements are therefore limited, although they may be important for certain offshore structures (Bjerrum, 1973). For loose non-cohesive materials, however, there is a possibility of complete failure.

If drainage is allowed to occur, then the above problems will be alleviated, but significant settlements can still take place. Even for permeable sands, if the dynamic loading is of comparatively large amplitude and short duration, then liquefaction can take place before the excess pore-water pressures can be dissipated.

Any form of dynamic loading can increase the pore-water pressure (repeated cycling, transient loading, etc.). Of greatest importance are storm wave loading, either directly by causing pressure fluctuations in sub-bottom sediments or indirectly via a structure, and earthquake loading, which is a problem both offshore and onshore. The main areas of concern offshore are connected with the foundations of gravity platforms, pipelines, sediment instability, and piling. It has also been reported that liquefaction is a potential problem for the artificial islands used for hydrocarbon exploration in the Beaufort Sea (Finn *et al.*, 1982). The majority of reported damage to offshore structures has occurred in the Gulf of Mexico, where storms have resulted in sediment slides and slumps. Even in the denser soils of the North Sea, dynamic loading can be a problem. Eide *et al.* (1979) reported on observed platform behaviour in the North Sea over the period 1973–1978. Of 13 platforms studied, only 3 had functioning pore-water pressure transducers installed in the base, of which 2 were founded on clay and 1 on sand. During storms with maximum wave forces up to 45 per cent of the design wave forces, pore-water pressure increases up to 20 kPa were recorded. The authors concluded that storm wave loading does generate excess pore water in both sand and clay, and that this behaviour must be taken into account during the design for displacement and stability for North Sea platforms.

In any analysis of the liquefaction potential of an area, the following basic steps are required. First, an estimate is required of the cyclic shear stresses to be expected at a particular location, together with a knowledge of the variation with depth. Second, the actual cyclic shear strength of the soils under investigation must be evaluated. Finally, these values are compared in order to compute a factor of safety (and its variation with depth).

Much of the early work was carried out with respect to the analysis and prediction of liquefaction during earthquakes, and studies of a relatively few past major earthquakes remain the primary source of data used to test methods of predicting the liquefaction potential of an area. The similarities between the cyclic shear stresses induced in soils by storm wave loading and earthquake loading (and the lack of a suitable alternative) have led researchers to use the same data base for both earthquake and storm wave loading.

It is with regard to the estimation of the cyclic strength of marine sands that this paper is concerned. Although much research has been undertaken into the dynamic behaviour of soils, there is still a large degree of uncertainty as to the response of an *in situ* deposit under a given dynamic load. The remainder of this paper describes current techniques for estimating the cyclic strength of soils, together with alternative methods that have been suggested. Finally, an investigation is described of work undertaken at the University of Newcastle-upon-Tyne into the use of shear wave velocity as a predictor of the cyclic strength of soils.

Present techniques for estimating the cyclic strength of sands

Use of undisturbed samples

Until recently most of the dynamic soil behaviour research for non-cohesive soils has involved the use of reconstituted samples. Liquefaction analyses that have been carried out for the offshore case using such samples include those by Christian *et al.* (1974), Lee and Focht (1975) and Clukey *et al.* (1980). These investigations have been useful in studying the effects of stress level, void ratio and other factors not connected with the soil fabric (although the effects of overconsolidation and vibration prestraining were studied by Lee and Focht, 1975). Some researchers (e.g. Mulilis *et al.*, 1975) have stressed the importance of the method of sample preparation, and thus sample structure, on the results of the dynamic testing. Differences in the method of sample preparation can affect the cyclic stresses giving rise to initial liquefaction, by up to 100 per cent at the same relative density. Similarly, the behaviour of reconstituted and of undisturbed samples can differ considerably, and the testing of undisturbed samples is likely to give an accurate reflection of *in situ* conditions only if samples of sufficiently high quality can be obtained.

As many authors have noted, obtaining 'undisturbed' samples is very difficult to accomplish. This is particularly true for non-cohesive soils in a marine environment. It has also been reported that the cyclic strength of a sample is more sensitive to disturbance than is the measurement of static strength (Espana *et al.,* 1978). Sample disturbance causes changes in the density and fabric of the sample. These changes can result from disturbance of the soil below the bottom of the bore-hole during drilling, from the insertion of a sample tube, from the ensuing transport and storage, from the sample extrusion and from preparation carried out prior to testing (Marsland and Windle, 1982). The disturbances include the following:

(1) Reduction in the *in situ* total stress on the sample, and changes from the *in situ* anisotropic state of stress to an isotropic state of stress, with a consequent possible change in the measured soil behaviour. It is also possible for the soil below the base of the borehole to swell as a result of stress release prior to sampling.

(2) In the offshore environment the presence of gas (either dissolved or free) is important. This is more so than in the onshore case, because of the large pore-water pressures caused by the hydrostatic pressure. The reduction in total stress on the sample will result in dissolved gas coming out of solution and expansion of free gas, with a consequent major disruption of the sample structure.

(3) The sampling process may cause the specimen to lose any strength it may have acquired from long-term loading under a sustained stress. Also, the sample may be stressed during the process of sampling. Loading of the sample is more of a problem in the offshore case, where the drilling vessel is often subject to wave motion and where heave compensation is required in an attempt to smooth out this motion.

(4) The sample may be subjected to vibration and impulse loads during the transportation and handling phases. In addition, migration of pore-water may cause changes in the sample properties.

(5) Specimen extrusion and laboratory preparation techniques can significantly affect the measured soil strength.

On land recent improved sampling techniques using freezing methods, block sampling and modifications to corers have made it possible to obtain good-quality samples for detailed investigations. However, with the exception of modifications to corers, it is unlikely that any of these techniques will have an application offshore, although freezing techniques may help to reduce many of the transport and handling problems. Even onshore, Broms (1978) considered that these improved techniques do not give samples of sufficiently high quality for the accurate determination of the cyclic strength of cohesionless soils.

In summary, the difficulties in obtaining good-quality samples of non-cohesive soil offshore appear insurmountable. The use of *in situ* techniques for the prediction of cyclic strength is to be preferred.

Use of SPT N-values

In an attempt to avoid the problems associated with the collection of undisturbed non-cohesive samples, and in order to reduce the time and cost of carrying out these tests, an alternative method of estimation was suggested by Seed and Idriss (1971) and developed by Seed *et al.* (1983), based on the use of N values derived from the standard penetrometer test (SPT). From a study of case histories of liquefaction during earthquakes, a relationship was developed between the cyclic shear stress ratio (defined as the applied shear stress divided by the vertical effective stress) and the modified penetration resistance (N_1). This modified penetration resistance is the measured blowcount value corrected to an overburden pressure of 1 ton/ft² (95.8 kPa).

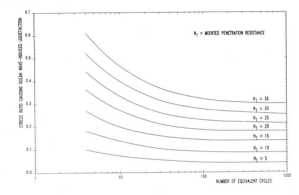

Fig. 1. Relationship between equivalent number of cycles and stress ratio required to cause ocean wave-induced liquefaction (After Nataraja and Singh, 1979).

For the case of ocean wave-induced liquefaction, Nataraja and Gill (1983) used a relationship developed by Seed (1976) between earthquake magnitude and number of equivalent stress cycles (at a stress level of 0.65 that of the maximum experienced during the earthquake), together with extrapolations for the number of stress cycles. This resulted in a chart, reproduced in Fig. 1, relating number of stress cycles, stress level giving rise to liquefaction, and the modified penetration resistance.

Charts such as Fig. 1 can thus be used to estimate cyclic strength from a knowledge of the N value; the cyclic strength can then be compared with the expected cyclic stresses for the design storm (or earthquake). Seed (1976) pointed out possible criticisms of the use of such charts. These included the observation that there is a wide scatter in the data used to compile the charts; that such charts, based on purely empirical data, may not include all the significant factors that can affect cyclic liquefaction; that penetration resistance may not be an appropriate index of the liquefaction characteristics; and that the SPT resistance of a soil is not always determined with reliability in the field and may vary depending on the particular test conditions.

In order to show why penetration resistance should be a good indicator of cyclic strength, Seed (1976) and Schmertmann (1978) listed the factors which affect both variables and showed that, qualitatively at least, such a relationship between them could be expected, since each factor affects the variables in the same sense. These factors are listed in Table 1.

TABLE 1 Factors affecting cyclic strength and penetration resistance

Factor	Effect on cyclic strength (factor of safety against liquefaction)	Effect on penetration resistance
Increased relative density	Increased	Increased
Increased stability of soil structure	Increased	Increased
Increased time under pressure	Increased	Probably increased
Increase in K_0 (coefficient of earth pressure at rest)	Increased	Increased
Prior seismic strains	Increased	Probably increased

In addition, Schmertmann (1978) noted other properties of the SPT, namely that the loading on a soil by an SPT test is a rapid pulsed loading and is thus essentially undrained, as is earthquake loading; and that there is a high percentage of shear wave energy introduced by the SPT (indeed, it is often used as an shear wave source) which matches the predominant shear wave energy of earthquakes.

Perhaps the greatest problem associated with the use of the SPT N values is that the field testing procedure varies a great deal from one company to another and the recorded N values can differ by factors of 2 or more, according to the particular technique used. Schmertmann (1976) concluded that 'the SPT represents, at present, such a poorly controlled and poorly understood test that its use for anything quantitative should be restricted to within individual organizations'. More stringent standardization is required before quoted N values can be compared from one site to another.

Examples of the uses of the SPT offshore were presented by Nataraja and Gill (1983), who used cyclic strength estimates derived from N values for assessing the liquefaction potential of sites of an ocean outfall,

a large diameter pipeline, a sewer outfall in Lake Ontario and an offshore gravity platform.

Use of the CPT

In the offshore environment the static cone penetrometer test (CPT) is often used instead of the SPT. The CPT gives a measure of the penetration resistance, q_c, as a penetrometer is pushed at a fixed rate into the ground. One advantage of this test is that a continuous profile with depth is obtained, making it possible to detect relatively thin layers of non-cohesive soil where liquefaction may be a problem. In order to use this measure of penetration resistance to determine the cyclic strength of a soil, it is necessary to convert the penetration resistance into an equivalent blowcount N value. However, the relationship between penetration resistance and N value varies with soil type. For the fairly uniform, fine or fine to medium, slightly silty to silty silica sands typically encountered in the North Sea, Semple and Johnston (1980) used the relationship

$$q_c = 1000N \quad (\text{for } N > 10; q_c \text{ in kPa}) \qquad (1)$$

However, they found that values of q_c estimated by this relationship were ± 50 per cent of the measured q_c value.

Thus errors involved in the use of the CPT in this way for estimating cyclic strength include not only all the errors associated with the relationship between SPT and cyclic strength detailed in the previous section, but also uncertainties in the conversion of q_c values to N values.

Consideration must be given to whether or not the CPT is sensitive to the same factors that affect the cyclic strength of soils. The test involves a slow shearing of soil with which large strains are associated. Cyclic strength involves the response of the soils to cyclic loads where only small strains occur (until the liquefaction condition is approached).

Alternative techniques for estimating the cyclic strength of sands

The techniques discussed in this section have been suggested as alternative methods for predicting the cylic strength of soils.

Piezometers

The rise in excess pore-water pressure due to cyclic loading of sediments has been observed with the use of piezometers under an offshore platform (Eide *et al.*, 1979). Piezometers have also been used in shallow water off the Mississippi delta to monitor increases in pore-water pressure during storms, and to investigate the relationship between the pore-water pressure in sediments and submarine slope stability (e.g. Williams *et al.*, 1981). It is believed that many of the submarine slides and slumps are triggered by loss of effective stress in storms (Suhayda and Prior, 1978).

By directly monitoring the build-up of pore-water pressure during storms, the piezometer probes are essentially acting as *in situ* tests for the cyclic strength of soils. If, for example, during a storm of known magnitude and duration, pore-water pressures are observed to increase by a certain amount, then an estimate can potentially be made of the storm loading conditions necessary to incur liquefaction (Richards *et al.*, 1975). The major advantage of this technique is that the effects of permeability and dissipation of excess pore-water pressure are taken into account. One disadvantage is the fact that the method is only applicable where sufficiently high loads are transmitted to the seabed for pore-water pressure increase to be significant (i.e. for the problem of wave loading in the coastal environment). It may not be suitable for seismic loading in deeper waters or for loading transmitted through structures. Readings may also need to be monitored for a long time period in order to observe the soil response during a large storm — Okusa and Uchida (1980) monitored pore-water pressure response for a period of just under 2 years but the area they studied did not suffer the expected large storms.

In view of the fact that a direct measurement of pore-water pressure build-up during cyclic loading is possible, the method would appear promising for certain situations.

Electrical Methods

The majority of work in relating electrical properties of soils to cyclic strength has been undertaken by Arulanandan and co-workers at the University of California. Arulanandan *et al.* (1981) presented a summary of the use of the electrical resistivity technique in assessing cyclic strength. They showed that the formation factor (the ratio between the soil resistivity and the pore-water resistivity), the shape factor (determined from plots of formation factor against porosity) and the anisotropy index (ratio between formation factors in the vertical and horizontal planes) are dependent on the physical properties of sand (grain size, grain shape, density) and the structure of that sand (packing, cementation). Since cyclic strength of sands also depends on these properties, they proposed that a knowledge of the electrical properties can be used to assess this strength. A relationship between the stress ratio giving rise to liquefaction and a combination of the electrical indices mentioned above (the electrical index) was developed. The estimation of cyclic strength by this method was stated to give similar results to estimations achieved by other methods.

The fact that electrical parameters of sands are related to factors that affect cyclic strength does imply that the method may be useful for estimating the resistance to liquefaction. In particular, both formation factor and cyclic strength are strongly influenced by sand density. However, the use of the 'electrical index', which is a combination of various electrical properties, compounds errors in measurement of the individual properties. For instance, the anisotropy index is related to sample packing. However, differences in methods of sample preparation can have a large effect on resistance to liquefaction (up to 100%), whereas the ratio between formation factors in the horizontal and vertical planes only shows variations of the order of a few per cent. A large degree of data scatter can be expected when such small variations are measured.

The cyclic strain approach and shear modulus

In a series of recent papers (Dobry and Ladd, 1980; Dobry *et al.*, 1980; Yokel *et al.*, 1980) a different approach has been suggested for the determination of liquefaction potential. In the majority of research on the subject of the prediction of liquefaction (as summarized, for example, by Seed, 1976) the cyclic strength and loading levels experienced in earthquakes have been expressed in terms of cyclic shear stresses. The cyclic strength measured in terms of stress level is dependent on many factors that are related to the soil structure. These include packing (investigated by different methods of sample preparation), cementation and overconsolidation. The importance of such factors is of the same order as that of sample density.

Dobry and Ladd (1980) proposed a cyclic strain approach. They suggested that earthquake loading be characterized by the cyclic strains rather than the cyclic stresses. Using this approach, they showed that there is a threshold cyclic shear strain below which there is no build-up of pore-water pressure. This threshold shear strain level theoretically varies with effective confining pressure (Yokel *et al.*, 1980), but over the pressure range of interest (about 20–200 kPa) it is approximately 10^{-4}. This level is supported by field evidence (Dobry and Ladd, 1980). The existence of a threshold level was suggested earlier by Stoll and Kald (1976), who expected it to depend on void ratio, confining stress and soil type.

For earthquake loading, shear stresses in the soil are calculated from an estimation of the peak ground accelerations (Seed and Idriss, 1971). Thus the shear modulus (G) is required in order that the shear strains can be calculated. At small strains this can be derived from a measurement of the shear wave velocity (V_s), by the following relation:

$$G = \rho V_s^2 \qquad (2)$$

where ρ is the bulk density of the soil. The approach outlined above predicts that pore-water pressure build-up is less likely in soils that are associated with a higher value of shear modulus — thus the risk of liquefaction is reduced. This is supported by the observation that factors that increase cyclic strength also increase the shear wave velocity and thus shear modulus (see the following section). Dobry and Ladd (1980) showed that the dependence of cyclic strength on the method of sample preparation is removed if cyclic strength is measured in terms of cyclic strains instead of cyclic stresses; i.e. the increase in stress level giving rise to liquefaction for certain methods of sample preparation corresponds with an associated increase in shear modulus.

Possible disadvantages of the method are:

(1) Given the existence of a threshold strain level, a knowledge of the shear modulus of the soil and total overburden pressure is used to predict the threshold peak ground acceleration that is necessary for the start of excess pore-water pressure development. This can then be compared with the accelerations of the design earthquake. Thus the method does not predict whether or not liquefaction will occur, but only whether the development of excess pore-water pressure is possible.

(2) The analysis to date has been for earthquake loading; the basic approach does not take number of loading cycles into consideration. The concept of a well-defined threshold strain level may not apply for the large number of cycles involved in ocean wave loading. Even a very small increase in excess pore-water pressure per cycle may be significant in the case of wave loading. Also, wave loads induce stresses on marine soils, with strain levels increasing as the soil weakens. That is, the loading on soils resulting from wave action is stress-controlled rather than strain-controlled.

In spite of these potential criticisms, the method appears promising in its use of shear modulus to take into account the effect of soil structure on cyclic strength.

Shear wave velocity as an indicator of cyclic strength

One of the most promising of the alternative methods which have been discussed is the use of shear modulus and the strain approach for estimating when pore-water pressure increase may be expected during un-drained cyclic loading. The results of a study using a similar line of approach is described in this section. It was considered that it would be interesting to test whether or not the shear wave velocity was directly related to the cyclic strength of soils (i.e. the stress level required to liquefy the sample in a given number of loading cycles) rather than to the possibility of pore-water pressure increase. A literature review of factors affecting the shear wave velocity was thus undertaken, with results summarized in Table 2.

There are many similarities between the loading caused by the transmission of shear wave energy and the cyclic loading that results in permanent defor-mations and the development of excess pore-water pressure. The main difference between the two cases is that of strain amplitude. Thus the experiments de-scribed in the following paragraphs essentially test whether the soil resistance to the very small amplitude cyclic shear stresses associated with shear wave propa-gation is proportional to the larger cyclic shear stresses that lead to permanent deformation of the structure. Such a possibility is supported by previous research work summarized by Richart (1978), showing that the shear modulus—strain amplitude relationships were of similar form for all sands.

The other potential advantages of the shear wave velocity over the SPT blowcount value are that it is a more accurate and repeatable measurement, and that it is possible to carry out laboratory investigations of relationships between the shear wave velocity and cyclic strength.

A laboratory programme was therefore pursued. Sand samples were reconstituted by a method of moist tamping in a triaxial cell (see Fig. 2). The usual endcaps of the cell were replaced by endcaps containing geo-physical transducers, including shear wave transducers. The shear wave transducers consisted of bender elements, similar to those described by Shirley and Hampton (1978). The shear wave velocity is calculated from a knowledge of the transit time of pulsed shear wave transmission and transducer separation. After the samples had been saturated and consolidated to the desired stress conditions and the shear wave velocity had been measured, the samples were cyclically loaded under drained conditions until they liquefied. Samples were constructed for a variety of sand types at a range of densities. All test sands were uniform sands in the fine to medium grain size range (which are the most prone to liquefaction). Their details are presented in Table 3, where e_{max} and e_{min} are the maximum and minimum void ratios respectively, D_{50} is the median grain size, D_{10} is the effective size, U is the uniformity coefficient, G_s is the specific gravity, Ro is the round-ness coefficient, and Sp is the sphericity coefficient. Roundness and sphericity were estimated from visual comparison charts after Powers (1953) and Rittenhouse (1943) respectively. Monterey and Toyoura sands are standard test sands of the USA and Japan respectively.

The results of a series of experiments are presented in Fig. 3. The vertical axis is plotted in terms of SR_{30}, the stress ratio giving rise to initial liquefaction in 30

TABLE 2 Factors influencing cyclic strength and S-wave velocity

Factor	Effect on cyclic strength of non-cohesive soils	Effect on S wave velocity
Increased effective confining pressure	Increased	Increased
Increased density	Increased	Increased
Increased grain size (for constant D_r)	Increased	Increased
Increased uniformity coefficient (for constant D_r)	?	?
Increased angularity	Increased	Increased
Soil structure (moist tamped relative to dry pluviated)	Increased	Increased
Vibration prestraining	Increased	Increased
Overconsolidation	Increased	No effect[a]
Ageing effects/cementation	Increased	Increased
Static shear stress	Increased	No effect[a]
Presence of small quantity of air	Increased	No effect
Temperature	Probably no effect	No effect

[a]Increased if mean effective confining pressure increased.

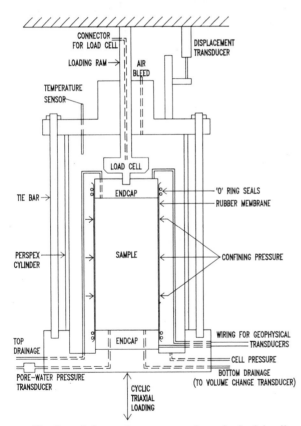

Fig. 2. Schematic representation of triaxial cell.

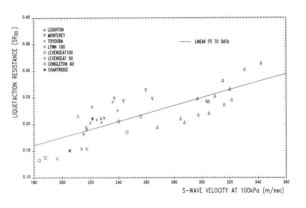

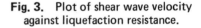

Fig. 3. Plot of shear wave velocity against liquefaction resistance.

loading cycles, which is a measure of the resistance of the sample to liquefaction. The stress ratio is the applied cyclic shear stress level normalized with respect to the initial confining pressure. Although there is a good deal of scatter (which results from the cyclic loading measurements rather than the shear-wave measurements — such scatter is not unusual in the field of the cyclic loading of soils), there is a definite trend of increasing shear wave velocity with increasing resistance to liquefaction. Such a relationship supports the idea that shear wave velocity can be used to predict liquefaction resistance.

In studies of the influence of effective confining pressure on the cyclic strength of soils, it is generally assumed that shear stress level and effective confining pressure are linearly related (e.g. Finn *et al.*, 1971; Lee and Focht, 1975). However, some researchers (Mulilis *et al.*, 1975; Castro and Poulos, 1976) have shown that the stress level increases with effective confining pressure at a rate less than directly proportional. In fact, if a power law is fitted to the data presented by Castro and Poulos, it is found that the power exponent of the effective pressure is of the order of 0.5–0.8. This is similar to that for the dependence of shear modulus on effective pressure where the power exponent

TABLE 3   Details of test sands used in this study

Test sand	Grain size parameters						Grain shape	
	e_{max}	e_{min}	D_{50} (mm)	D_{10} (mm)	U	G_s	Ro	Sp
Monterey	0.851	0.557	0.47	0.33	1.48	2.637	3.5	0.71
Toyoura	0.987	0.610	0.24	0.15	1.77	2.641	2.0	0.67
Leighton	0.801	0.515	0.50	0.37	1.45	2.649	4.0	0.73
Chartridge	1.026	0.638	0.12	0.076	1.68	2.651	2.0	0.75
Levenseat 100	1.200	0.718	0.080	0.054	1.67	2.666	1.5	0.61
Levenseat 50	0.939	0.572	0.23	0.14	1.71	2.654	2.0	0.65
Lynn 100	1.012	0.665	0.13	0.10	1.39	2.649	2.0	0.69
Congleton 60	0.751	0.469	0.25	0.16	1.66	2.649	4.0	0.77

is about 0.5 (e.g. Richart, 1978). Such a relationship is worthy of further study. If the shear modulus—cyclic strength relationship were shown to be independent of effective pressure, then it would be better to use shear modulus (easily obtainable from the shear wave velocity by equation 2), rather than the shear wave velocity, as a direct predictor of cyclic strength.

Summary

The present techniques for estimating the cyclic strength of soils have been presented, together with a discussion of their disadvantages. Of the alternative procedures which have been suggested, the use of acoustic techniques (particularly the measurement of shear wave velocity) would seem to be advantageous. A laboratory investigation has shown that there is a relationship between cyclic strength and shear wave velocity over a range of sand types and densities. A review of other factors which influence cyclic strength and shear wave velocity, together with the work of Dobry and colleagues, would seem to indicate that these variables also respond in a similar manner to variations in sample structure. It is thought that the use of shear wave velocity (or shear modulus) would prove a useful predictor of the cyclic strength of sands. The subject deserves further investigation.

Acknowledgements

The work described in this paper was undertaken at the University of Newcastle-upon-Tyne and was financed under the Science and Engineering Research Council's Marine Technology Programme. Thanks are due to Dr E. James, of Fugro Ltd, for providing the source of the quotation used in the introduction.

References

Arulanandan, K., Harvey, S. J. and Chak, J. S., 1981. Electrical characterization of soil for *in situ* measurement of liquefaction potential. *Int. Conf. Recent Advances in Geotech. Earthquake Eng. and Soil Dynamics.*, St. Louis, Missouri, Vol. 3, pp. 1223–1229.

Bjerrum, L., 1973. Geotechnical problems involved in foundations of structures in the North Sea. *Geotechnique, 23* (3), 319–358.

Broms, B., 1978. European experience in soil sampling and its influence on dynamic laboratory testing. In: *Soil Sampling and its Importance to Dynamic Laboratory Testing.* Preprint 3440, ASCE, pp. 1–86.

Castro, G. and Poulos, J., 1976. Factors affecting liquefaction and cyclic mobility. In: *Liquefaction Problems in Geotechnical Engineering.* Preprint 2752, ASCE, pp. 105–137.

Christian, J. T., Taylor, P. K., Yen, J. K. C. and Erali, D. R., 1974. Large diameter underwater pipeline for nuclear power plant designed against soil liquefaction. Pap. No. 2094. *Proc. Offshore Tech. Conf.,* pp. 597–606.

Clukey, E., Cacchione, D. A. and Nelson, C. H., 1980. Liquefaction potential of the Yukon prodelta, Bering Sea. Pap. No. 3773, *Proc. Offshore Tech. Conf.* pp. 315–325.

Dobry, R. and Ladd, R. S. 1980. Discussion on 'Soil liquefaction and cyclic mobility evaluation for level ground during earthquakes' by Seed, H. B. (1979). *J. Geotech. Engng Div., Proc. ASCE,* **106** (GT6), 720–724.

Dobry, R., Powell, D. J., Yokel, F. Y. and Ladd, R. S., 1980. Liquefaction potential of saturated sand — the stiffness method. *7th World Conf. Earthquake Engng,* Istanbul, Vol. 3, pp. 25–32.

Eide, O., Anderson, K. H. and Lunne, T., 1979. Observed foundation behaviour of concrete gravity platforms installed in the North Sea, 1973–1978. Pap. No. 78, *2nd Int. Conf. Behaviour of Offshore Structures,* London, pp. 435–456.

Espana, C., Chaney, R. C. and Duffy, D., 1978. Cyclic strengths compared from two sampling methods. In: *Soil Sampling and its Importance to Dynamic Laboratory Testing*. Preprint 3440, ASCE, pp. 287–319.

Finn, W. D. L., Iai, S. and Ishihara, K., 1982. Performance of artificial offshore islands under wave and earthquake loading: field data analysis. *Proc. Offshore Tech. Conf.*, Paper 4220, pp. 661–671.

Finn, W. D. L., Pickering, D. J. and Bransby, P. L., 1971. Sand liquefaction in triaxial and simple shear tests. *J. Soil Mech. Fndn. Engng Div., Proc. ASCE*, 97 (SM4), 639–659.

Lee, K. L. and Focht, J. A. Jr., 1975. Liquefaction potential at the Ekofisk tank in the North Sea. *J. Geotech. Engng Div., Proc. ASCE*, 101 (GT1), 1–18.

Marsland, A. and Windle, D., 1982. Developments in offshore site investigation. Pap. No. 2.7, *Oceanology International*, Brighton.

Mulilis, J. P., Chan, C. K. and Seed, H. B., 1975. The effect of method of sample preparation on the cyclic stress-strain behaviour of sands. Rep. No. EERC 75-18, College of Engineering, University of California, Berkeley.

Nataraja, M. S. and Gill, H. S., 1983. Ocean wave-induced liquefaction analysis. *J. Geotech. Engng. Div., Proc. ASCE*, 109 (GT4), 573–590.

Okusa, S. and Uchida, A., 1980. Pore-water pressure change in submarine sediments due to waves. *Mar. Geotech.*, 4 (2), 145–161.

Powers, M. C., 1953. A new roundness scale for sedimentary particles. *J. Sediment. Petrol.*, 23, 117–119.

Richards, A. F., Øien, K., Keller, G. H. and Lai, J. Y., 1975. Differential piezometer probe for an *in situ* measurement of sea-floor pore-pressure. *Geotechnique*, 25 (2), 229–238.

Richart, F. E., 1978. Field and laboratory measurement of dynamic soil properties. In: B. Prange (ed.), *Dynamical Methods in Soil and Rock Mechanics*, Vol. 1. Balkema, Rotterdam, pp. 3–36.

Rittenhouse, G., 1943. A visual method of estimating two-dimensional sphericity. *J. Sediment. Petrol.*, 13, 80–81.

Schmertmann, J. H., 1976. Measurement of *in situ* shear strength. In: *In situ Measurement of Soil Properties*, Vol. 2. ASCE, pp. 57–138.

Schmertmann, J. H., 1978. Use of the SPT to measure dynamic soil properties? – Yes, but . . .! In: *Dynamic Geotechnical Testing*. ASTM, STP 654, pp. 341–355.

Seed, H. B., 1976. Evaluation of soil liquefaction effects on level ground during earthquakes. In: *Liquefaction Problems in Geotechnical Engineering*. Preprint 2752, ASCE, pp. 1–104.

Seed, H. B. and Idriss, I. M., 1971. Simplified procedure for evaluating soil liquefaction potential. *J. Soil Mech. Fndn Engng Div., Proc. ASCE*, 97 (SM9), 1249–1273.

Seed, H. B., Idriss, I. M. and Arango, I., 1983. Evaluation of liquefaction potential using field performance data. *J. Geotech. Engng. Div., Proc. ASCE*, 109 (GT3), 458–482.

Semple, R. M. and Johnston, J. W., 1980. Performance of Stingray in soil sampling and *in situ* testing. In: D. Ardus (ed.), *Offshore Site Investigation*. Graham and Trotman, London, pp. 169–182.

Shirley, D. J. and Hampton, L. D. 1978. Shear-wave measurements in laboratory sediments. *J. Acoust. Soc. Am.*, 63 (2), 607–613.

Stoll, R. D. and Kald, L., 1976. The threshold of dilation under cyclic loading. In: *Liquefaction Problems in Geotechnical Engineering*. Preprint 2752, ASCE, pp. 343–358.

Suhayda, J. N. and Prior, D. B., 1978. Explanation of submarine landslide morphology by stability analysis and rheological models. *Proc. Offshore Tech. Conf.*, Paper 3171, pp. 1075–1082.

Williams, G. N., Dunlap, W. A. and Hansen, B., 1981. Storm induced bottom sediment data: Seaswab II results. *Proc. Offshore Tech. Conf.*, Paper 3974, pp. 211–221.

Yokel, F. Y., Dobry, R., Ladd, R. S. and Powell, D. T., 1980. Liquefaction of sands during earthquakes – the cyclic strain approach. *Proc. Int. Symp. on Soils under Cyclic and Transient Loading*, Swansea, Vol. 2, pp. 571–580.

14

Methods of Predicting the Deformation of the Seabed due to Cyclic Loading

Ronald C. Chaney

*Department of Environmental Resources Engineering, Humboldt State University,
Arcata, California, USA*

The stability of underwater slopes and seafloor sediments is of major importance for offshore structures such as pipelines, jacket platforms and gravity platforms. In this paper laboratory simulation studies of marine slopes under cyclic loading are analyzed. Results indicate that the development of slope movement is dependent on (1) soil strength, (2) bottom pressure, (3) depth of unstable zone and pore-water pressure generation. To predict slope performance requires (a) quantifying the soil response to cyclic and static loading (total or effective stress analysis), and (b) selection of the appropriate slope stability technique (limit equilibrium or finite element method). Tables are presented comparing total and effective stress analysis as well as the various methods of slope stability analysis. On the basis of this study it was concluded that the permanent deformation procedure utilizing the finite element method is the most general.

Introduction

The stability of underwater slopes and seafloor sediments is of major importance for offshore structures such as pipelines, jacket platforms and gravity platforms. These slopes are subjected to storm wave loading, and in many instances may also be excited by earthquake motions. In this paper only the methods of evaluating the initial stability or instability of the marine slope will be discussed.

An insight into the behaviour of a marine slope under wave loading can be gained from wave tank studies. These studies have been performed by Henkel (1970), Singh (1971), Tsui (1972) and Doyle (1973). A summary of the results of these studies is presented in Table 1. In general, the studies show that the effect of waves on a seabed depend on the strength of the material and the amount of pressure induced at the mudline.

To evaluate the stability of the seabed, there are at present three procedures available. These methods are (1) the factor of safety approach, (2) the strain potential approach and (3) the permanent deformation approach. The factor of safety approach is based on the principle of limiting equilibrium, in which the strength available is compared with the applied stress. The difficulty with this procedure is that the stability or instability of the slope is depicted by a single number which does not convey to the analyst any feeling of the amount of movement involved. In response to this drawback, Seed (1966) proposed using a 'strain potential' procedure in which the strain of an individual element or slice comprising the slope could be estimated. The drawback with this procedure arises in the basic assumption that each element or slice comprising the slope acts independently of every other slice or element. The third procedure, the 'permanent deformation method' originally developed by Lee (1974), built upon the 'strain potential' procedure. In this refinement the various elements comprising the slope being investigated are combined together to act as an aggregate rather than as individual parts.

In the following sections the effect of waves on seabed sediments will be discussed along with analytical methods of modeling the soil response. These sections will be followed by a review of methods of evaluating slope performance.

TABLE 1   Qualitative behavior of marine slopes under wave loading

	Depth[a]	Soil strength[a]	Bottom pressures[a]	General comments
Orbital movement (vertical)	Decreases faster than horizontal with depth	Decreases	Increases	(1) Remolding of soil in unstable zone a gradual process
Depth of unstable zone	NA	Decreases	Increases	(2) After movement stops, pore tubes develop to relieve excess pore-water pressure
Orbital movement (horizontal)	Decreases	Decreases	Increases	(3) Depth of unstable zone achieved quickly
Downslope	NA	Decreases	Increases	(4) Zones of materials of higher strength within the unstable zone are translated intact.

[a]Parameters increasing.

Effect of waves on seabed sediments

The primary difficulty with wave or seismic loading is that the induced stresses are transient, each peak lasting from only a fraction of a second to a few seconds. The response of the slope to this type of loading is different from the response under a sustained load. One possible response is that the soil progressively loses most of its strength or 'liquefies' and flows under its own weight, which results in large permanent deformations. Another alternative is that permanent soil deformations occur only during the short period of time in which an induced stress peak exceeds some threshold strength. In this case movement caused by the several peaks of stress during a storm or earthquake may not be serious, provided that large deformations have not occurred and the soil strength after the cyclic loading is still adequate to resist the applied static gravity loads.

The responses of clays and of sands to cyclic loading are very different and will be treated separately in the following sections.

Behavior of soft clay

Under cyclic loading a soft clay exhibits two predominant changes in its physical behavior. These changes are (1) a decrease in its stiffness (modulus) and (2) an increase in the amount of energy being consumed per cycle of loading (damping). This behavior is shown graphically in Fig. 1 for a clay material from the Gulf of Mexico as it undergoes a constant stress cyclic loading test. In this plot the stiffness or modulus can be represented as the slope of a straight line drawn through the two end points of a single hysteresis loop. The modulus is observed to decrease with increasing cycles of loading. In addition, on each cycle of loading the hysteresis loop becomes larger and larger, which indicates that the amount of energy being consumed is increasing. A summary of the information contained in the hysteresis loops can be presented graphically as plots of the modulus (G) and the damping ratio (D) against the shearing strain (Fig. 2).

The post-static strength of the soft clay is also influenced by the magnitude and duration of the cyclic loading. A graph depicting the decrease of the static strength after loading is presented in Fig. 3. A review of this figure shows that below a critical strain level ($\epsilon_p/\epsilon_{fs} < 0.5$) there is a negligible decrease in strength, while above this level there is a substantial decrease in strength (upwards of 75 per cent).

Behavior of sands

If a relatively loose (relative density = 70 per cent), saturated sandy soil is subjected to cyclic loads, it tends to compact and decrease in volume. If drainage is prohibited during the cyclic loading, there is a subsequent build-up in pore-water pressure until it is equal to the overburden pressure. The effective stress becomes zero and the soil loses its shear strength and develops a condition in which it no longer can support shear loads (liquefied condition). A typical test record of a cyclic loading test on a sand material under anisotropic loading conditions is shown in Fig. 4. A review of Fig. 4 shows that as the cyclic loading is occurring there is a corresponding build-up in the pore-water pressure and a subsequent axial strain. If this condition is of a large extent, and the pore-water pressure is not relieved, it can result in a lateral movement due to a loss in shearing resistance.

The factors that influence cyclic loading induced liquefaction are (a) soil type, (b) relative density or void ratio, (c) initial confining pressure and (d) nature of cyclic loading (magnitude and duration).

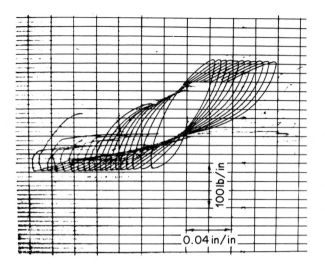

Fig. 1.   Cyclic triaxial test of Gulf of Mexico clay.

Methods of modeling soil behavior

The effects of cyclic loading are extremely dependent upon the soil and nature of cyclic deformation, as

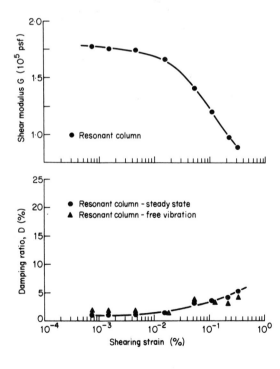

Fig. 2.   Results of cyclic triaxial test on Gulf of Mexico clay.

shown in the previous section. The potential for pore-water pressure generation in either clays or sands governs their behavior under cyclic loading. Two analytical approaches have been developed to model this behavior:

(1) Total stress–strain approaches, in which the cyclic strain or stress conditions of interest are simulated on laboratory specimens and behavior inferred directly.

(2) Effective stress–strain approaches, in which one attempts to determine analytically the pore-water pressures generated and their influence on the soil behavior. Laboratory tests on the soils are utilized at a more fundamental level to determine pore-water pressure generation characteristics.

The effective stress approaches are more intuitively satisfactory from a fundamental standpoint. However, in present offshore practice it is not common that enough information is known about either the *in situ* stress condition or the stress/strain pore-water pressure characteristics of the material under cyclic loading conditions to enable effective implementation of this approach. Research and development of field, laboratory and analytical tools will change this position in the future.

The total stress–strain approaches for cyclic loading problems have largely evolved from work on earthquake response related problems. In general, there have evolved two methodologies of using the total stress–strain approach. The first method involves the development of parameters for input into a stress–strain matrix such as employed in a typical finite element simulation. In general, the basic model of the soil consists of the following elements: (a) initial backbone curve (frequently that of Ramberg–Osgood); (b) criteria to describe the hysteretic behavior (normally Masing criteria are employed); (c) formulation to degrade the backbone curve under either constant strain or stress cyclic loading. This technique will be discussed in greater detail in a later section of this paper.

The second approach is a graphical methodology as depicted in Fig. 5. In this technique laboratory cyclic loading tests are conducted on specimens representative of the slope being evaluated. Typically the cylic triaxial apparatus is employed. A schematic of this operation is shown in Fig. 5(a–c). Results from the test program are then replotted on logarithmic axes as ratio between cyclic shear stresses and static normal stresses ($\Delta\sigma_d/2\sigma_{3c}$) against the number of cycles of load application (N) for a given strain level, as shown in Fig. 5(d). The number of equivalent cycles of load application can be estimated on the basis of Miner's criteria, as shown by Lee and Focht (1975). Knowing the equivalent number of cycles, N, and using Fig. 5(d), a level of cyclic stress, $\Delta\sigma_d$, can be estimated. Then knowing the cyclic stress,

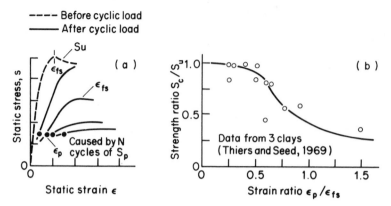

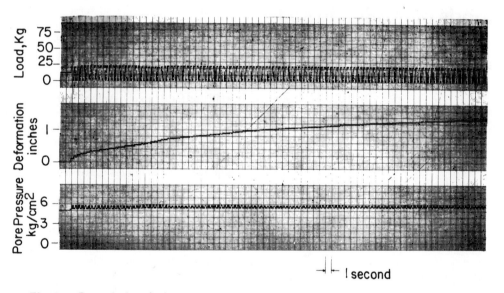

Fig. 3. Static strength after cyclic loading (Lee and Focht, 1976). S_u, static strength in an undrained test performed on an undisturbed sample prior to any cyclic loading; S_c, static strength in an undrained test following cyclic loading; S_p, cyclic strength amplitude; ϵ_{fs}; strain to failure in an undrained test performed on an undisturbed sample prior to any cyclic loading; ϵ_{fc}, strain to failure in a static test following cyclic loading; ϵ_p, cyclic strain amplitude; ϵ_c, cumulative strain.

Fig. 4. Record of typical non-reversing stress test on loose sand: $K_c = 2.0$; $e = 0.87$.

$\Delta\sigma_d$, and the initial stress condition, σ_{1c}, σ_{3c}, of the slice or element comprising the slope, a stress τ_{ff} can be determined. The stress τ_{ff} is the maximum shear stress resulting from the cyclic loading corresponding to a normal stress, σ_{fc}, on the failure plane of the element or slice. Results from a series of such analyses at different strain levels can be subsequently plotted as τ_{ff} against σ_{fc}, as shown in Fig. 5. In use, the imposed cyclic loading stresses combined with the initial static stress condition of the element or slice under study are superimposed on Fig. 5. The resulting strain of the the element or slice can then be read directly from the figure. A summary of the input data requirements

and the subsequent advantages/disadvantages of both the effective and total stress methodologies is presented in Table 2.

Two categories of numerical models have been developed to assess the initial stability and movement of sediments. The majority of these models were initially developed for terrestrial applications (i.e. design of earth dams) but have been, or have potential to be, employed to advantage in the marine environment. These models are (1) limit equilibrium and (2) finite element. In the following sections each of these models will be discussed.

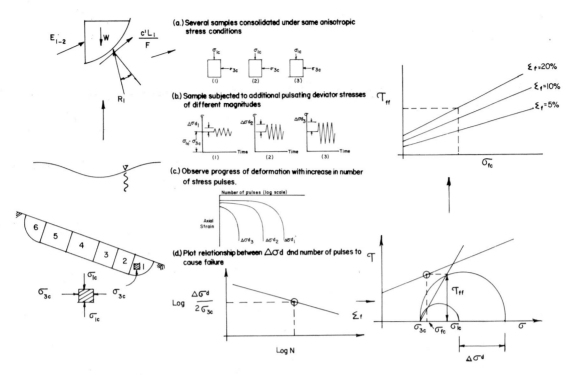

Fig. 5. Methodology to determine elemental strain potential or factor of safety.

Evaluation of slope instability

Limit Equilibrium Methods

Whenever a mass of soil has an inclined surface, the potential exists for part of the soil mass to slide from a higher location to a lower location. The initiation of sliding will occur if driving forces developed in the soil exceed the corresponding resisting forces of the soil. In theory the above problem is simple but certain practical considerations make precise stability analysis of slopes difficult in practice. The primary difficulties are: (a) identification of the failure surface; (b) variation of shear strength in time and space; and (c) identification and quantification of the loading in time and space. There are a number of techniques available for performing limit equilibrium analysis. The various techniques differ primarily in the manner in which they handle the slope geometry, the geometry of potential failure surfaces and utilization of either total or effective stress analysis, and satisfy static equilibrium conditions. The primary methods are (1) infinite slope analysis, (2) the Swedish circle method and (3) the method of slices. The final evaluation of the stability or instability of the slope is made normally by the use of appropriate safety factors.

Plane Failure Surface

Submarine slopes are commonly fairly gentle, uniform and homogeneous over considerable horizontal distances. Thus in many cases 'infinite slope' limiting equilibrium methods are applicable. One disadvantage with this type of analysis is that it can only handle earthquake accelerations (Morgenstern, 1967) and not storm wave loading.

Circular Failure Surface

One of the first procedures proposed for evaluating seafloor stability under wave loading was described by Henkel (1970). The analytical model described by Henkel is a total stress analysis (neglecting soil pore-water pressures) based on the principle of limiting equilibrium and employing the assumption of a circular failure surface. The driving moment initiating a slide in this model is produced by the gravity forces introduced by a sloping bottom profile and the couple developed by the differential bottom pressures produced by surface waves. If the driving moment is sufficient to overcome the resisting moment developed by the shear strength of the soil along the circular failure arc, large displacements may be expected to occur. Henkel employed this model to show that the overturning moment produced by large waves in the Gulf of Mexico could exceed the resisting moment provided by some of the soft sediments offshore from the Mississippi delta (Henkel, 1970). Bea (1971) used this same model to investigate the failures of two jacket

TABLE 3 Characteristics of total and effective stress analysis

TOTAL STRESS ANALYSIS
Input data requirements
(1) Bulk unit weight
(2) Stress–strain total stress parameter
(3) Slope geometry
(4) Stratification
Advantages
(1) Concept is simple
(2) Does not require knowledge of *in situ* initial pore pressure distribution
Disadvantages
(1) Requires accurate simulation of loading in laboratory
(2) Does not give a fundamental understanding of phenomena
(3) Requires extensive laboratory test program

EFFECTIVE STRESS ANALYSIS
Input data requirements
(1) Intitial pore-water pressure distribution
(2) Bulk unit weights
(3) Stress–strain effective stress parameters
(4) Slope geometry
(5) Stratification
Advantages
(1) Gives a fundamental understanding of phenomena
(2) Requires only limited laboratory test program to determine characteristics of pore-water pressure generation. This assumes availability of effective stress–strain relation
Disadvantages
(1) Requires knowledge of *in situ* initial pore-water pressure distribution
(2) No universally accepted effective stress–strain relation is at present available.

pile platforms in the Gulf of Mexico (South Pass block 70). Bea showed that the model correctly predicted the initiation of movement in the sediments. The evaluation of this stability is accomplished through the use of a factor of safety. Finn and Lee (1979) presented results from a more general effective stress method based on Sarma's model (Sarma, 1973), which, in addition to wave loading, also included earthquake loading and excess pore pressures.

Finite Element Methods

The finite element method has been used extensively for the evaluation of slope stability. This use has been primarily for terrestrial soil but it has also been used for marine soils. The basic methodology utilized is based on the original work by Wilson (1965). The method itself is based on the determination of the resulting deformations of a frame under an imposed load. The deformations are based on the minimum

potential energy. The relationship between the deformations $\{q_j\}$ and the loads $\{Q_i\}$ is given by the following expressions:

$$\{q_j\} = [k]^{-1} Q_i$$
$$k = \int_v [A]^T [C] [A] \, dv$$
$$\{\xi_i\} = [A_{ij}] \{q_j\};$$
$$\{\sigma_i\} = [C_{ij}] \{\xi_j\};$$
$$C_{ij} = \begin{bmatrix} (3B+4G)/3 & (3B-2G)/3 & 0 \\ & (3B+4G)/3 & 0 \\ & & G \end{bmatrix};$$

where
$[k]$ = stiffness matrix;
$[A_{ij}]$ = geometric matrix;
$[C_{ij}]$ = stress strain matrix;
$\{\xi_i\}$ = strain vector;
$\{\sigma_i\}$ = stress vector;
B = bulk modulus;
G = shear modulus;

In the use of the finite element method three distinct ways of quantifying the initial stability of slopes have been developed. The first two methods utilize element calculations in which the slope is modeled as a series of individual elements (Seed, 1966). The first method uses a calculated factor of safety for each element. In contrast, the second method is based on an evaluation of the potential strain experienced by each element. The third method involves predicting the overall resulting permanent deformations.

Factor of Safety Approach

In this method the factor of safety is defined as the ratio between the cyclic strength and the induced dynamic stress in each element of the slope. One of the difficulties in this approach is that a different factor of safety is calculated for each element and no single overall definition of the factor of safety for the entire slope is readily determined. This difficulty is overcome by applying judgement or an averaging process to the results. Consideration of the mechanisms of potential failure and the existence of a large number of safe elements may indicate a safe slope, whereas a large number of failure elements may suggest that the slope is unsafe.

Potential Strain Approach

Knowing the factor of safety at each element based on failure defined by a specified strain in a laboratory test, it is possible to evaluate the equivalent cyclic strain at each element. These strains are strains that would develop in a laboratory test specimen which is

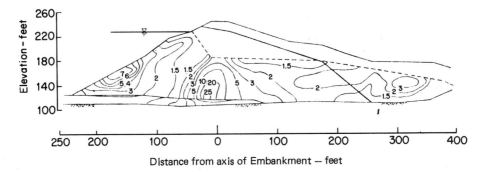

Fig. 6.   Strain potential approach utilizing finite element method (Makdisi *et al.,* 1978).

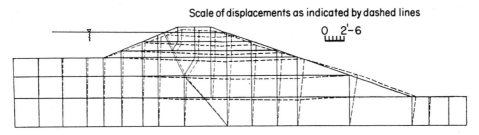

Fig. 7.   Permanent deformation approach utilizing finite element method (Chaney, 1979).

subjected to the same static and seismic stresses imposed on a field element. However, an element in the field cannot strain like the test specimen, because it is constrained by the deformations of adjacent elements. Thus, with reference to the field loading case, the cyclic strains are referred to as potential strains. Judgement is again required to interpret the meaning of strain potential in terms of the overall dynamic stability of the slope. An example of this method is shown in Fig. 6 for an earth embankment (dike).

Permanent Deformation Approach

Displacement-type analysis can be an important method in evaluating wave loading effects on submarine slopes. Wright (1976) developed a static finite element model which included both wave forces and gravity stresses, and utilized a hyperbolic stress—strain relationship. In an analysis of the South Pass block 70 failure, Wright showed that lateral downslope displacements were more than twice the corresponding deflections upslope. This is a direct result of the downward component of gravity stresses. Wright also showed that the deterioration of stress—strain parameters due to cyclic loading led to a drastic increase in displacements.

The principal limitation of Wright's static displacement model is that it does not consider the accumulation of deformations due to passage of number of waves. This problem has been approached by Schapery and Dunlap (1978), modeling the soil as a linearly visco-

elastic material. Their analysis also included the effect of energy adsorption of the seafloor on the wave characteristics.

Lee (1974), Serff (1976) and Chaney (1979, 1980) have developed more rigorous methods of estimating the permanent deformations experienced by earth structures during a dynamic loading. These methods follow the basic Seed approach for calculating the factor of safety and strain potential for individual elements (Seed, 1966). Differences between the two methods depend on procedures used to calculate deformations from the strain potential values. The procedure initially developed by Lee (1974) and later refined by Chaney (1979, 1980) involved the concept that permanent seismic deformations of a slope may be computed by evaluating dynamically induced softened slope stiffness values for soil elements with the resultant settling of the slope to a new condition being compatible with pseudo or apparent stress—strain properties of the soils comprising the slope. An example of a permanent deformation method for an earth embankment (dike) is shown in Fig. 7.

Summary and conclusions

The purpose of this paper has been to present a background and critical review of the methods for predicting the behavior of a marine slope under cyclic loading. The problem was approached by first looking at the

TABLE 3   Summary of methods to evaluate initiation of instability

Method	Approach	Type of failure surface	Type of stress analysis handled	Type of external cyclic loading handled	Method of evaluating failure	Comment
Limit equilibrium	Plane failure surface	Plane	(a) Total (b) Effective	Seismic	Factor of safety	Morgenstern (1967)
	Circular failure surface	Circular	(a) Total (b) Effective	(a) Seismic (b) Wave	Factor of safety	Henkel (1970); Finn and Lee (1979)
Finite element	Factor of safety	Not required	Total	(a) Seismic (b) Wave	Factor of safety	Seed (1966)
	Potential strain	Not required	Total	(a) Seismic (b) Wave	Strain potential of individual elements	Seed (1966)
	Permanent deformation	Not required	(a) Total (b) Effective	(a) Seismic (b) Wave	Deformation	Lee (1974); Wright (1976); Serff (1976); Chaney (1979, 1980)

behavior of a laboratory simulation of a marine slope under wave loading as reported by various investigators. It was observed from these various studies that the behavior of an unstable zone which is involved in a slope movement is dependent on various parameters. These parameters are soil strength, bottom pressures, depth of unstable zone and pore-water pressure generation.

To predict the performance of a marine slope knowing the cyclic loading depends therefore on (a) quantifying the soil response to cyclic and static loading (total or effective stress analysis) and (b) selection of an appropriate slope stability analysis technique (limit equilibrium method or finite element method). The selection of a slope stability analysis technique is dependent upon: (1) the configuration of the failure surface; (2) the nature of cyclic loading; and (3) the required method of evaluating failure (factor of safety, potential strain or permanent deformation). A summary of the various methods to evaluate instability is presented in Table 3. The most general methodology currently available based on this review is the permanent deformation procedure.

References

Bea, R. G., 1971. How sea-floor slides affect offshore structures. *Oil Gas J.*, 19 November, 88–92.

Chaney, R. C., 1979. Earthquake induced deformations of earth dams. *Proc. US National Conf. Earthquake Engineering.* Earthquake Engineering Research Institute, Stanford, California, pp. 632–642.

Chaney, R. C., 1980. Seismically induced deformations in earth dams. *7th World Conf. Earthquake Engineering, Istanbul, Turkey*, Vol. 31, pp. 483–486.

Doyle, E. H., 1973. Soil-wave tank studies of marine soil instability. *Proc. 5th Annual Offshore Technology Conf.*, Vol. 2, pp. 753–766.

Finn, W. D. L. and Lee, M. K. W., 1979. Seafloor stability under seismic and wave loading. *Proc. Soil Dynamics In The Marine Environment.* ASCE, Reprint 3604.

Henkel, D. J., 1970. The role of waves in causing submarine landslides. *Geotechnique*, **20** (1), 75–80.

Lee, K. L., 1974. Earthquake induced permanent deformations of embankments. UCLA-ENG-7498, University of California, Los Angeles.

Lee, K. L. and Focht, J. A., Jr., 1975. Cyclic testing of soil for ocean wave loading problems. *Proc. 7th Annual Offshore Technology Conf., Houston*, Vol. 1, pp. 343–354.

Lee, K. L. and Focht, J. A., 1976. Cyclic testing of soil for ocean wave loading problems. *Mar. Geotech.*, **1**, (4), 305–325.

Makdisi, F. I., Seed, H. B. and Idriss, I. M., 1978. Analysis of Chabot Dam during the 1906 earthquake. *Proc. Earthquake Engineering and Soil Dynamics.* ASCE, Vol. II, pp. 569–587.

Morgenstern, N. R., 1967. Submarine slumping and the initiation of turbidity currents. In: A. F. Richards (ed.), University of Illinois Press, Urbana, Ill., pp. 189–220.

Sarma, S. K., 1973. Stability analysis of embankments and slopes. *Geotechnique*, **23**(3), 423–433.

Schapery, R. A. and Dunlap, W. A., 1978. Prediction of storm-induced sea bottom movement and platform forces. *Proc. 10th Offshore Technology Conf., Houston*, Vol. 3, pp. 1789–1796.

Seed, H. B., 1966. A method for earthquake resistant design of earth dams. *J. Soil Mech. Foundns Div., ASCE.*, Paper 4616, 13–41.

Serff, N., 1976. Earthquake-induced deformations of Earth Dams. Ph.D. Thesis, University of California, Berkeley.

Singh, H., 1971. The Behaviour of Normally Consolidated and Heavily Overconsolidated Clays At Low Effective Stress. Ph.D. Thesis, Cornell University, Ithaca, New York.

Tsui, K. K., 1972. Stability of Submarine Slopes. Ph.D. Thesis, Queen's University, Kingston, Ontario.

Wilson, E. L., 1965. Structural analysis of axisymmetric solids. *J. Am. Inst. Aeronaut. Astronaut.*, **3**, 2269–2274.

Wright, S. G., 1976. Analysis for wave induced sea-floor movements. *Proc. 8th Annual Offshore Technology Conf., Houston*, Vol. 1, pp. 41–52.

15

Geotechnical Hazards Associated with Leg Penetration of Jack-up Rigs

R. Kee and B. W. Ims
Fugro Ltd., Hemel Hempstead, UK

This paper describes the geotechnical problems of locating jack-up rigs on the seabed, whether on mat supports or footings. The potential hazards which may lead to leg penetration into the seabed are outlined, being punch-through failure, seabed scour, gas pockets, soil collapse and the effects of cyclic/horizontal leg loads.

Introduction

In recent years there has been a large increase in offshore oil and gas exploration activities. About 60% of the work is being performed with jack-up drilling rigs, of which there are about 300 already operating worldwide.

Jack-up rigs basically are of two types, depending on their foundation: (a) mat-supported, (b) individual footings (spud-cans). Mat-supported rigs have a large foundation bearing area and therefore impose relatively low bearing pressures on the seabed, which results in limited penetration. Mats are usually A-shaped and are suitable for use when the seabed comprises very soft clay soils. However, mat-supported rigs require that the seabed be fairly level and their stability is sensitive to factors such as subsoil variability, lateral sliding resistance and seabed instability. Rigs supported on individual footings are more numerous and much more widely used. This paper presents aspects of geotechnical hazards that may be encountered when the legs of these rigs penetrate into the seabed.

Typical rig

Figure 1(a) shows a typical profile of a jack-up rig supported on individual legs, and the principal dimensions are shown in Fig. 1(b). The rigs are supported on three or more legs. The overall length of the legs is up to 130–140 m range and therefore the water depth in which a rig can operate will depend on the extent of the penetration of its leg into the seabed and the air gap required at a given location. Typical configurations and dimensions of circular footings of various jack-up legs are shown in Fig. 2.

Jack-up procedure

When a rig arrives at a drilling location, the legs are lowered until contact is made with the seabed and nominal penetration of the legs achieved. After the position of the rig is checked, the legs are jacked down into the seabed continually against the weight of the rig. The legs will penetrate into the seabed until sufficient bearing resistance is encountered by the footings, when the hull will start to rise clear of the water surface.

Before a rig is put into operation it is normal practice to impose a temporary preload on the legs by pumping sea-water into compartments in the hull of the rig. The preload is to allow for environmental and live loads that the rig might experience during operation so as to prevent additional leg penetration. The magnitude of preload that is generally used depends on the rig design and anticipated additional environmental and live loads, and may vary between 20 and 70% of the maximum normal vertical load of the rig.

Before the preload is added, the hull is raised clear of

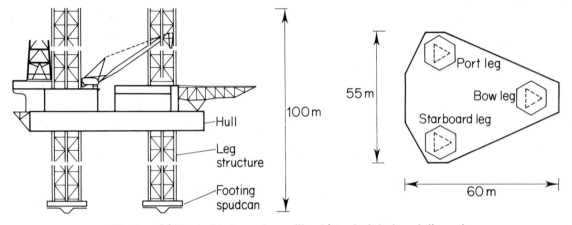

Fig. 1. (a) Typical jack-up rig profile; (b) typical deck and dimensions.

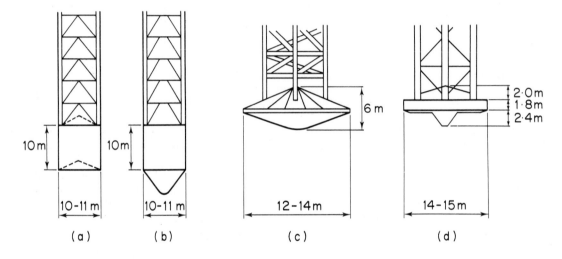

Fig. 2. Typical configurations and dimensions of footings.

the water to maintain a minimum air gap consistent with weather conditions and operational constraints. This is an important consideration in order to minimize the effect of punch–through failure of one or more legs resulting in sudden additional penetration which might cause substantial damage to the leg structure or rig. Punch-through failure is one of the hazards that is discussed later. It is caused by a footing temporarily resting over a relatively thin strong soil layer underlain by a weaker layer. The latter fails when its bearing capacity is exceeded by the pressure imposed by the footing, owing to an increase in leg load.

After preloading and when all conditions are deemed satisfactory, the preload is removed by dumping the ballast and the hull is jacked up to the required air gap.

Leg penetration

A rig may be adequate from the point of view of exploration requirements, but to determine whether it is suitable for a particular location it is important that the likely magnitude of leg penetration be capable of being assessed. In this context, the suitability of the rig will depend on whether the available length of the legs is sufficient, given the water depth and minimum air gap requirements at the location.

To evaluate the magnitude of leg penetration the following data should be known: (a) dimensions and magnitude of loads acting on a leg footing; (b) sequence and engineering properties of the strata underlying the seabed; (c) footing shape and size.

Item (a) can be evaluated quite accurately. Leg loads could be up to 4500 ton, and assuming a 14 m diameter

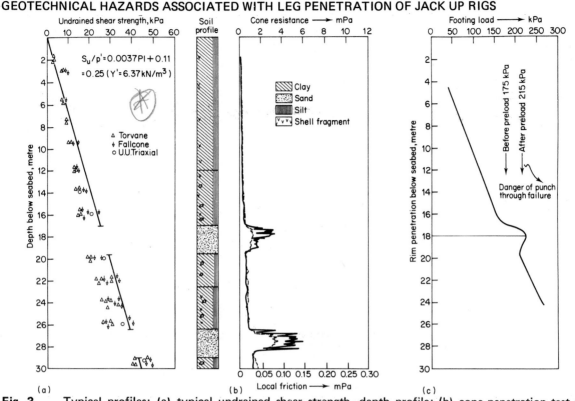

Fig. 3.   Typical profiles: (a) typical undrained shear strength—depth profile; (b) cone penetration test result; (c) footing load/penetration below seabed (footing diameter 14 m) of the type shown in Fig. 2(d).

footing the bearing would be just under 300 kPa. In typical normally consolidated clays leg penetration could exceed 30 m below mudline during preload.

Item (b) should be obtained from boreholes with associated undisturbed sampling and laboratory testing to determine the shear strength profile together with other relevant properties for a better understanding of the soil characteristics. Figure 3(a) shows typical soil data requirements. In the authors' opinion the best solution is to supplement and enhance the borehole data with an adjacent static cone penetration test (CPT). The latter provides a continuous profile of the soil resistance and would reflect strength variations which might be missed in the borehole. Figure 3(b) shows a typical CPT profile. Hence, for evaluation of leg penetration, knowledge of the soil conditions is a prerequisite. In areas where no soil data are available a soil investigation should then be performed in advance of the rig's arrival. Where soil data are available in the area but not at the specific location and the weather conditions are predictable, a soil boring could be performed from the rig after initial jack-up but before the application of the preload. This is done to check that there is no danger of a punch-through failure during or after preloading.

In many areas of the world where jack-ups are used the strata underlying the seabed comprise normally consolidated clays. In these conditions classical bearing capacity theory can be applied to evaluate the magnitude of leg penetration. The footings will penetrate into the clay so long as the pressure imposed by the footings is greater than the ultimate bearing capacity of the underlying clay. The method presented by Skempton (1951) is generally used for determining the ultimate bearing capacity of shallow foundations in clays:

$$Q = A(CN_c + \gamma'D) \qquad (1)$$

where Q = ultimate bearing capacity of footing; A = cross-sectional area of footing (largest diameter for a spud-can); C = average undrained cohesion to significant depth below footing; D = depth of footing; B = footing width (largest diameter in the case of a spud-can); γ' = submerged density of soil; N_c = bearing capacity factor, varying between 6.2 and 9 as D/B varies between 0 and 4.

This method recommends that the average shear strength over a depth of $2B/3$ below the footing be used if the shear strength at this depth does not vary by more than ± 50% of the average value. Other authors suggest that the average shear within a depth of $0.5B$ below the footing is more appropriate.

Another method frequently used is that presented by Davis and Booker (1973) for linearly increasing shear strength of the soil with depth.

The authors' experience, based on about 25 case histories, with footing types as shown in Figs. 2(c, d) is that equation (1) overestimates the bearing capacity and, hence, underestimates the magnitude of leg penetration in a normally consolidated or lightly over-consolidated clay.

The authors have found that in a soft clay the following simple approach gives a good correlation between observed and predicted leg penetration:

$$Q = 8 \cdot A \cdot C_0 \quad \text{or} \quad Q = 6 \cdot A \cdot C \qquad (2)$$

whichever is the lesser, where C_0 = strength immediately beneath the largest part of the footing; C = average shear strength determined over a depth of $B/2$ beneath the footing. In either case the shear strength should be determined from triaxial tests on 'good quality undisturbed' samples.

The reasons for the underpredictions obtained when using equation (1) could be that: (a) the assumption that the conical shape of the spud-can can be idealized as a flat footing is invalid; (b) the continuous failure of the soil during footing penetration results in a lower mobilized soil strength as measured by conventional tests; (c) soil strength anisotropy is a significant factor; (d) the compressibility of the soft clay may negate the conventional depth effect ($\gamma'D$), as suggested by Vesic (1963); (e) soil falling into the holes created by the penetrating footings increases the vertical pressure at the undersurface of the footing.

Punch-through hazard

Punch-through hazard occurs in stratified soils when there is a relatively thin layer of cemented material, stiff clay or sand underlain by weaker soil. The leg footings penetrate to the strong stratum under preload. If the footing pressure of a leg exceeds the combined bearing capacity of the two-layered system, then a punch-through failure will occur. Sudden and large additional penetration accompanied with intolerable tilting could be disastrous and many a rig has been severely damaged because of this and some have been lost as a direct result of it. Therefore the value of soil information before rig placement cannot be over-emphasized, as this will also influence rig selection. A rig with footing pressure that can operate safely under these circumstances should be selected.

If a potential punch-through hazard is known to exist and it is essential to drill at a particular location, under favourable conditions and controlled procedures this hazard could be overcome. One technique would be to jack down the legs to the top of the layer by jetting water through outlets which are incorporated in the underside of the footings. However, this operation is far from being foolproof and rig damage has been known.

To determine the combined bearing capacity of a two-layered system comprising a stronger layer of finite thickness overlying a soft layer, the following techniques have been applied successfully.

For the case of a stiff clay layer overlying a soft clay, the method given by Brown and Meyerhof (1969) is relevant. The method assumes that shear punching occurs around the footing perimeter; thus, for a circular footing:

$$q = \frac{3HC_1}{B} + 6C_2 \qquad (3)$$

where q = ultimate unit bearing capacity of footing; C_1 = average undrained cohesion of upper stiff clay; H = thickness of stiff clay layer below footing; C_2 = undrained cohesion of underlying soft clay. This largely empirical method based on small-scale experimental results has been confirmed by numerical techniques (Griffiths, 1982).

Where a layer of sand overlies soft clay, Vesic (1970) suggests a generalised method whereby

$$q = q_2 \exp\left(\frac{4H}{3B}\right) \qquad (4)$$

where q_2 = bearing capacity of a footing of the same size resting on top of the soft clay (see equation 2).

Vesic also gave a simple expression for critical depth of the sand layer beyond which the underlying soft clay will have little effect on the bearing capacity:

$$H/B = 3/4 \ln (q_1/q_2) \qquad (5)$$

where q_1 = bearing capacity of sand layer, assuming infinite thickness. If typical values are used, the critical H is $1.5-2B$ for a loose sand layer.

Other hazards

Scour

In general, when the seabed consists of either sand or dense cohesionless silt to a significant depth, leg penetration will be limited. This is due to the relatively high bearing capacity of these materials compared with that of a typical footing pressure.

In this instance scour of the seafloor could pose a danger to the stability of a rig by erosion of the foundation material. To guard against such a problem, the footings could be made to penetrate below the anticipated scour depth by jetting. Alternatively, gravel bags or anti-scour netting could be placed around the footing for protection.

Presence of shallow gas pockets

In many areas of the world gas pockets under pressure have been found in sand layers within normally consolidated clay deposits at depths of 20–30 cm below mudline. Such pockets have been encountered in soil investigation boreholes. When a footing penetrates into such a stratum, the release of gas under pressure from these pockets could be disastrous. Quite apart from other consequences of gas escape, the release in pressure might cause liquefaction of the sand, resulting in complete loss of strength and support of the footing.

When geophysical surveys are being performed at a proposed rig location, they should concentrate on the possibility of detecting the presence of shallow gas pockets. If any doubt exists, there is no substitute for investigation by boreholes and cone testing. The possible presence of gas in the sediments can be determined by detailed chemical analyses of soil samples obtained using recently developed ambient pressure samplers or from comparative measurements of shear and compression wave velocities on undisturbed samples in the laboratory.

Collapse of adjacent soil over footings

Inspection reports by divers at some locations in soft normally consolidated clays have indicated that there is a progressive infilling of soil over the footings as they penetrate below mudline. This is used to help in justifying the application of equation (2) for predicting leg penetration in a soft clay.

However, it can be imagined that when the clay is firmer, 'clean' holes are formed as the legs penetrate. Subsequent collapse of soil over the footings would add sudden load on them, and when precarious soil conditions exist underneath, this could be a source of danger.

Time-dependent leg penetration in clay

When a rig is expected to be at a location for some time, additional penetration of the legs would occur due to consolidation of clay underneath the footings. Normally this is not expected to disturb the stability of the rig, as the rate and magnitude of settlement are relatively low, in general. The level of the hull is usually monitored and the effect of differential settlement of the legs can be corrected by raising the hull at the appropriate leg for compensation.

Operation and environment loads

During its working life the rig foundation will be subjected to loads different from those imposed during preloading. Drilling operations will impose both cyclic and occasional shock loads. Wind and wave action will impose cyclic horizontal loads, the line of action of which will result in rotational moments being applied to the footing (Fig. 4).

Cyclic loading has the effect of reducing soil strength and, in granular soils at shallow penetrations, possibly

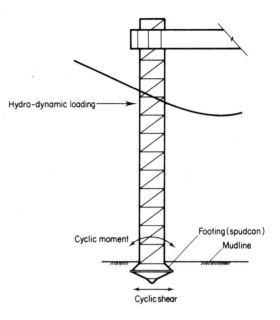

Fig. 4.   Environmental loadings.

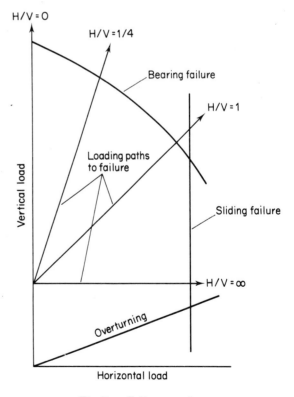

Fig. 5.   Failure envelope.

causing washout beneath the edges of the footing. Horizontal loads and moments have the effect of reducing the bearing capacity of a foundation (Fig. 5). The net effect is to have a foundation subjected to a loading different from that imposed during the 'check' preloading, bearing in a soil whose strength may be lower than originally measured.

Hence, in extreme conditions foundation failures resulting in additional and sudden leg penetrations could and have been known to occur.

Techniques to evaluate jack-up footing behaviour under these loading conditions are based on observed laboratory behaviour. Confirmation of the accuracy of these techniques by back-analysis of field behaviour is limited by the poorly defined environmental loads. Research is currently being carried out to develop monitoring equipment to improve the data bank on this aspect by a number of companies and universities throughout the world.

Conclusions

Rig failures are usually not reported in detail, if at all. Foundation failures range from unexpected additional leg penetration causing only minor damage to dramatic collapse and loss of the rig. With the expected increase in the numbers of rigs that will be operating in the future, it is as well that potential hazards are borne in mind so that appropriate precautions may be taken to avoid mishaps.

Potential geotechnical hazards associated with jack-up rigs supported on individual legs have been discussed. These include: (1) punch-through failure; (2) scour of the seabed; (3) presence of shallow gas pockets; (4) collapse of soil over footings causing sudden increase in load; (5) combined effect of cyclic and horizontal loads during the lifetime of the rig at location.

The value of comprehensive geotechnical investigation combined with geophysical and oceanographic surveys has been emphasised. Potential problems can be identified and risks avoided. Geotechnical methods are available for analyses; but experience and the exercise of engineering judgement are necessary for good effect.

References

Brown, J. D. and Meyerhof, G. G., 1969. Experimental study of bearing capacity in layered clays. *Proc. 7th Int. Conf. Soil Mechanics and Foundation Engineering, Mexico City,* Vol. 2, pp. 45–51.

Davis, E. H. and Booker, J. R., 1973. The effect of increasing strength with depth on the bearing capacity. *Geotechnique,* 23 (4), 551–563.

Griffiths, D. V., 1982. Computation of bearing capacity on layered soils. *Proc. 4th Int. Conf. Numerical Methods in Geomechanics, Edmonton.*

Skempton, A. W., 1951. Bearing capacity of clays. *Building Research Congress, London,* Div. 1, pp. 180–189.

Vesic, A. S., 1963. Bearing capacity of deep foundations in sand. *High. Res. Rec.,* **39**, 112–153.

Appendix to Parts I and II

Sub-seabed Studies Workshop

A. R. Biddle* and N. G. T. Fannin† (Chairmen)
*McClelland Engineering Ltd., Harrow, UK
†British Geological Survey, Nottingham, UK

Government funded seafloor surveys and studies

The role of government-funded geological and marine exploration offshore has been discussed.

The main stimulus for recent government seabed mapping of the Continental Shelf in many areas of the world has been the search for oil and gas reserves offshore. Large government budgets are now involved. The objectives and constraints of such work have been discussed at length, including differences in phrasing legislation and guidance regulations can affect the extent of seabed surveys and studies carried out by exploration licence holders. In the UK, the Department of Energy Guidance Notes have been as brief as possible, consistent with giving adequate guidance.

In the USA, the USGS has been given the brief to ensure that oil and gas exploration permits are only issued for areas where the seabed is geologically safe. This approach has proved difficult to implement as the geologist's time-scale is orders of magnitude different to that of commercial oil development, and risks are impossible to quantify for many hazards, e.g. earthquake triggered mechanisms. Using the geologist's judgement often results in large tracts of seabed being withdrawn from lease sale due to doubts over the degree of risk associated with many suspect features of the seabed. The need for a better quantified risk assessment, demands detailed analysis of seabed features by geotechnical engineers. Such studies have been carried out by oil companies for development sites but the extent is necessarily limited.

There are only a few isolated cases of geotechnical engineers having been involved or commissioned to evaluate the significance of seabed hazard features in government funded work in the USA or UK. Sufficiently multi-disciplined teams have been used in seabed studies offshore Israel, The Netherlands and Japan.

There was recognition by many participants at the Conference that their studies would have been more conclusive and meaningful if an expert from another discipline had been involved as well. In many cases there were teams of scientists without any engineering input for quantitative analysis and risk assessment. There was a considerable body of opinion in favour of the involvement of broad-based coordinators from industry to take the lead role in government sponsored seabed studies of the type run by the USGS.

The discussion then turned to detailed technical topics, as given below.

Liquefaction

A short presentation was given by Delft Hydraulics Laboratory, The Netherlands, on the subject of 'Wave-Induced Seepage Flow Around Pipelines'. Cases involving fully-buried and partially-buried pipelines have been studied. Wave-induced pressure on the seabed creates cyclic forces and changes in bouyancy during the oscillatory flow.

At 17 km a length of pipeline nearshore Netherlands has settled over 1 m into the seabed and has become completely covered. This is not thought to be due to liquefaction but due to sand-bed movement. Most of the burial is known to occur in severe storms.

Current engineering practice to assess the risk of liquefaction involves measuring soil voids ratio, estimating shear stresses due to the environmental exciting forces, and then estimating the level of pore pressures that are set up in the soil using data from laboratory cyclic shear tests and resonant column tests on extracted samples concerned. Methods such as those of Yamamoto or Ishihara are then used to predict the thickness of surface soil that may liquefy, and depending on the engineering judgement the foundation loads of offshore structures are often transferred to soils below this depth using skirts or piles. However, it is recognised that too little is known about the subject of liquefaction, particularly earthquake excited phenomena. Long term monitoring programmes are required by governments in earthquake areas to not only record the seismographs but also the ground effects caused in a more scientific manner than before. For instance, the measurement of seabed pore pressure during earthquakes is required to better predict the pore pressure levels that can be produced and to understand the activating level causing submarine slope failures.

Many effects of earthquakes such as seabed sand 'boils' disappear within a few days of the event. Large effects such as slope failures can be examined in detail as they appear to retain their geometry on a geological timescale.

There was a call from the Conference participants for standardisation of measurement of soil properties in the interests of compiling an earthquake data base throughout the world. Examples were given of imprecise methods such as the so called 'Standard Penetration Test' being used as base data in China and Japan. The SPT has many variations throughout the world and this renders it unacceptable in scientific comparisons. In geotechnical circles, the international conferences such as ESOPT are focussing attention on these problems. The use of new methods, such as measurement of shear wave velocity in soils, may be more applicable to earthquake susceptibility than traditional mechanical tests.

Seabed erosion features, scour and bed movement

In the discussion of liquefaction, several other topics were raised that are best reported under this separate heading.

The point was made that due to ignorance, a large number of seabed features and hazards were being treated as liquefaction phenomena, when they were most probably due to bed movement under wave action. The scour potential of large waves around offshore structures may not be fully appreciated. Seabed particle transport due to waves can occur right out to the Continental Shelf edge at depths up to 200 m. Reflected waves from the structure cause a reversal in seabed particle motion and a nett translation along the sides of the obstruction. Research is required into this aspect to relate field scour observations to wave effects via instrumentation of the seabed and model studies.

Submarine slope stability

Participants agreed that seabed slope failures should be accurately mapped, measured and sampled in a similar manner to terrestrial practice to permit engineering analysis and thereby an understanding of the likely failure criteria. It should be standard practice during seafloor exploration surveys to measure the slope accurately and to sample comprehensively the soils in the upper geological units to a depth of at least 40 m. It was apparent from the discussion that seabed sampling using dart corers, gravity corers and vibrocorers was only 'scratching the surface'. Most slope failures were on a massive scale and obtaining representative samples of the soils affected would require use of geotechnical drilling vessels. Dynamically positioned drill ships are now available in Europe which can operate in water depths down to 1000 m, which permits soil sampling of the Continental Shelf margin slopes in many areas of the world.

It was evident from the discussion that slope stability analysis requires the involvement of experienced geotechnical engineers in the seafloor survey and study teams. The exact solutions sought by scientists are not possible and the analytical process requires sensitivity studies using a selection of parameters coupled with engineering judgement to bound the problem.

It was concluded by IUTAM participants that the workshop session had been extremely valuable in bringing together experts from the scientific, academic and engineering disciplines. Much had been gained from a frank discussion in an informal atmosphere.

Part III

Sediment Transport

Section 5 — Wave- and Current-induced Transport
Section 6 — Storm Impact and Forecasting

Section 5 — Wave- and Current-induced Transport

Introduction

D. M. McDowell (Chairman)
Simon Engineering Laboratories, University of Manchester, UK

Three papers presented here are concerned with theoretical and laboratory studies of sediment transport due to wave action, and the other is concerned with observations of sediment transport in the field.

Research into a particular subject usually follows the same general course. Curiosity about some observed event leads to measurements and attempts to explain them in terms of the physical parameters that are thought to be relevant. Failure to explain satisfactorily leads to proliferation of measurements and theories, which are often successful in specific situations. This leads to the realisation that a general theory is needed. A general theory may only emerge slowly after a considerable amount of effort, but once established, it is invaluable in providing a better framework for solution of specific problems.

The subject of sediment transport is one of great difficulty because of the variety of relevant parameters. It was inevitable that early work in theoretical and laboratory studies should concentrate on steady flow over grains with uniform properties, and that real problems would have to be solved by means of empirical relationships based on crude measurements and limited experience.

Until very recently, nearly all serious research on sediment transport had been on steady flow acting on non-cohesive grains. Some attempts have been made to account for mixed non-cohesive sediments of varied shape and size, and the resultant bed forms have been included in studies. Within the past two decades there has been work on cohesive sediments beginning with tentative measurements of gross parameters. Only in the last few years has there been widening interest in the action of waves in causing sediment transport.

There is now a pressing need for a generalised theory of sediment transport due to unsteady flows. Such a theory would incorporate the rate of development of bed forms, the modification of fluid properties by material in suspension, and the growth and decay of turbulence in unsteady flows. One vital element is that, in unsteady flows, instantaneous conditions are a function of the past history of both the flow field and the sediment motion. Until these points are faced and overcome, generalised solutions will not be obtained. It is often argued that in many situations, steady flow solutions give an acceptable answer. There is some truth in this, just as there is in the statement that truly steady flow seldom exists in natural situations.

It is one thing to state the goals to be sought in a generalised theory; it is quite another to strive to reach them. One of the main reasons for the lack of progress has been the difficulty of obtaining measurements and the near impossibility of analysing so many varied and varying parameters. University workers, in particular, have not had the resources to undertake this work on the scale needed. Now, however, the subject has been transformed by modern instrumentation and the power of data acquisition and analysis. Nevertheless the complexity of the problem is such that its solution will only evolve gradually by many stages.

Ahilan and Sleath (1984) describe here work on one of the stages towards a generalised solution. Their work was confined to laminar flow induced by wave action over a bed of uniform grains. Sediment transport over a flat bed occurs under two extreme conditions; when motion is only just initiated by the flow, and when it is so intense that bed forms are washed out. The authors chose to study the case of intense motion of bed material.

The choice of laminar flows has obvious advantages in that standard analytical solutions exist for fluid behaviour in several situations. It is therefore possible to work towards an acceptable solution from sound physical principles. When sediment transport has to be taken into account, however, the solution is less simple. The authors assumed that it occurred by material being carried in suspension in the fluid at concentrations that are related to the boundary shear stress. The shear stress and fluid viscosity are affected in turn by the concentration of particles entrained in the fluid. All the relationships between particle concentration and fluid properties that were used were semi-empirical and were devised for conditions of steady flow (Bagnold, 1954, 1956). They are non-linear functions of concentration, so the full set of equations had to be solved numerically.

In order to achieve high rates of transport over a flat bed under laminar flow, it was necessary to use a light-weight bed material. Nylon with a submerged relative density of 0.137 was used. Experiments were conducted with the objective of observing velocity distributions within the moving layer of sediment. The methods used were inexpensive and ingenious and appeared to give reliable results. They would have been simpler to achieve and potentially much more detailed if a Laser Doppler anemometer had been used.

The main finding was that measured results could only be made to compare with theory if the viscosity was increased by a factor of about six times. This is close to 5.9, which had been suggested by Kajiura for eddy viscosity in the inner boundary layer of flow over a rough bed. The authors drew attention to the need for more experiments to resolve this question.

This highlights one of the fundamental problems with work on laminar flows. Solution of the equations of fluid motion depend on the assumption that flow really is laminar. Such flows cannot sustain particles in suspension in the fluid and if they are realised, the only particles in suspension would be there as a result of collisions with, or within, the bed. However, we know that flows over rough beds at Reynolds numbers near the critical are anything but laminar. Vortices are shed into the flow, often in short bursts of activity. These vortices may be irrotational, except in regions of high shear, but nevertheless they look like turbulence and register as such on instruments that record one component of velocity. They can be distinguished by measuring the Reynolds stress, which can only be done by simultaneous measurement of two components of velocity at a point in space.

We need to know more about the influence of particles in suspension on effective viscosity and shear stress.

Bearing in mind the limitations of the apparatus, this was a well-conceived and conducted study. Its objectives were necessarily modest but they resulted in a sharpening of some of the questions that need to be answered.

Kemp and Simons (1984) have tackled the problem of interaction between waves and currents in the absence of sediment particles and commented briefly on more recent experiments in which lightweight materials were brought into suspension by waves. They used a Laser Doppler anemometer to measure velocities and turbulence levels and an on-line minicomputer to analyse the large amounts of experimental data. They estimated Reynolds stresses by repeating measurements under identical conditions with the anemometer aligned first at 45° to one side of the main flow direction and then at 45° to the other side. Of course, one cannot measure Reynolds stress in this manner; the theory on which the method is based requires that the two sets of measurements should be simultaneous (Kemp and Simons, 1982). Nevertheless, the fact that the experiments were based on an ensemble of a large number of waves probably resulted in quite good estimates. Experiments were conducted with smooth and rough beds, with steady flow, with waves alone and with waves travelling both with and against the flow. The range of conditions was such that steady flows were fully turbulent while waves on their own were laminar.

The roughness consisted of fixed triangular elements with height and spacing within the ratios found in natural sand ripples.

This project was directed towards a vital part of generalised sediment transport; the interaction of waves and currents travelling in line over rigid roughness elements of a particular type. It became possible to address it seriously only with the advent of the LDA and minicomputer.

Elements missing from this project were the influence that sediment motion would have had on fluid properties, and the interaction that takes place between flow and bed-form. The use of triangular roughness elements achieved the purpose of inducing flow separation and vortices. However, it is now well established that bed ripples and dunes adjust their length and height to every change in flow condition and that the patterns of separation and vortex shedding are an integral part of these adjustments. The results of the experiments are of great value, but they fall short of providing information about shear stresses and wave attenuation factors under similar flow conditions but with an active sediment bed. No doubt such studies will follow.

One further comment on this project is called for. The waves used in it were not capable of generating turbulence in the absence of a current. It is not clear how this might affect the case of combined waves and currents. There is likely to be a scale effect when comparing some of the results with larger waves, but its range has yet to be explored.

Prida Thimakorn (1984) describes a theoretical

study of the entrainment of sediment by wave action. He based his parameters on experience of fine sediment brought into a bay by river flow. On contact with saline water, the sediment flocculates and acquires a fall velocity comparable to fine sand. The analysis does not take into account the rheological properties of clay in suspension; it would be equally applicable to fine sands. Nevertheless the experimental results were obtained using clay-sized particles.

The problem was approached by assuming that a relationship for the rate of transport of sediment by uni-directional flow could be modified by replacing the shear stresses due to currents by equivalent shear stresses due to waves. A standard dispersion equation was used to define concentration at any height above the bed. Transformation of this equation resulted in an analytical expression for sediment concentration at any elevation in terms of particle fall velocity, an eddy diffusivity and a boundary value of concentration.

To solve this expression, the eddy diffusivity had to be expressed in suitable form. The author used a function based on the work of Rotta (1972). The final expression included a function of particle fall velocity with time. This function was not detailed in the paper and was presumably based on experiment.

The equation for particle concentration described in this paper was tested by applying it to three sets of wave data. Predicted and measured initial bottom concentrations compared well. Details of the experiments are given elsewhere (Thimakorn, 1980).

This paper is a good example of the application of analytical methods to a real problem. Steps in the analysis were based on semi-empirical expressions for the rate of sediment entrainment and for the coefficient of diffusivity. Initial comparisons with experimental results are encouraging.

Tetsuo Yanagi (1984) describes a series of detailed observations of distribution of sediments contaminated by various pollutants and demonstrates that they could be explained by reference to the long-term tidal residual flow rather than by the stronger tidal currents. He began with the case of heavy metal contamination of surface sediments in the Seto Inland Sea in Western Japan. This sea is some 500 km long by 50 km wide and has a depth of about 30 m. Its tides are dominated by the M2 tidal component. A comparison was made between M2 tidal current ellipses, presumably measured in a large tidal model of the Seto Sea, measured tidal residual surface currents and heavy metal concentrations in surface sediments in several parts of this sea. It is concluded only that the heavy metal concentration in the surface sediments is strongly influenced by the tidal residual flow.

The author compares the dispersion relationships for effluents and for heavy metals and shows that the exponent for dispersion of heavy metals in surface sediments was lower than that for effluents. It can be inferred that much of the heavy metal dispersion took place after attachment to sediment.

The author next discusses the three-dimensional structure of tidal residual flows in the light of observations made with three current meters placed near the bed at locations separated by several hundred metres in Kasado Bay. He reproduces a graph of vorticity and divergence compared with tidal rise and fall based on the three sets of measurements. These showed that there was a residual counter-clockwise circulation near the bed. Reference was made to the distribution of tidal residual flows shown in a tidal model of the bay. The grain size distribution of sediments strongly supports the existence of vertical residual currents in addition to the established horizontal circulation. In conclusion, the author describes the results of experiments in a hydraulic model of Seto Inland Sea which shows the effect of closing two entrances on the residual circulation. This was supplemented by experiments in a rectangular basin in which the effect of a training wall on residual circulation was examined.

This is a thought-provoking paper from the point of view of both the problem under discussion and the techniques that were used for its study. It adds significantly to the fund of knowledge of sediment transport under circumstances that are very important to Japan and are very similar to situations in many other parts of the world that are not yet under such strong economic pressure.

References

Ahilan, R. V. and Sleath, J. F. A., 1984. Wave-induced bed load transport. This volume, pp. 183–190.

Bagnold, R. A., 1954. Experiments on a gravity-free dispersion of large solid spheres in a Newtonian fluid under shear. *Proc. Roy. Soc., Ser. A*, **225**, 49–63.

Bagnold, R. A., 1956. The flow of cohesionless grains in fluids. *Phil. Trans. Roy. Soc. Ser. A*, **249**, 235–297.

Kemp, P. H. and Simons, R. R., 1982. The interaction between waves and turbulent current: waves propagating with the current. *J. Fluid Mech.*, **116**, 227–250.

Kemp, P. H. and Simons, R. R., 1984. Sediment transport due to waves and tidal currents. This volume, pp. 197–206.

Rotta, J. C., 1972. *Turbulent Stromumgen*. Teubner, Stuttgart.

Thimakorn, P., 1980. An experiment on clay suspension under water waves. *Proc. 17th Coastal Engineering Conf.*

Thimakorn, P., 1984. Resuspension of clays beneath waves. This volume, pp. 191–196.

Yanagi, T. 1984. Sediment transport by tidal residual flow in bays. This volume, pp. 207–214.

16

Wave-induced Bed Load Transport

R. V. Ahilan and J. F. A. Sleath
Engineering Department, Cambridge University, UK

A theory is developed for the velocity and concentration distribution within a moving layer of sediment in oscillatory flow. The bed is assumed flat and consequently all of the sediment moves as bed load. Tests have been carried out in an oscillatory-flow water tunnel. The velocity distribution within the moving layer of sediment was measured with a ciné camera and also with a cross-correlation device. Agreement between the original laminar flow theory and the experimental results is found to be poor, but better agreement is obtained if the flow is assumed turbulent with eddy viscosity constant across the bed layer.

Introduction

The way in which sand is moved round by the sea is of great importance to the coastal engineer. The erosion of beaches, the siltation of ports, the undermining of offshore structures are all examples of problems caused by sediment transport.

This paper is concerned with the transport of sediment by waves over flat beds. Flat beds occur particularly with relatively coarse sediment, when only small amounts of sediment are in motion and also when there is intense sediment transport — for example, in the vicinity of breaking waves. Because the bed is flat, all of the sediment is transported as bed load. In addition, we shall consider only the case of pure oscillatory flow. This may be regarded as a first step towards the treatment of the more complex problem of sediment transport by combined waves and currents.

There have already been a number of studies of sediment transport in oscillatory flow. Manohar (1955), Vincent (1957), Kalkanis (1964) and Abou Seida (1965) made measurements with trays set into the bed. This method allows only the mean sediment transport rate during the course of a half-wave cycle to be determined. More recently Sleath (1978), Shibayama and Horikawa (1980) and Horikawa *et al.* (1982) have used optical techniques which also give information on the way in which the transport rate varies with time. Most of the results available at the present time are for relatively low transport rates. The object of the present study is to make measurements at very high sediment transport rates and, if possible, to develop a theoretical model. This paper reports some of the preliminary results obtained so far.

Theory

The main difficulty in any theoretical treatment of this problem is what assumption to make about the relationship between shear stress and velocity gradient in the moving layer of sediment. Measurements have been made by Bagnold (1954) for steady flow but it is not clear how applicable these results are to the case of oscillatory flow, with which we are concerned here. According to Savage and McKeown (1983), there is even some doubt whether Bagnold's results are wholly correct for steady flow. The only way to proceed is to start by assuming that existing steady flow relations are applicable to oscillatory flow and then compare the resulting theoretical curves with experiment. This should allow us to revise the initial assumptions and improve the theory.

We consider the case of a semi-infinite bed of sediment with a flat, horizontal surface. The flow is oscillatory and parallel to the surface. At large distances from the bed the velocity is given by

$$u = U_0 \cos \omega t \qquad (1)$$

Let us assume that the flow is strong enough to move the surface layers of sediment. Since, at any given height, the flow is in the horizontal direction, convective acceleration terms in the equations of motion may be neglected and the momentum equation reduces to

$$\rho_m \frac{\partial u}{\partial t} = -\frac{\partial p}{\partial x} + \frac{\partial \tau}{\partial y} \qquad (2)$$

In this equation u is the velocity of the fluid/sediment mixture, p is the pressure, τ is shear stress, x is measured in the direction of oscillation and y is measured vertically up from the level at which sediment just remains stationary throughout the cycle. Thus the condition at the lower boundary is

$$u = 0 \quad \text{at} \quad y = 0 \qquad (3)$$

The density, ρ_m, is that of the fluid–sediment mixture. If C is the volumetric concentration of sediment at any given point,

$$\rho_m = \rho(1 - C) + \rho_s C \qquad (4)$$

where ρ is the fluid density and ρ_s that of the dry sediment.

Bagnold (1954) suggested that the shear stress might be divided into two parts

$$\tau = T + \tau^1 \qquad (5)$$

where T is the shear stress due to encounters between grains and τ^1 is the stress due to distortion of the intergranular fluid. For laminar flow he found

$$T = 2.2\lambda^{3/2}\mu \frac{\partial u}{\partial y} \qquad (6)$$

The linear concentration, λ, is related to the volumetric concentration, C, by

$$\lambda = \frac{1}{(C^*/C)^{1/3} - 1} \qquad (7)$$

where C^* is the maximum possible value of C where all grains are in static contact. Bagnold (1956) suggested a value of 0.65 for C^* for reasonably rounded and uniform grains.

Bagnold (1956) also suggested that τ^1 might be obtained from Einstein's relation for the viscosity of a fluid containing suspended sediment. Thus

$$\tau^1 = \left(1 + \frac{5C}{2}\right)\mu \frac{\partial u}{\partial y} \qquad (8)$$

In order to make use of these expressions we need some relation linking the concentration, C, to the flow conditions. Bagnold (1954) suggested that

$$T = P \tan \alpha \qquad (9)$$

where P is the normal stress and $\tan \alpha = 0.75$ in laminar flow. If the flow is steady and the distribution of concentration has reached an equilibrium, the normal stress will be equal to the weight of sediment above. Thus, at height y,

$$P = (\rho_s - \rho)g \int_y^\infty C \, dy \qquad (10)$$

In oscillatory flow the normal stress due to encounters between grains need not necessarily be in equilibrium with the weight of sediment above at each instant of the cycle. There may be instants, when the velocity gradient is low, when the normal stress is below that given by equation (10) and the grains of sediment tend to settle towards the bed. Correspondingly, there will be other parts of the cycle when the normal stress is greater and the grains are pushed back away from the bed. However, even under these conditions, it is clear that once the flow has settled down, the average value of the normal stress during a half-cycle must equal the average weight of sediment above. Thus, substituting from equation (10) into equation (9) and differentiating,

$$\frac{\partial \bar{T}}{\partial y} = -(\rho_s - \rho)g \, \bar{C} \tan \alpha \qquad (11)$$

where the overbar indicates the time-mean, averaged over a half-cycle.

Finally, it will be assumed that the boundary layer is sufficiently thin for the pressure gradient $\partial p/\partial x$ to be unchanged across it. Thus, making use of equation (1) and the fact that $\partial \tau/\partial y$ in equation (2) is zero outside the boundary layer,

$$\frac{\partial p}{\partial x} = \rho U_0 \omega \sin \omega t \qquad (12)$$

A solution may now be obtained to equation (2), making use of equations (5)–(8), (11) and (12). The simplest procedure is to make use of a small-perturbation technique. The concentration, C, may be expressed as a power series

$$C = \bar{C}(y) + \epsilon C_1(y, t) + \epsilon^2 C_2(y, t) + \ldots \qquad (13)$$

where ϵ is a small parameter which, for present purposes, may be taken equal to W/U_0. If the fall velocity, W, of

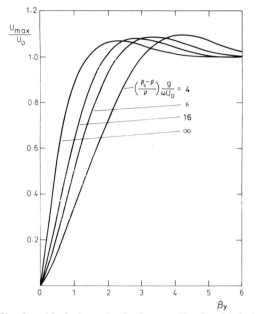

Fig. 1. Variation of velocity amplitude with height.

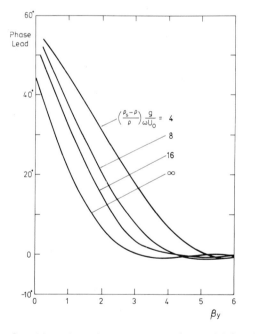

Fig. 2. Variation of zero-crossing phase with height.

the sediment is small compared with U_0, the variation in concentration at any given point during the course of the cycle will also be small. Thus a first approximation is obtained by taking C to be independent of time.

Since the equations are linear in velocity and shear stress, both u and τ may be expressed as functions of y times $\exp(i\omega t)$ and the equations reduced to ordinary differential equations in y. However, because of the

non-linearity of equations (6) and (7) it is still necessary to obtain the solution numerically. This was done on the Cambridge University IBM 3081 computer, the 'shooting and matching' technique suggested by Dahlquist and Bjork (1974) being used. The equations are solved as an initial-value problem with estimates for unknown conditions at the inner boundary. These initial estimates are then improved by comparison of the known and computed results at the outer boundary, and the process is repeated until a satisfactory solution is obtained.

The theoretical curves for the maximum velocity, U_{max}, at any given height and the zero-crossing phase relative to that in the freestream are shown in Figs 1 and 2. Only the results for the first-order solution are shown. The parameter $1/\beta$ is the Stokes thickness, $(2\nu/\omega)^{1/2}$, and $(\rho_s - \rho)g/\rho\omega U_0$ is inversely related to the intensity of sediment movement. When $(\rho_s - \rho)g/\rho\omega U_0$ is infinite, there is no sediment transport and consequently the solution reduces to that for a clear fluid oscillating above a smooth plate. We see that as $(\rho_s - \rho)g/\rho\omega U_0$ decreases, and the intensity of sediment transport increases, the thickness of the boundary layer increases, as might be expected. Figure 2 also shows that the phase lead of the fluid–sediment mixture in the vicinity of $y = 0$ increases with increasing sediment transport.

Figures 3 and 4 show the concentration and sediment transport rates predicted by the theory. The results are very much as might have been expected. As $(\rho_s - \rho)g/\rho\omega U_0$ decreases, the concentration at any given value of y increases and so does the sediment transport rate, Cu_{max}. It will be noted that maximum sediment transport rate occurs some way below the top of the moving layer of sediment, because, although velocity falls off, concentration increases with depth.

Experimental layout and apparatus

Experiments have been carried out in an oscillatory-flow water tunnel consisting essentially of a U-tube in which the water was caused to oscillate at, or near, its resonant frequency by a paddle in one the the rising arms. The paddle was driven by a DC electric motor with feedback control via a connecting rod and crank. The horizontal working section of the tunnel was of length 1.83 m and square cross-section, with internal width and height equal to 0.45 m. For the tests reported below, the thickness of the bed of sediment was 0.14 m.

The sediment consisted of nylon pellets of specific gravity 1.137 and equivalent sphere diameter 4.0 mm. The bottom of the water tunnel was carefully roughened to prevent slipping of the sediment.

The principal object of the tests was to observe

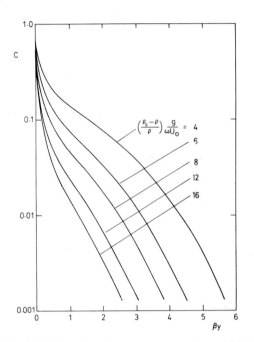

Fig. 3. Variation of sediment concentration with height.

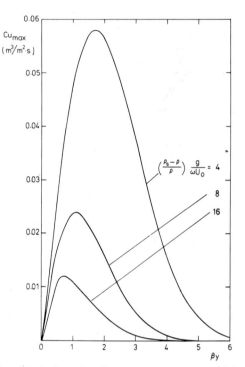

Fig. 4. Variation of sediment transport rate with height.

velocity distributions within the moving layer of sediment for purposes of comparison with the theoretical solution described in the preceding section. Although the theory outlined above has treated the sediment—fluid mixture as a single continuum, it is probable that there will be some slippage between grains and fluid at certain points in the cycle. Even if the flow of fluid relative to the grains is slight, the velocity at a given point in the fluid will not necessarily be the same as either the mean velocity of the fluid or that of the grains. This discrepancy will be particularly marked when, as in the present tests, the concentration of sediment is high. Consequently, it is not desirable to make use of an instrument, such as a laser Doppler anemometer, which provides only a measurement of velocity at a fixed point in the fluid. Instead it was decided to measure the velocity of the grains of sediment directly. This was done in two different ways.

The first made use of two photo-transistors mounted side by side and directed through the Perspex sidewall of the tunnel at the sediment within. Each photo-transistor was screened so that light only reached it from an area of the visible mass of sediment approximately 3.2 mm in diameter. The middle of the field of view of one photo-transistor was separated from that of the other by a horizontal distance of 3.7 mm. As grains of sediment moved through the field of view, the varying intensity of light which they reflected caused the output from the photo-transistor to fluctuate. Cross-correlation of the signals from the two photo-transistors thus allowed the time taken for a grain to

pass the 3.7 mm from the middle of the field of view to one to the middle of the field of view of the other to be determined.

The normal procedure when this device was used was for the apparatus to be set in motion and allowed to attain equilibrium. The signal from each photo-transistor was then recorded on magnetic tape at fixed heights. The output from a phase marker attached to the flywheel of the motor was also recorded. At the end of the test the recording was played back into a computer via an analogue-to-digital converter. Cross-correlation of the signals from the two photo-transistors was then carried out for segments of record corresponding to approximately 1/29 of the cycle.

To provide a check on the results obtained in this way a ciné camera running at 64 frames/s was used to film the motion of the grains of sediment through the sidewall of the tunnel. The film was played back through an analysing projector and the motion of individual grains of sediment was recorded. This method was straightforward but laborious.

Test results and discussion

Figure 5 shows an example of the recorded output from the two photo-transistors. In this test the intensity of the flow was sufficiently high for the bed to appear plane. The period of oscillation was 3.67 s and the

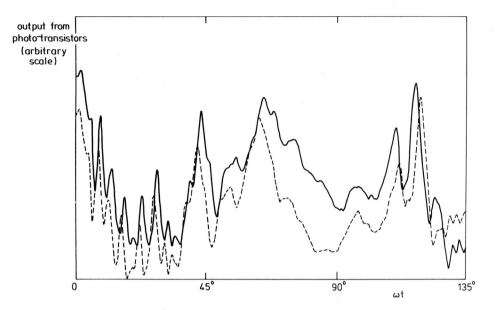

Fig. 5. Example of the output from the two phototransistors (T = 3.67 s).

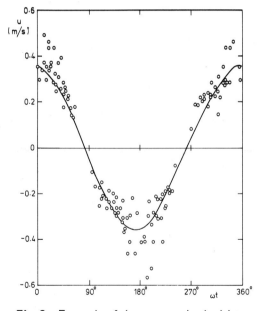

Fig. 6. Example of the measured velocities at a given height.

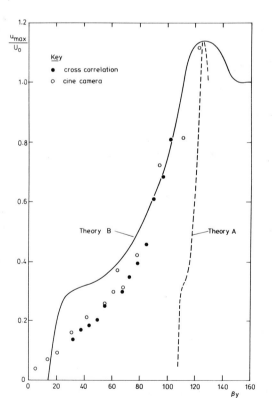

Fig. 7. Variation of velocity amplitude with height (T = 3.67 s, u_0 = 0.45 m/s): ●, cross correlation; ○ ciné camera.

record shown in Fig. 5 covers a time interval of 1.38 s. The variation in signal intensity as individual grains of sediment or groups of grains pass through the field of view of the two photo-transistors is clearly marked. In the vicinity of $\omega t = 0°$ the velocity is maximum and each record shows closely spaced peaks as grains are swept rapidly by. On the other hand, near $\omega t = 90°$, where the velocity passes through zero, the peaks are more widely separated.

An example of the velocities calculated from cross-correlation of signals such as those of Fig. 5 is shown in Fig. 6. These results were obtained from seven recorded cycles at a point near the top of the moving layer of sediment. The reason why there are few experimental points in the vicinity of zero velocity is that, when the flow reverses, grains of sediment passing through the field of view of one photo-transistor are not necessarily carried past the other. Also shown in Fig. 6 is the mean curve through the experimental points.

Similar results to those of Fig. 6 are obtained at other heights within the moving layer of sediment. Figure 7 shows how the amplitude of the mean velocity varies with depth. Also shown in this figure are the results obtained from analysis of the ciné film. The agreement between the two sets of results is encouraging.

On the other hand, the agreement between theory and experiment is poor. The curve labelled 'Theory A' in Fig. 7 is that given by the theoretical solution discussed above. In order to facilitate comparison, the orgin of the theoretical curve has been shifted to make the theoretical maximum coincide with the maximum estimated from the measured velocities. It is clear that the actual thickness of the boundary layer is much greater than that predicted by the theory. Figure 7 shows some thickening of the boundary layer with increasing sediment transport but it would seem that it is not nearly enough.

The reason for this discrepancy probably lies in the assumed relation between shear stress and velocity gradient. According to Bagnold (1954), equation (6) only applied if the parameter $(\rho_s/\mu)\lambda^{1/2}D^2\,\partial u/\partial y$ is less than about 40. This condition is well satisfied in the present tests. However, Savage and McKeown (1983) found that even below this limit the shear stress could be significantly greater than that given by equation (6) at high concentrations. They suggested that under such conditions the flow was 'turbulent-like'. The scatter of the measurements in Fig. 6 would seem to support this hypothesis. Kalkanis (1964), Sleath (1970) and others have found that a good approximation to the velocity distribution in turbulent oscillatory boundary layers is obtained if the viscosity, ν, is replaced by an eddy viscosity, ϵ, which does not vary with height or phase. Under these circumstances the velocity distribution is unchanged except that the parameter βy is multiplied by a constant factor. The curves referred to as 'Theory B' in Figs. 7 and 8 are the same as the curves obtained from the theoretical solution discussed above, except that the values of βy have been scaled by a constant factor equal to 6.0. Although the agreement with the velocity amplitude measurement in Fig. 7 is still not particularly good, the agreement with the measurements of phase in Fig. 8 is close.

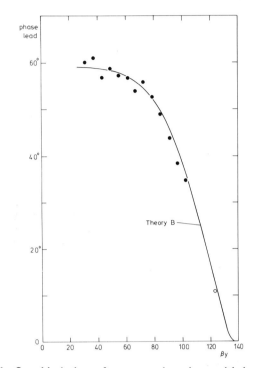

Fig. 8. Variation of zero-crossing phase with height (T = 3.67 s, u_0 = 0.45m/s).

The factor 6.0 was arbitrarily chosen to fit the experimental results. The same value was used for both the amplitude and the phase measurements. Since the characteristics of the turbulence in a moving layer of sediment are likely to be different from those in a clear fluid, it is difficult to justify this value by reference to published velocity measurements. However, it may be noted that the expression for eddy viscosity suggested by Kajiura (1968) for the inner layer over a rough bed would give a scaling factor of 5.9 for the present test conditions if the Nikuradse roughness length is taken equal to grain size, so the factor of 6.0 is certainly not unreasonable.

There are, of course, other possible explanations for the discrepancy between Theory A and experiment. The shape and texture of the grains used in these tests are different from those used by Bagnold. Also, the finite grain size of the sediment might have some effect on the smaller turbulent eddies. There is clearly a need for more tests to investigate these possibilities.

The origin of y for the Theory B curve has also been shifted to align the maximum velocity with that estimated from the measurements. The fact that the theoretical velocity falls to zero at a slightly higher level than was observed in the experiment may be explained as follows. The theory assumes that the level at which no sediment movement takes place is at a constant height $y = 0$. It was observed that, in

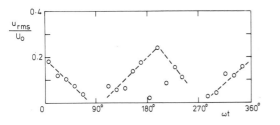

Fig. 9. Example of the way in which the root mean square fluctuation in velocity varies during the course of the cycle ($T = 3.67$ s, $u_0 = 0.45$ m/s).

reality, grains settle out when the velocity gradient is low, and so the no-motion level rises, whereas when the velocity gradient is high, erosion occurs and grains move lower down. In other words, the mean zero level is slightly above the level at which some motion may be detected at given instants during the cycle.

So far we have discussed only the mean velocity. It is clear from Fig. 6 that the actual velocity fluctuates around this mean. Figure 9 shows how the root mean square fluctuation in velocity, u_{rms}, varies during the course of the cycle. This quantity was obtained by dividing the cycle into 25 equal time intervals and then calculating the mean and root mean square deviation from the mean for each interval separately. Some of these time intervals contained very few velocity measurements and this is one reason for the considerable experimental scatter in Fig. 9. Nevertheless, there is a clear tendency for u_{rms} to increase when the mean velocity increases, and vice versa, as might be expected.

Conclusions

Measurements have been made of the velocity distribution in the moving layer of sediment produced by oscillatory flow over a flat bed in a water tunnel. The measurements, which were made with aid of a cross-correlation device and a ciné camera, all show similar trends. Near the bottom of the moving layer the variation with height of the amplitude and phase of the velocity is slow, but the rate of change increases higher up. This is because the concentration of sediment increases with depth and consequently the effective viscosity also increases.

A theory making use of Bagnold's (1954) results for shear stresses in laminar flow shows poor agreement

with the measurements. However, better agreement is found if the flow is assumed to be turbulent and the kinematic viscosity, v, of the fluid is replaced by an eddy viscosity, ϵ, which is constant across the moving layer.

The work reported here represents the preliminary result of a continuing investigation.

References

Abou Seida, M. M., 1965. Bed load function due to wave action. Rep. No. HEL-2-11, Hydraulic Engineering Laboratory, University of California, Berkeley.

Bagnold, R. A., 1954. Experiments on a gravity free dispersion of large solid spheres in a Newtonian fluid under shear. *Proc. Roy. Soc., Ser. A,* **225**, 49–63.

Bagnold, R. A., 1956. The flow of cohesionless grains in fluids. *Phil. Trans. Roy. Soc., Ser. A,* **249**, 235–297.

Dahlquist, G. and Bjork, A., 1974. *Numerical Methods.* Prentice-Hall, Englewood Cliffs, N.J.

Horikawa, K., Watanabe, A. and Katori, S., 1982. Sediment transport under sheet flow condition. *Proc. 18th Conference on Coastal Engineering,* Cape Town.

Kajiura, K., 1968. A model of the bottom boundary layer in water waves. *Bull. Earthquake Res. Inst.,* **46**, 75–123.

Kalkanis, G., 1964. Transportation of bed material due to wave action. Tech. Mem. No. 2, Coastal Engineering Research Center, US Army Corps of Engineers.

Manohar, M., 1955. Mechanics of bottom sediment movement due to wave action. Tech. Mem. No. 75, Beach Erosion Board, US Army Corps of Engineers.

Savage, S. B. and McKeown, S., 1983. Shear stresses developed during rapid shear of concentrated suspensions of large spherical particles between concentric cyclinders. *J. Fluid Mech.,* **127**, 453–472.

Shibayama, T. and Horikawa, K., 1980. Bed load measurements and prediction of two-dimensional beach transformation due to waves. *Coastal Engng Jap.,* **23**, 179–190.

Sleath, J. F. A., 1970. Velocity measurements close to the bed in a wave tank. *J. Fluid Mech.,* **42**, 111–123.

Sleath, J. F. A., 1978. Measurements of bed load in oscillating flow. *J. Waterway Port Coastal Ocean Div. ASCE,* **104** (WW3), 291–307.

Vincent, G. E., 1957. Contribution to the study of sediment transport on a horizontal bed due to wave action. *Proc. 6th Conference on Coastal Engineering,* Miami, pp. 326–355.

17

Resuspension of Clays Under Waves

Prida Thimakorn [1]

Asian Institute of Technology, Bangkok, Thailand

Owing to their complex form, very small size and relatively low density, clays freshly deposited in a bay are vulnerable to resuspension when agitated by waves. The model of resuspension of clays under wave motion is presented from the point of view of (a) entrainment of the bottom clay at rest into the active semi-fluid layer; (b) diffusion of the near-bottom high-concentration mixture into the lower-concentration field in the upper layer. The near-bottom concentration at the initiation of the wave is estimated from the rate of entrainment. The diffusion equation containing the vertical convection term including the terminal fall velocity of clay is derived to obtain the transient concentration profile. The value of the vertical diffusion coefficient is derived from the energy equation of turbulence where energy production is balanced by dissipation. Then the diffusion coefficient is approximated directly from wave parameters. The model is executed by computing the initial near-bottom concentration and the transient concentration profile using experimental wave data form. It was found that the estimated initial concentration in the bottom region is compatible with those obtained from the measured data and then calculated transient concentration profiles show good agreement with the measured profiles.

Introduction

Freshly deposited clays in a bay connected with a large river where very fine suspended sediment is transported is resuspended by wave action. Field studies show that those resuspended materials are transported again by tidal current, resulting in additional transport of clay in the estuary (Thimakorn and Das Gupta, 1978). In the event that this should happen during a flood tide, a portion of the materials will be redeposited in the lower region of the river during the slack tide between the flood and ebb tides. The wave condition in the bay possesses a transient character, depending upon the local wind speed, which shows its variation within a relatively short period. This phenomenon is common in tropical regions where a large land mass surrounds a bay. Here the sea is relatively calm in the morning and the waves arise in the afternoon (when the land temperature is higher than that of water) and last for about 4–6 h until the evening. This means that agitation of the clays takes place within a relatively short time and the transient condition of resuspension

is apparent. Estimation of the sediment concentration in the wave field resulting from wave agitation is important in determining long-term net sediment deposition at different locations in an estuary in connection with the influence of tidal current and river flow.

A comprehensive model of clay resuspension in a wave field is constructed for the shallow and intermediate waves. The clay bottom is flat and smooth, since the freshly deposited clay is in the semi-fluid state and it cannot maintain any vertical dimension of the bed in the form of either ripples or dunes.

The model

The model of clay resuspension under the action of waves is characterized by two continuing processes. One is the bottom agitation of the waves which brings the materials into the wave field. The other is transmission of the higher-concentration materials in the near bottom into the lower concentration of the upper layer (the diffusion model). Figure 1 shows the definition sketch of this model.

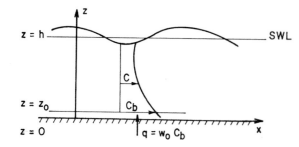

Fig. 1.   Definition of the concentration field under wave motion.

The shear model

The shear model is adopted to determine the near-bottom concentration of clay being agitated by waves. From Fig. 1, at a reference near-bottom layer $z = z_0$, the rate of entrainment, q, is balanced by the near-bottom concentration, C_b, as:

$$q = w_0 C_b \tag{1}$$

The entrainment rate, q, is estimated from the analogy of bed-load transport over a flat bed, as suggested by Vincent (1958). Applying the classic bed-load transport equation of Du Boys (1879) for unidirectional flow

$$q_B = \frac{\Delta V D'}{2\tau_c^2} \tau_0(\tau_0 - \tau_c) \tag{2}$$

Modifying equation (2) for waves, the entrainment rate is

$$q = \frac{\hat{U}\delta_w}{2\tau_{w,c}^2} \tau_w(\tau_w - \tau_{w,c}) \tag{3}$$

The values of $\hat{U}$, δ_w and τ_w are obtained (after Johnson, 1965) as follows:

$$\hat{U} = \frac{\pi H}{T} \frac{1}{\sinh \dfrac{2\pi h}{L}} \tag{4}$$

$$\tau_w = \tfrac{1}{2} f_w \rho \hat{U}^2 \tag{5}$$

$$\delta_w = \left(\frac{\pi}{4}\nu T\right)^{1/2} \tag{6}$$

The value of the critical shear stress is obtained from experiment. Combining equations (1) and (3) gives the near-bottom concentration, C_b, when the value of the terminal fall velocity is known from experiment (Thimakorn, 1980).

The diffusion model

The diffusion model is constructed to calculate the continuing supply of the sediment from the lower layer of high concentration to the upper layer of initially lower concentration. The diffusion equation in the vertical direction of the transient concentration of clay having terminal fall velocity is

$$\frac{\partial C}{\partial t} = w_0 \frac{\partial C}{\partial z} + \epsilon \frac{\partial^2 C}{\partial z^2} \tag{7}$$

since there is initially some concentration of the suspended clay in the water and in the course of the sedimentation is diffused into the upper layer. We have:

$$\left. \begin{array}{l} \text{initial condition} \quad C(z, 0) = \phi(z) \\[4pt] \text{boundary condition} \ C(0, t) = \psi(t) \end{array} \right\} \tag{8}$$

Equation (7) is the second-order differential equation with convection term. This convection term is eliminated by introducing

$$C = \Gamma \exp - (w_0 z/2\epsilon) \exp - (w_0^2 t/4\epsilon) \tag{9}$$

Differentiating C from equation (9) w.r.t. t, z and z^2, the original differential equation is transformed to

$$\frac{\partial \Gamma}{\partial t} = \epsilon \frac{\partial^2 \Gamma}{\partial z^2} \tag{10}$$

The initial condition and the boundary conditions are modified to

$$\left. \begin{array}{l} \Gamma(x, 0) = \phi_1(z) \\[4pt] \Gamma(0, t) = \psi_1(t) \end{array} \right\} \tag{11}$$

Let $F(z, t)$ be the solution of equation (10) such that the original problem of the change in the strength of Γ is a direct function of time only, so that

$$F(z, t - \tau) - F(z, t - \Gamma - \partial t) \simeq \frac{\partial F}{\partial t} \phi(F) \, d\tau \tag{12}$$

$$F(z, t) = \int_0^t \psi'\tau \frac{\partial F}{\partial t}(z, t - \tau) \, d\tau \tag{13}$$

Introducing the Laplace transform,

$$\bar{\Gamma}(z, p) = \int_0^\infty \exp(-pt)\,\Gamma(z, t)\, dt \tag{14}$$

Now with the Laplace operator we have

$$\begin{aligned} \mathcal{L}(\Gamma t, p) &= \int_0^\infty \exp(-pt)\,\Gamma t \, dt \\ &= \exp(-tp)\,\Gamma(z, t) \Big|_0^\infty + p\,\Gamma(z, p) \\ &= -\Phi'(z) + p\,\bar{\Gamma}(z, p) \end{aligned} \tag{15}$$

Equation (14) is reduced to an ordinary equation of second order as

$$-\Phi'(z) + p\,\bar{\Gamma}(z,p) = \epsilon\frac{\partial^2\Gamma(z,p)}{\partial z^2} \qquad (16)$$

with boundary conditions

$$\left.\begin{array}{l}\bar{\Gamma}(0,p) = f(p) = \displaystyle\int_0^\infty \exp(-pt)\,\psi'(t)\,dt\\[2mm]\text{and}\qquad\qquad \bar{\Gamma}(\infty,p) < \infty\end{array}\right\} \qquad (17)$$

For $\bar{\Gamma}(0,p)$

$$\bar{\Gamma} = f(p)\exp\left\{-\left(\frac{p}{\epsilon}z\right)^{1/2}\right\}$$

$$\bar{\Gamma} = \int_0^\infty \frac{\Phi'(\xi)}{\epsilon} G(z,\xi)\,d\xi$$

Thus the solution of equation (16) is

$$\bar{\Gamma}(z,p) = f(p)\exp\left(-\frac{p}{\epsilon}z\right)^{1/2} - \int_0^\infty \frac{\Phi'(\xi)}{\epsilon}G(z,\xi)\,d\xi \quad (18)$$

where $f(p) = 1/p$ and

$$G(z,\xi) = -\frac{\exp(-q\xi)\sinh qz}{q} \qquad q = \left(\frac{p}{\epsilon}\right)^{1/2}$$

$$= -\frac{\exp(-q\xi)}{q}\left\{\frac{\exp(qz) - \exp(-qz)}{2}\right\}$$

$$= -\frac{\exp(-q^2\xi z)}{2q} + \frac{\exp(q^2\xi z)}{2q}$$

$$= -\frac{\exp\left(\frac{p}{\epsilon}z\right)}{2(p/\epsilon)^{1/2}} + \frac{\exp\left(\frac{p}{\epsilon}\xi z\right)}{2(p/\epsilon)^{1/2}}$$

so equation (18) can be written as

$$\bar{\Gamma}(z,p) = \frac{1}{p}\exp\left\{-\left(\frac{p}{\epsilon}z\right)^{1/2}\right\}$$

$$-\int_0^\infty \frac{\Phi(\xi)}{\epsilon}\left[-\frac{\left(\exp-\frac{p}{\epsilon}\xi z\right)}{2(p/\epsilon)^{1/2}} + \frac{\left(\exp\frac{p}{\epsilon}\xi\right)}{2(p/\epsilon)^{1/2}}\right]d\xi \ (19)$$

In order to obtain Γ, using the inverse Laplace transform: the first term in the right-hand side of equation (19) is

$$\frac{1}{p}\exp\left(\frac{p}{\epsilon}z\right)^{1/2} = 1 - \mathrm{erf}\left\{\frac{z}{2(\epsilon t)^{1/2}}\right\}$$

$$= \frac{2}{\pi}\int_{\frac{z}{2(\epsilon t)^{1/2}}}^\infty \exp(-\eta^2)\,d\eta$$

The terms

$$-\frac{\exp\left(-\frac{p}{\epsilon}\xi z\right)}{2(p/\epsilon)^{1/2}}$$

and

$$\frac{\exp\left(\frac{p}{\epsilon}\xi z\right)}{2(p/\epsilon)^{1/2}}$$

can be written as

$$-\frac{\cos 2(\xi z t)^{1/2}}{2(\pi t)^{1/2}}$$

and

$$\frac{\sin 2(\xi z t)^{1/2}}{2(\pi t)^{1/2}};$$

Therefore we can write

$$\Gamma = \frac{2}{\pi}\int_{\frac{z}{2(\epsilon t)^{1/2}}}^\infty \exp(-\eta^2)\,d\eta$$
$$-\int_0^\infty \frac{\phi(\xi)}{\epsilon}\left\{-\frac{\cos 2(\xi z t)^{1/2}}{2(\pi t)^{1/2}} + \frac{\sin 2(\xi z t)^{1/2}}{2(\pi t)^{1/2}}\right\}dt \ (20)$$

which gives the solution

$$\Gamma = \frac{z}{2\sqrt{\pi k}}\int_0^t \frac{\exp\left\{-\frac{z^2}{4\epsilon(t-\tau)}\right\}}{(t-\tau)^{3/2}}\psi'(\tau)\,d\tau\int_0^\infty \frac{-\Phi'(\xi)}{2(\pi k t)^{1/2}}$$

$$\times\left[\exp\left\{-\left(\frac{\xi-z}{4\epsilon t}\right)^2\right\} - \exp\left\{-\left(\frac{\xi+z}{4\epsilon t}\right)^2\right\}\right]d\xi \ (21)$$

Let $\lambda = z/[2\{\epsilon(t-\tau)\}^{1/2}]$; the final solution of equation (10) is

$$\Gamma = \frac{2}{\pi}C_0\exp\left(\frac{w_0^2 t}{4}\right)\left(\int_0^\infty - \int_0^\alpha\right)\exp\left(-\lambda^2 - \frac{\epsilon^2}{\lambda^2}\right)d\lambda$$
$$+ \int_0^\infty \frac{\Phi'(w_0 t)}{2(\pi\epsilon t)^{1/2}}\left[\exp\left\{-\left(\frac{w_0 t - z}{4t}\right)^2\right\}\right.$$
$$\left. - \exp\left\{-\left(\frac{w_0 t + z}{4\epsilon t}\right)^2\right\}\right]dt \qquad (22)$$

Substituting equation (22) into equation (9), the solution of equation (7) is

$$\frac{C(z,t)}{C_0} = \frac{1}{2}\left\{\mathrm{erfc}\left(\frac{z - w_0 t}{2(\epsilon t)^{1/2}}\right) + \exp\left(\frac{w_0 z}{\epsilon}\right)\mathrm{erfc}\left(\frac{z + w_0 t}{2(\epsilon t)^{1/2}}\right)\right\}$$
$$+ \int_0^\infty -\frac{\Phi(w_0 t)}{2(\pi t)^{1/2}}\exp\left\{-\left(\frac{w_0 t - z}{4\epsilon t}\right)^2\right\} - \exp\left\{-\left(\frac{w_0 t + z}{4\epsilon t}\right)^2\right\}$$
$$(23)$$

Model execution

The model is tested using experimental wave data reported by Thimakorn (1980). First the entrainment rate of the sediment is calculated from equation (3). A value of terminal fall velocity of $w_0 = 0.215$ cm/s is used. It was derived from the experimental data on clay concentration in a wave field obtained during its transient state by plotting the value $\ln\left\{(\bar{C}_{eq} - \bar{C}_t)/(\bar{C}_{eq} - \bar{C}_0)\right\}$ with time t to obtain its slope, $\beta' = \beta w_0/h$, when $\beta = \bar{C}/\bar{C}_b$. This value of w_0 can only be used and applied in the wave field in terms of 'apparent fall velocity', which is different from that terminal fall velocity defined by Stokes' law. The initial bottom concentration obtained from the calculation, C_{b0}, and that obtained from the measurement, $C_{b0,m}$, is compared, as shown by the fitting of the $45°$ line showing the correlation coefficient $R = 0.953$ (Fig. 2).

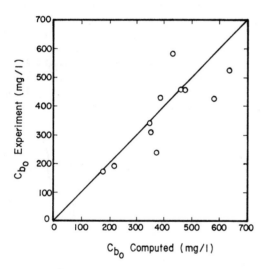

Fig. 2. Comparison between the computed and the experimental values of the initial bottom concentration.

This diffusion model is tested by applying equation (23) to the three sets of wave data. The eddy diffusivity, ϵ, is evaluated from the concept of energy in a turbulent field. It starts from the two-dimensional energy equation of turbulence:

$$\frac{D}{Dt}\frac{\bar{q}^2}{2} = -\left\{\overline{u'^2}\frac{\partial \bar{u}}{\partial x} + \overline{u'w'}\frac{\partial \bar{u}}{\partial z} + \overline{u'w'}\frac{\partial \bar{w}}{\partial x} - \overline{w'^2}\frac{\partial \bar{w}}{\partial z}\right\}$$
$$- \left\{\overline{u'\frac{\partial q^2/2}{\partial x}} + \overline{w'\frac{\partial q^2/2}{\partial z}}\right\}$$
$$- \frac{1}{\rho}\left\{\overline{\frac{\partial u'p'}{\partial x}} + \overline{\frac{\partial w'p'}{\partial z}}\right\}$$

$$- \nu\left\{\frac{\partial}{\partial x}\overline{\left(u'\frac{\partial u'}{\partial x}\right)} + \frac{\partial}{\partial z}\overline{\left(u'\frac{\partial u'}{\partial z}\right)}\right.$$
$$+ \frac{\partial}{\partial x}\overline{\left(w'\frac{\partial w'}{\partial x'}\right)} + \frac{\partial}{\partial z}\overline{\left(w'\frac{\partial w'}{\partial z}\right)}$$
$$\left. - \nu\left\{\overline{\left(\frac{\partial u'}{\partial x}\right)^2} + \overline{\left(\frac{\partial u'}{\partial z}\right)^2} + \overline{\left(\frac{\partial w'}{\partial x}\right)^2} + \overline{\left(\frac{\partial w'}{\partial z}\right)^2}\right\}\right.$$
$$\tag{24}$$

Rotta (1972) noted that for flow over a relatively flat bed the energy production and dissipation are balanced. Assuming that the values of $\bar{u}$ and $\bar{w}$ are approximated by the potential flow theory for waves, then

$$\frac{\partial \bar{u}}{\partial x} = -\frac{\partial \bar{w}}{\partial z}; \quad \frac{\partial \bar{w}}{\partial x} = \frac{\partial \bar{u}}{\partial z} \tag{25}$$

The production term in equation (23) changes to

$$-\left\{-\overline{u'^2}\frac{\partial \bar{w}}{\partial z} + \overline{u'w'}\frac{\partial \bar{u}}{\partial z} + \overline{w'u'}\frac{\partial \bar{u}}{\partial z} - \overline{w'^2}\frac{\partial \bar{w}}{\partial z}\right\}$$

The dissipation term is replaced by

$$E = \frac{A}{L}(\bar{q}^2)^{3/2} \tag{26}$$

The eddy viscosity is defined by

$$\epsilon = B(\bar{q}^2)^{1/2}L \tag{27}$$

Introduce the vorticity diffusivity, ϵil, which is defined as

$$-\overline{u_i u_j} = -\epsilon il\overline{\frac{\partial u_i}{\partial x_j}} + \overline{\frac{\partial u_j}{\partial x_i}} \tag{28}$$

Expanding equation (28) into four components, using the orthogonal property of $\bar{u}$ and $\bar{w}$ and substituting into the production term, one has

$$2\epsilon_{11}\left\{\left(\frac{\partial \bar{w}}{\partial z}\right)^2 + \left(\frac{\partial \bar{u}}{\partial z}\right)^2\right\} + 2\epsilon_{33}\left\{\left(\frac{\partial \bar{u}}{\partial z}\right)^2 + \left(\frac{\partial \bar{w}}{\partial z}\right)^2\right\}$$
$$= 4\epsilon_{11}\left\{\left(\frac{\partial \bar{u}}{\partial z}\right)^2 + \left(\frac{\partial \bar{w}}{\partial z}\right)^2\right\} \tag{29}$$

From the property of linear wave

$$\bar{u} = a\sigma\frac{\cosh kz}{\sinh kh}\cos(kx - \sigma t) \tag{30}$$

$$\bar{w} = a\sigma\frac{\sinh kz}{\sinh kh}\sin(kx - \sigma t) \tag{31}$$

The energy production term in equation (29), when balanced with the dissipation term in equations (26) and (27):

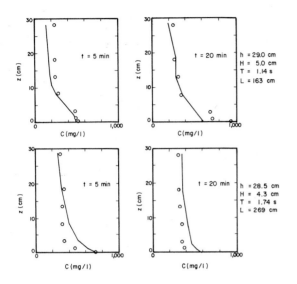

Fig. 3. Comparison between the estimated concentration profile and the measured value: (———) computed profile; (○) flume data.

$$\bar{q}^2 = \frac{4cBL^2k^2}{A}(\bar{u}^2 + \bar{w}^2) \qquad (32)$$

$$C = \frac{2}{\pi}$$

By assuming that the mixing length, L, equals the vertical axis, b, of the orbital ellipse and using the value of $A = 0.124$ and $B = 0.396$ (after Rotta, 1972), the final expression of ϵ is

$$\epsilon = 5.56 \frac{H^2}{LT} \frac{\sinh^2 kz}{\sinh^3 kh} \qquad (33)$$

From comparison of the estimated values and the measured data shown in Fig. 3 it will be noted that this model is capable of estimating the transient concentration profile of a clay—water mixture under shallow water and intermediate waves.

Conclusion

The method developed to estimate the concentration of clay under water waves due to resuspension introduces the two important factors in the entrainment process and the diffusion field. They are the terminal fall velocity, w_0, and the vertical diffusion coefficient, ϵ. It has to be borne in mind that for any solid—fluid flow system — in particular, when the vertical dynamics of the system's behaviour are evaluated — all necessary vertical components in the process must be taken care of. In this model, even though clay particles do not exhibit an actual terminal fall velocity similar to that of

sand when Stokes' law is appropriate, there exists an apparent fall velocity which cannot be obtained by ordinary method such as hydrometer analysis. Details of the determination of w_0 have been provided by Thimakorn (1980). A new method of estimating the vertical diffusion coefficient, ϵ, from direct evaluation of turbulence in waves is demonstrated. It shows that direct evaluation of ϵ from wave parameters is possible, and when it is applied, together with the value of w_0, a good estimation of the resuspension, even at transient state, is possible.

Acknowledgements

The author wishes to acknowledge the inspiration of this work from Professor H. Kikawa, of Tokyo Institute of Technology; continuing encouragement and support was received by the author from Professor H. Shi-igai and Dr H. Nishimura, of Tskuba University. Professor K. Horikawa and Dr A. Watanabe, of the University of Tokyo, assisted the author in overcoming many difficulties during the course of the work. Dr M. Sawamoto, of Tokyo Institute of Technology, was also of great assistance. Finally, the author had the full support of the Asian Institute of Technology and, in particular, Dr Suphat Vongvisessomjai, Chairman, Division of Water Resources Engineering.

Symbols

a	wave amplitude
b	vertical axis of ellipse
d'	thickness of active layer of sediment at the seabed
f_w	bottom friction under wave motion
h	water depth to still water level
k	wave number
p	pressure
q	rate of sediment entrainment; energy of turbulence
q_B	bed-load transport rate
t	time
u	horizontal velocity
$\bar{u}$	average horizontal velocity
u'	horizontal turbulent velocity
w	vertical velocity
$\bar{w}$	average vertical velocity
w'	vertical turbulent velocity
w_0	terminal fall velocity
x	horizontal axis
z	vertical axis
z_0	reference level near seabed
A	constant in turbulent energy dissipation
B	constant in eddy viscosity
C	concentration of clay—water mixture (mg/l)
C_b	concentration at $z = z_0$
C_{b0}	near-bottom concentration at the initiation of the motion.

$\bar{C}_0$	average initial concentration
$\bar{C}_t$	average concentration at time t
$\bar{C}_{eq}$	average equilibrium concentration
L	wavelength
$\hat{U}$	maximum velocity of water particle at seabed
ΔV	velocity of active sediment layer
ϵ	eddy diffusivity
δ_w	thickness of boundary layer under waves
ν	kinematic viscosity
ρ	density
τ	shear stress at the seabed in unidirectional flow
τ_c	critical shear stress
τ_w	bottom shear stress under waves
τ_{wc}	critical shear stress under waves
σ	wave frequency
ϵ	turbulent energy dissipation
L	mixing length

References

Du Boys, M. P. , 1879. Le Rhon et les Rivières à lit Affouillable. *Mem. Doc. Pout Chauss., Ser. 5,* **XVIII**.

Jonnson, I. G., 1965. Wave boundary layers and friction factors. *Proc. 10th Coastal Engineering Conf.*

Rotta, J. C., 1972. *Turbulent Stromumgen,* Teubener, Stuttgart.

Thimakorn, P. and Das Gupta, A., 1978. Concentration of suspended clay in tidal estuary. *Proc. 1st Congr. APD-IAHR Int. Conf. Water Resources Engineering*

Thimakorn, P., 1980. An experiment on clay suspension under water waves. *Proc. 17th Coastal Engineering Conf.*

Vincent, G. E. 1958. Concentration to the study of sediment transport due to wave action. *Proc. 6th Coastal Engineering Conf.*

18

Sediment Transport due to Waves and Tidal Currents

P. H. Kemp and R. R. Simons
Civil Engineering Department, University College, London, UK

The paper reviews existing methods for the prediction of sediment transport under the action of combined waves and currents, and presents the results of an experimental programme aimed at verifying some of the main assumptions implicit in such methods. A logarithmic layer is confirmed to exist in the turbulent oscillatory boundary layer, and bed shear stresses and roughness lengths are compared favourably with theoretical estimates. Field observations of sediment transport under combined waves and tidal currents are shown to be predicted with the use of an appropriate power law.

Introduction

One of the long-established maxims of the coastal engineer is that 'waves disturb, currents transport'. In spite of considerable study and research, there is still some way to go before the physics of this process is understood and can be expressed in quantitative terms. However, it is accepted that waves may entrain significant amounts of sediment when a current of comparable magnitude may be too weak to initiate sediment motion.

The complex nature of loose boundary hydraulics is related to the continual interaction between the fluid and the sediment, and hence, to the bed geometry. Even in purely tidal situations in the absence of waves, the ability to predict siltation to within a factor of 2 is considered reasonable by Thorn (1981). When waves are superimposed on currents, it is clear from field observations that the increase in sediment transport is dramatic. Owen and Thorn (1978a, b) in a nearshore field study compared the suspended sediment transport for tidal currents alone with measurements made during periods of wave action, and found that the waves increased the transport to up to 40 times that in the current-only case. Kana and Ward (1980) similarly found increases by a factor of 10 in sediment load, and an increase by a factor of 60 in sediment transport, when the increase in extent of the active zone and

the greater height to which the sand was suspended were taken into account.

Brief review of the main factors

The shear stress at the bed is one of the dominant factors in all sediment transport calculations. The mean shear stress is most commonly used, although there is both a distribution of shear stress on the bed and a distribution of shear stress value at which the sediment will move. In addition, the large-scale turbulence structures known as ejections and sweeps, associated with the bursting process, account for the major contribution to the Reynolds stress.

The many sediment transport equations for unidirectional flow have been reviewed and compared by various authors. White *et al.* (1975) compared eight theories relating to unidirectional flow on the basis of 1000 flume and 260 field measurements. For the better theories the predicted transport rates were between 0.25 and 4 times the observed rates for about 85 per cent of the data. Ackers and White (1980) made a further review when proposing a new theory. Graaf and Overeem (1979) compare formulae on the basis of their overall trends and sensitivity to changes in the value of the parameters. They point out that these formulae have been adapted and not designed for coastal conditions, and they do not see any significant advances in this area until the general physics is better understood.

Bed friction is related to the nature of the bed material and to the geometry of the bed. It must be remembered that so far as the bed material is concerned, this is not always non-cohesive. Mantz (1980) points out that grain surface interaction effects begin to affect silica grains for median diameters less than $100\,\mu$m. Different effects are produced in pure water, sea-water and soft water. Whereas sea-water and hard water increase cohesion, pure water induces repulsion and soft water makes the grains cohesionless. The latter effect was exploited by Mantz in his experiments. In addition, mud has a differential response to bed shear stress, depending on the depth of erosion: see Miles (1981) and Thorn (1981).

Bed roughness can include grain roughness, ripple roughness and a component associated with the movement or suspension of the bed material. When ripples are present, their influence on sediment transport will depend on whether or not separation occurs. Separation may not occur if the ripples were formed under earlier wave conditions (Soulsby *et al.*, 1983). The occurrence of bed forms can be related to a modified Shields parameter, as discussed by Sleath (1978), and in the form proposed by Brebner (1980). The ripple steepness, η/λ, depends on the shear stress. Here η is the ripple height and λ is the ripple wavelength. There is general agreement that ripple steepness lies in the range 0.13–0.22 for values of Shields parameter, θ', less than 0.4, where θ' is given by

$$\theta' = \tfrac{1}{2}(a\omega)^2 f_{\mathrm{w}}/[(s-1)gD]$$

as quoted by Nielsen (1981). The friction factor, f_{w}, is based on Jonsson (1965, 1966) or Swart (1974). For $\theta' > 0.4$ ripples start to flatten out, and for $\theta' > 1$ the bed is virtually plane.

The effect of asymmetrical ripples on the direction of sediment transport was studied by Inman and Bowen (1962), Inman and Tunstall (1972) and Bijker *et al.* (1976). Inman and Tunstall found that for $\eta/\lambda < 0.2$ sediment moved in a direction from the steep to the flatter face of the ripple. This was observed both in the laboratory and in the field.

Ikeda (1980) comments on a tenfold increase in sediment transport when ripples change from two- to three-dimensional shape, although the mean shear stress remained almost constant. The change was associated with turbulent bursts and spiral flow.

The combination of waves and currents is frequently met in tidal channels. Even in the absence of waves the situation is complicated. The flow is frequently unsteady,

non-uniform and spatially variable, although for certain phases of the tide the flow may be quasi-steady. In addition, a wide range of sediments may be present, although some areas of the bed may consist of inerodible material. In this situation mathematical models are no better than the extent to which the simplified assumptions happen to fit the particular circumstances.

The diffusion coefficient, the Reynolds stress and the kinetic energy are larger in decelerating flow than in accelerating flow (Russell and Kemp, 1977; Anwar and Atkins, 1980). Field measurements by Soulsby and Dyer (1981) suggest that during accelerating flow the sediment is responding to turbulence produced at an earlier point in time. Also, in tidal flow the fall velocity of the particles causes a time lag of suspended solids concentration with respect to the depth-mean water velocity, as shown by Thorn (1981).

Sediment transport due to waves and currents

Methods which are currently used for calculating sediment transport for combined wave and current conditions can be used to identify areas where uncertainties exist and research is required.

Broadly speaking, it is necessary to define the combined shear stress at the bed. This must be derived from assumptions or measurements of the separate wave- and current-induced velocities, and their resultant value. This involves a choice of wave theory and an assumption of, say, a logarithmic current velocity profile. Implicit in these definitions is an assessment of a bed roughness or friction factor. The formulation must then define a diffusion coefficient which can be used in conjunction with a separately determined value of sediment concentration in the bed layer, to give a sediment concentration distribution with height. On the basis of the assumed current velocity distribution, the total sediment transport can then be obtained by integration over the depth. There are variations on this approach but it is presented in order to highlight the uncertainties involved.

The general approach outlined above may be illustrated by reference to the equation developed by Bijker (1967, 1971, 1980). Central to his method is the computation of the total bed shear stress, τ_{r}, in terms of the shear stress, τ_{c}, due to the current alone, in the form

$$\frac{\tau_{\mathrm{r}}}{\tau_{\mathrm{c}}} = \left(1 + \tfrac{1}{2}\xi^2 \frac{U_0^2}{V^2}\right)$$

where U_0 is the amplitude of the orbital velocity at the bed and V is the mean current velocity. The coefficient ξ can be evaluated from the friction factors of Jonsson

(1965, 1966) or Swart (1974), expressed as f_w, from which

$$\xi = C_h(f_w/2g)^{1/2}$$

where C_h is the Chezy coefficient. The friction coefficient based on Swart's formula is

$$f_w = \exp\left\{5.213(k/a)^{0.194} - 5.977\right\}$$

and k is the roughness value. The sediment concentration at a height z is then

$$C = C_0 \exp(-wz/\epsilon)$$

Bijker (1971) adapts a sediment transport formula due to Frijlink for unidirectional flow, viz.

$$S_b = 5D_{50}\left(\frac{\mu - \tau}{\rho}\right)^{1/2} \exp\left(-0.27\Delta D \cdot \frac{\rho g}{\mu\tau}\right)$$

where D is the grain diameter, μ is a ripple coefficient and Δ is the relative density of the bed material. By substituting τ_r for τ, and after expressing τ_c as $g(V^2/C_h^2)$, he obtains

$$S_b = 5DV\frac{g^{1/2}}{C_h}$$

$$\times \exp\left[-0.27\Delta DC_h^2\bigg/\left\{\mu \cdot V^2\left(1 + \frac{1}{2}\left(\xi\frac{U_0}{V}\right)^2\right)\right\}\right]$$

The authors of all existing sediment transport formulations generally conclude that there is a need for more information from experimentation.

Wave and current interaction

The prediction of sediment motion relies on a knowledge of the fluid motion. Clearly it is of interest to know whether a logarithmic velocity profile persists when waves are superimposed on a current. In addition, the conditions under which flow reversal occurs near the bed and the extent to which wave motion may modify the current-induced turbulence in the boundary layer are important in relation to sediment motion. Interest also centres around the predicted increase in apparent roughness under the combined fluid motion. The research programme pursued by the authors was designed to look at the interaction between waves and currents in the absence of sediment, in order to define the mean velocity components, the structure of the turbulence and the shear stresses. The study proceeded from experiments on waves alone to waves propagating with the current and against the current. In all three cases the tests were carried out in the first instance with a smooth bed and subsequently with a rough bed consisting of two-dimensional triangular slats. One of the main areas of interest was the height to which the water was disturbed above the bed when acted on by

waves alone and the comparable situation when a current was superimposed on the waves.

Since the characteristics of the current were measured independently, it was possible to deduce whether there had been any interaction between the waves and the current, and also to infer what might happen to the distribution of the sediment which it was assumed would be put into suspension in the two cases.

Comparatively little previous work had been carried out on the interaction between waves and currents, particularly near the bed. Grant and Madsen (1979) produced a theoretical analysis of combined wave and current flow over a rough boundary, predicting an increase in apparent bed roughness and shear stress when waves were superimposed on the current. A similar theory has been presented by Christoffersen (1980). Bakker and van Doorn (1978) also found an apparent increase in bed roughness. George and Sleath (1979) described the cycle of vortex formation and ejection around spherical roughness elements in the presence of a weak current. The stronger downstream vortex was found to induce a weak reverse mean current just above the roughness elements. This is consistent with the observations of Inman and Bowen (1963) and Bijker *et al.* (1976), who both reported enhanced upstream sediment transport when a weak current was superimposed on waves. The present authors have carried out an extensive investigation on the subject of wave and current interaction, and a report in more detail may be found in Kemp and Simons (1982, 1983).

Experimental investigation of wave – current interaction

The apparatus used by the authors in their investigation is described in Kemp and Simons (1982). The work was carried out in an open channel, monochromatic waves being used. Two bed conditions were used, as mentioned above. With the smooth bed the current alone was well into the turbulent regimen, whereas the waves alone were laminar. With the two-dimensional roughness elements the current was fully rough turbulent. The larger waves by Jonsson's criterion were at the top of the transitional range, but it was felt that the sharp-crested triangular elements would induce earlier transition to turbulence than he predicted. Spectral analysis was carried out on the waves, and this eliminated any suggestion that any of the effects observed might have been due to the presence of parasitic waves resulting from imperfect wave generation. The height and spacing of the roughness also came within the range of possible ratios between height and length found in natural sand ripples. Fluid velocities were measured with a laser-Doppler anemometer, and the analysis of the turbulent and wave-induced velocities was carried out with an on-line minicomputer. The computer was programmed to produce ensemble average velocities, Reynolds stresses and wave elevation data. The cycle was sampled at 200 separate phase positions, with up to 250 observations at each position. Measurements were made at up to 30 points in the vertical.

The flow depth was kept constant at 200 mm, and one overall flow rate was used throughout each test. For currents flowing with the waves the mean centreline velocity was 184 mm/s. For waves opposing the current the equivalent velocity was 111 mm/s. Four different wave heights were used, with a wave period of 1 s.

When scaled on suitable flow parameters, a logarithmic mean velocity profile was found for the currents alone, following the universal law of the wall for both rough and smooth boundaries, with von Karman's constants of 0.36 and 0.4, respectively. Turbulent and Reynolds stress measurements in the wall layer were in good agreement with previous research (Laufer, 1950; Grass, 1971).

For the waves the range of a_{bm}/k_s lay between 0.46 and 0.85, where a_{bm} is the orbital amplitude of a particle at the outer edge of the wave boundary layer and k_s is the Nikuradse equivalent sand roughness. Wave surface profiles and wave-induced periodic velocities corresponded very closely with Stokes' second-order theory and with the theory of Brink-Kjaer and Jonsson (1975).

The test programme included measurement of wave height attenuation along the channel and determination of the rough boundary friction factor, f_w. The wave attenuation was found to be greatly increased by the addition of an opposing current and reduced by a following current. The wave friction factors were determined for waves alone from Kamphuis (1978) and compared with those derived from Kajiura (1968) and Jonsson and Carlsen (1976). The experimental results lay between those predicted by these two theories, but suggested that f_w continues to rise for decreasing values of bed excursion to roughness height ratio (a_{bm}/k_s), rather than taking the constant value of 0.24 reported in Jonsson and Carlsen (1976) and illustrated in Jonsson (1978), or alternatively the value of 0.28 based on Bagnold's (1946) experiments, as quoted by Nielsen (1979).

Combined wave and current test results

For waves propagating against the current the periodic velocities outside the rough bed boundary layer, of about two roughness heights, and up to the surface were in good agreement (i.e. within 2%) with the theory of Brink-Kjaer and Jonsson (1975) for second-order waves on a linear shear current. Velocities predicted by neglecting

the current and simply applying second-order wave theory to the measured wave length and height would underestimate results by 15%.

Within the layer two roughness heights above the rough bed apex level, periodic velocities were greater than the profile predicted. Values increased to a maximum at bed apex level, where values 50% in excess of theory were measured. This behaviour was induced by vortices shed from the rough bed, and is similar to that experienced in the tests on waves alone.

When waves propagated in the same direction as the current, the periodic velocities showed similar behaviour to that described above. However, the tests over both rough and smooth beds revealed a marked reduction in periodic velocity maxima and minima near the bed in the layer corresponding to the logarithmic region of high turbulent activity in the mean velocity profile.

Considering now the mean velocities over both rough and smooth boundaries, it was found that the effect of superimposing a wave onto a current flowing either with or against the waves is to cause a change in velocity in the upper layer of an opposite sense to the direction of wave propagation and of greater magnitude than the measured wave-induced mass transport. In the present case this could be interpreted as an increase in mass transport near the water surface; in this layer the change in velocity clearly depends on the relative directions of wave and current.

Near the bed it was found that the changes in mean velocity were independent of current/wave direction. Over the smooth bed the effect of superimposing waves of increasing height is to increase the near-bed mean velocities, whereas over the rough bed the velocities measured over a roughness apex are progressively decreased, with the reduction extending up into the logarithmic layer. Measurements over the roughness trough of the bed show an increase in return (against the current) flow between the elements as the wave height is increased. The duration of this return flow increases with wave height, thus allowing the formation of 'upstream' vortices, which induce downstream velocities in the roughness trough. It is of interest to note that flow reversal took place near the smooth bed for all the waves tested. Over the rough bed the reversal layer was thicker, varying between 3 mm and 15 mm above apex level.

Most of the current sediment transport theories for the combined wave and current situation have assumed that the mean velocity profile remains logarithmic after the waves have been superimposed. The authors' results as depicted in Fig. 1 confirm this. In the case of the rough boundary, however, the mean shear stress deduced from the slope of the logarithmic profile is significantly higher than that for the current alone. Figure 2 also shows that the roughness length scale, z_0 (the intercept

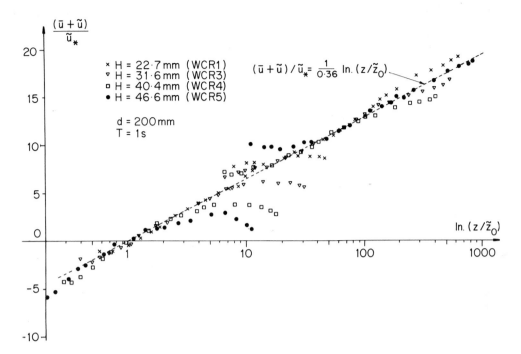

Fig. 1. Normalised semi-logarithmic velocity profiles measured over a rough-bed apex, at phases corresponding to wave crest and trough.

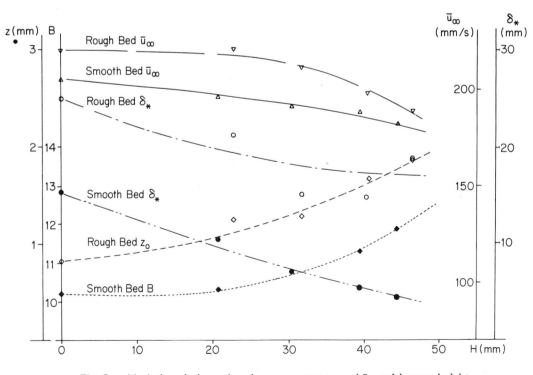

Fig. 2. Variations in boundary-layer parameters and $\bar{u}_{\infty}$ with wave height.

of the log line with ... 0), increases with wave height up to twice the current only value for the largest waves propagating with the current. In the case opposing the wave the largest wave tested gave threefold increase in shear stress and a sixfold increase in roughness length.

From the velocity profiles close to the smooth bed it is possible to calculate the maximum and minimum stresses τ_{max} and τ_{min} induced periodically by the waves and currents. For the codirectional waves and current the maximum bed shear is approximately double the mean over the cycle estimate for the larger waves. It is also apparent that the large periodic variations in $\tilde{u}^*$ close to the bed are not transmitted out into the logarithmic region, although the value of $\tilde{u}^*$ was found to vary from $\bar{u}^*$ in this layer by 20% in the larger waves and 10% in the smaller ones. This indicates a wave-induced turbulence not found in tests on waves alone.

From the foregoing results it is seen that a linear superposition of wave and current velocities will not accurately predict those of the combined flow field.

The turbulence characteristics for the smooth boundary indicate that the horizontal turbulence component varies through the wave cycle. Close to the bed, where wave action induces flow reversal, there are peaks in r.m.s. at the phases of zero velocity, the larger disturbances occurring during the deceleration phase. Further from the bed, with wave and current flowing in the same direction, the combined flow intensities fall below the values for current alone. This bears out the results of the mean velocity profiles, that the addition of waves reduces the thickness of the boundary layer. For this smooth bed case it appears that there is no significant increase in turbulence intensity caused by the addition of the wave.

In the case of the rough boundary both the horizontal and vertical turbulence intensities fluctuate considerably through the wave cycle, particularly within two roughness heights above the bed apex, a layer 10 mm thick. Here there is a maximum intensity at a phase corresponding to maximum deceleration, with a smaller peak during the acceleration phase. With the current flowing against the direction of wave propagation, the maximum occurs as the wave crest approaches and the water level is rising. In the experiments with the current in the direction of wave travel and in which a stronger current velocity was present, there was only one peak near the bed, corresponding to the decelerating phase, in that case after the crest had passed.

The superposition of even the smallest wave tested caused a dramatic increase in turbulence over the value for current flow alone. Figure 3 emphasizes the way in which intensities increase with wave height, illustrated with respect to the vertical component. The rate of increase with wave height of turbulence intensities

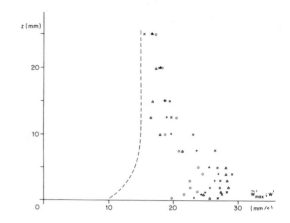

Fig. 3. Rates of increase in turbulence relative to vertical wave height. w'; $\tilde{w}'_{max}$ vs. z within 25 mm of bed roughness apex. Current alone: w' (———). Wave plus current: $\tilde{w}'_{max}$ (○) WCR1, (+) WCR3, (x) WCR4, (△) WCR5.

near the boundary was reduced by the addition of the current.

The variation in Reynolds stress through the wave cycle for the rough bed confirmed the results of the horizontal turbulence. At the phase corresponding to maximum deceleration there was a sharp reversal in Reynolds stress to a value 10 times that for current alone. During the acceleration phase there was a relatively mild peak, though with a positive value five times that for current alone.

For the rough boundary the wave-induced turbulent boundary layer is far thicker with the combined flow, and of two or three times the intensity of the current alone.

The distribution of eddy viscosity with depth was computed from the mean bed shear stress and mean velocity profile, the relationship

$$\epsilon = \frac{\bar{\tau}_b}{\rho \ d\bar{u}/dz}$$

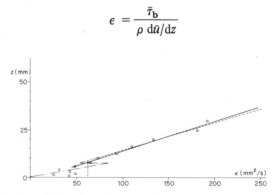

Fig. 4. Eddy-viscosity distribution over rough bed: run WCR3. (———) Christoffersen (1980); (—·—) Grant and Madsen (1979); (———) present data. $\epsilon = \bar{\tau}_b (\rho d\bar{u}/dy)$ for wave and current.

as assumed by Christoffersen (1980) being used. Figure 4, which shows a typical eddy viscosity distribution over the rough bed, suggests that the assumption of Grant and Madsen (1979) that there is a boundary layer of two regions with a linearly varying eddy viscosity of different slope in each is reasonable for most of the outer layer, although in the wave layer near the bed the distirbution is scattered for the larger waves and suggests a constant value for eddy viscosity. In the outer current layer the slope $d\epsilon/dz$ increased with increasing wave height.

Discussion

Much of the discussion is included in the text, and only some of the more general points are summarized here.

Changes in wave attenuation rate were different for the currents of opposite directions, which indicates a direct interaction effect.

The logarithmic description of the mean velocity profile was found to be valid for waves in both directions, and throughout the wave cycle. However, longer shallow-water waves might influence the logarithmic layer.

Near the rough bed mean velocities are decreased in magnitude by the presence of waves in both cases.

Variations in turbulence intensity during a wave cycle near the rough bed were different for the two tests. This was due to the different magnitudes of the current velocities relative to the wave velocities. For the weaker opposing current, flow reversal against the current took place every cycle, causing the generation and ejection of an upstream vortex in addition to the usual downstream vortex. This behaviour was similar to that for waves alone. With the stronger following current, the upstream vortex was far weaker, if formed at all, leading to the formation and weaker ejection of only one vortex each wave cycle.

The presence of wave-induced vortices in the layer above the rough bed causes a drop in mean velocity, and a large increase in apparent bed roughness and shear stress as determined from the mean velocity profiles. This was found for waves propagating both with and against the current, and is as predicted by the theories of both Grant and Madsen (1979) and Christoffersen (1980). The measured values appear to be in better agreement with the latter.

Turbulence intensities averaged over the wave cycle, as shown in Fig. 3, are increased by the presence of waves irrespective of current direction, the increase being within two roughness heights of the bed.

For all combined wave and current tests, including smooth and rough beds, flow reversal was experienced near the bed.

Possible implications in relation to sediment transport

The results of the research outlined above should enable some of the existing assumptions, such as that of the logarithmic velocity distributions, the thickness of the apparent roughness and the eddy diffusivity, to be used with added confidence in sediment transport theory. The results concerning flow reversal at the bed confirm and extend the understanding of the circumstances and reasons for this effect. The increases in shear stress and turbulence may also help to explain or interpret observations of sediment motion.

The entrainment of material from the bed can be considered to show a significant increase under the combined action of waves and currents, but the distribution of turbulence intensities suggests that the zone of diffusion would not increase unless the sediment were further diffused under the action of spilling breakers or of the turbulence in the current.

As an illustration of the use of predicted shear stress in this regard, two possible approaches are chosen for the prediction of sediment transport under combined waves and currents, and these are applied to some possible coastal conditions. Both involve an initial calculation of the bed shear stress, subsequently proceeding to relate these values to sediment motion.

The first method uses Christoffersen's (1980) theory to obtain the maximum bed shear stress for a given current, flow depth, wave height period and boundary roughness. The theory has been found to give results which correspond favourably with the authors' data for wave–current interactions. The final equations to be solved in this approach are set out in the Appendix. While there have been many suggestions for the relationship between maximum bed shear stress and the resulting sediment transport, for the purpose of this calculation to the transport rate has been taken as proportional to the bed shear stress raised to the power 3/2, as used by Engelund and Hansen (1967) and as modified for coastal use by Sayao and Kamphuis (1982).

The second method is that due to Bijker (1967, 1971, 1980), in which the total bed shear stress is computed from the relative wave and current velocities by use of the factor ξ. Sediment movement is then derived by use of the modified Frijlink formula, which relates sediment flux to bed shear stress and bed form parameters, on the basis of the equations given earlier in this paper.

In the equations for the latter approach, as in some other publications to do with wave–current interactions, it is not immediately clear whether the values of wave length and height to be used are those for the waves before the current is superimposed or as they would be measured in the combined flow. However,

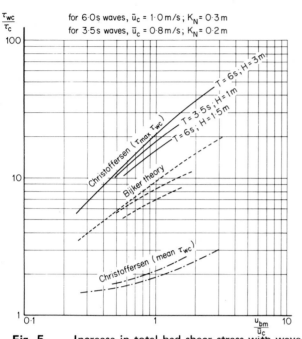

Fig. 5. Increase in total bed shear stress with wave-induced velocity: a comparison of results using Christoffersen and Bijker theories.

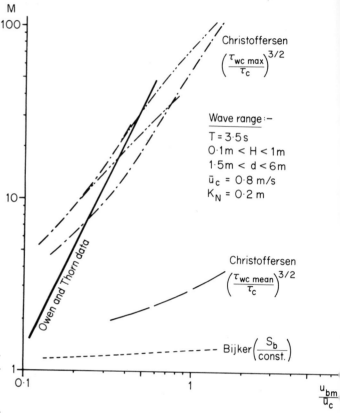

Fig. 6. Sediment flux multiplier, *M*, for waves on a tidal current; a comparison of theories with field data.

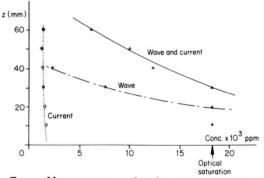

Fig. 7. Measurements of polystyrene concentration over a rough bed for combined wave and current, wave alone and current alone.

trial calculations comparing the bed shear stress values derived from the initial unmodified wave parameters with those using heights and lengths predicted by the theory of Brink-Kjaer and Jonsson (1975) for the combined flow revealed that there was no significant difference between the two for the likely range of variables.

Figure 5 shows the predicted increase in bed shear stress for the combined waves and currents over that for current alone for three possible coastal conditions, with waves of constant height and period propagating into shallow water. It is seen that Christoffersen's method gives values approximately double those derived from the Bijker approach. It should be noted, however, that Bijker considers average shear stresses over the wave cycle, whereas Christoffersen's method yields a maximum value. If the mean stress is calculated from the Christoffersen theory, it is found to be well below that of Bijker.

Owen and Thorn (1978a, b) published field data, measured off Maplin Sands, UK relating the sediment flux multiplier (a factor by which the superposition of waves increases the sediment transport caused by the current alone) to the wave-induced bed velocity. Increases by factors of up to 40 were reported. This information has been transposed onto Fig. 6 to be compared with the two methods of transport prediction under consideration. It is seen that the results obtained

by raising the maximum shear stress from the Christoffersen theory to the 3/2 power coincide reasonably well with the field data. The set of theoretical curves encompass the observed range of wave heights and water depths for a particular wave period, current and bed roughness. However, Bijker's method lies along the bottom of the graph and bears little resemblance to the data. This is probably because this approach is based on bed load transport, whereas at Maplin the sediment flux was predominantly in suspension. Use of the mean instead of the maximum shear stress derived from Christoffersen's theory similarly underestimated the sediment flux.

The Owen and Thorn data show that sediment flux is proportional to U^4 (i.e. τ^2), where suspension could be attributed to U^3 (i.e. $\tau^{3/2}$) and the transport to U (i.e. $\tau^{1/2}$). However, it is not the authors' intention in this paper to define the relationship, but to indicate the possibility of evaluating sediment transport from the shear stress derived from wave–current theory.

From these somewhat arbitrary calculations it does seem that a reasonable estimate can be made of likely sediment transport rates under waves and currents using maximum bed shear stresses and relating these to transport rates by an appropriate power law.

Continuing research

Separate experiments are currently being carried out in a standing wave channel and an oscillating water tunnel, using lightweight bed materials, in order to observe whether the inferences made from the clear-water study are borne out by comparable changes in the distribution of the sediment in suspension. Preliminary results are shown in Fig. 7.

References

Ackers, P. and White, W. R., 1980. Bed material transport: a theory for total load and its verification. *Int. Symp. River Sedimentation*, Beijing.

Anwar, H. O. and Atkins, R., 1980. Turbulence measurements in simulated tidal flow. *J. Hydraul. Div. ASCE*, (HY8), 1273–1289.

Bakker, W. T. and Van Boorn, Th., 1978. Near bottom velocities in waves with a current. *Proc. 16th Conf. Coastal Engineering*, Hamburg, Paper 110.

Bagnold, R. A., 1946. Motion of waves in shallow water: interaction between waves and sand bottoms. *Proc. Roy. Soc., Ser. A*, **187**, 1–55.

Bijker, E. W., 1967. Some considerations about scales for coastal models with moveable bed. *Delft Hydraul. Lab. Publ.* 50.

Bijker, E. W., 1971. Longshore transport computations. *J. Waterways and Harbours ASCE*, (WW4), 687–701.

Bijker, E. W., Hijum, E. V. and Vellinger, P., 1976. Sand transport by waves. *Proc. 15th Conf. Coastal Engineering* Hawaii, Vol. 2, pp. 1149–1167.

Bijker, E. W., 1980. Sedimentation in channels and trenches. *Proc. 17th Conf. Coastal Engineering*, Sydney.

Brebner, A., 1980. Sand bed-form lengths under oscillatory flow. *Proc. 17th Conf. Coastal Engineering*, Sydney, pp. 1340–1341.

Brink-Kjaer, O. and Jonsson, I. G., 1975. Radiation stress and energy flux in water waves on a shear current. *Inst. Hydrodyn. Hydraul. Engng Tech.* Univ. Denmark, Prog. Rep. 36, pp. 27–32.

Christoffersen, J. B. 1980. A simple turbulence model for a three-dimensional wave motion on a rough bed. *Internal Rep.* No. 1, Inst. Hydrodyn. Hydraul., Tech Univ. Denmark.

Englund, F. and Hansen, E., 1967. *A Monograph on Sediment Transport in Alluvial Streams.* Teknisk Vorlag, Copenhagen.

George, C. B. and Sleath, J. F. A., 1979. Measurements of combined oscillatory and steady flow over a rough bed. *J. Hydraul Res.*, **17**, 303–313.

Graaf, J. van de and Overeem, J. van, 1979. Evaluation of sediment transport formulae in coastal engineering practice. *Coastal Engng*, **3**, 1–32.

Grant, W. D. and Madsen, O. S., 1979. Combined wave and current interaction with a rough bottom. *J. Geophys. Res.*, **84**, 1797–1808.

Grass, A. J., 1971. Structural features of turbulent flow over smooth and rough boundaries. *J. Fluid Mech.*, **50**, 233–255.

Ikeda, S., 1980. Suspended sediment on sand ripples. *Proc. 3rd Int. Symp. Stochastic Hydraulics*, Tokyo, pp. 599–608.

Inman, D. L. and Bowen, A. J., 1962. Flume experiments on sand transport by waves and current. *Proc. 8th Coastal Engineering Conf.*, Mexico, pp. 137–150.

Inman, D. L. and Tunstall, E. B., 1972. Phase dependent roughness control of sand movement. *Proc. 13th Coastal Engineering Conf.*, Vancouver, pp. 1155–1171.

Jonsson, I. G. 1965. Friction factor diagram for oscillatory boundary layers. *Prog. Rep. 10, Coastal Engng Lab.*, Tech. Univ. Denmark.

Jonsson. I. G., 1966. Wave boundary layers and friction factors. *Proc. 10th Conf. Coastal Engineering*, Tokyo.

Jonsson, I. G., 1978. A new approach to oscillatory rough turbulent boundary layers. *Inst. Hydrodyn. Hydraul. Engng. Tech. Univ. Denmark,* Paper 17.

Jonsson, I. G. and Carlsen, N. A., 1976. Experimental and theoretical investigation in an oscillatory turbulent boundary layer. *J. Hydraul. Res.*, **14**, 45–60.

Kajiura, K., 1968. A model of the bottom boundary layer in water waves. *Bull. Earthquake Res. Insat.*, **46**, 75–123.

Kamphuis, J. W., 1978. Attenuation of gravity waves by bottom friction. *Coastal Engng*, **2**, 111–118.

Kana, T. W. and Ward, L. G., 1980. Nearshore suspended sediment load during storm and post-storm conditions. *Proc. 17th Conf. Coastal Engineering*, Sydney pp. 1158–1173.

Kemp, P. H. and Simons, R. R., 1982. The interaction between waves and a turbulent current: waves propagating with the current. *J. Fluid Mech.* **116**, 227–250.

Kemp, P. H. and Simons, R. R., 1983. The interaction of waves and a turbulent current: waves propagating against the current. *J. Fluid Mech.*, **130**, 73–89.

Laufer, J., 1950. Some recent measurements in a two-dimensional turbulent channel. *J. Aeronaut. Sci.*, **20**, 277–287.

Mantz, P. A., 1980. Laboratory flume experiments on the transport of cohesionless silica silts by water streams. *Proc. Instn Civ. Engrs*, 2 (69), 977–944.

Miles, G. V., 1981. Sediment Transport Models for Estuaries. *Hydraulics Res. Stn. Publn*, Wallingford, UK.

Nielsen, P., 1979. *Some basic concepts of wave sediment transport*. Ser. Pap. No. 20, Inst. Hydrodyn. Hydraul., Tech. Univ. Denmark.

Nielsen, P., 1981. Dynamics and geometry of wave generated ripples. *J. Geophys. Res.*, **86** (C7), 6467–6472.

Owen, M. W. and Thorn, M. F. C., 1971a. Effect of waves on sand transport by currents. *Proc. 16th Conf. Coastal Engineering, Hamburg.*

Owen, M. W. and Thorn, M. F. C., 1978b. Sand Transport in Waves and Currents. *Hydraul. Res. Stn, Annual Rep.* HMSO, London.

Russell, J. V. and Kemp, P. H., 1977. A suggestion of interaction of suspended sediment with turbulence in the Thames Estuary. *Proc. 6th Autralasian Hydraulics and Fluid Mechanics Conf.*, Adelaide.

Sayao, O. F. and Kamphuis, J. W., 1982. Littoral sand transport. Review of the state of the art. *C. E. Res. Rep. 78*, Queens University, Ontario.

Sleath, J. F. A., 1978. Measurements of bed load in oscillatory flow. *Proc. ASCE, Waterways Port Coastal Ocean Div.*, **104** (WW4), 291–307.

Soulsby, R. L. and Dyer, K. R., 1981. The form of the near-bed velocity profile in a tidally accelerating flow. *J. Geophys. Res.*, **86** (C9), 8067–8074.

Soulsby, R. L., Davies, A. G. and Wilkinson, R. H., 1983. The detailed processes of sediment transport by tidal currents and by surface waves. *Inst. Oceanogr Sci. Rep. 152.*

Swart, D. H., 1974. Offshore Sediment Transport and Equilibrium Beach Profiles. *Pub. No. 131, Delft Hydraul. Lab.*

Thorn, M. F. C., 1981. Physical processes of siltation in tidal currents. *Proc. Conf. Hydraulic Modelling in Maritime Engineering.* Institution of Civil Engineers. London, pp. 65–75.

White, W. R., Milli, H. and Crabbe, A. D., 1975. Sediment transport theories: a review. *Proc. Instn Civ. Engrs*, **59**, Pt 2, 265–292.

Appendix

In order to calculate the boundary shear stresses under combined wave current action by the Christofferssen (1950) simplified method, the following equations have to be solved, by use of an interactive procedure:

$$f_c = 2\kappa^2 \frac{1}{\left(\ln \dfrac{30h}{e \cdot K_N}\right)^2}$$

$$f_{cw} = 2\beta^{2/3} \left(\frac{K_N}{a_{bm}} \cdot \frac{\omega_a}{\omega_r}\right)^{2/3} m^{*1/3}$$

$$\frac{K_A}{K_N} = 30\beta \left(\frac{2m^*}{f_{cw}}\right)^{1/2} \exp\left\{-\kappa\left(\frac{2\sigma}{f_{cw}}\right)^{1/2}\right\}$$

$$f_s = \frac{2}{\left\{\left(\dfrac{2}{f_c}\right)^{1/2} - \dfrac{1}{\kappa}\ln\left(\dfrac{K_A}{K_N}\right)\right\}^2}$$

where f_c is the friction factor for current; f_{cw} is the friction factor for combined wave and current; f_s is the friction factor for average wave and current; β is an empirical constant $= 0.0747$; m^* and σ relate the relative magnitudes and directions of current and wave; K_A is the apparent bed roughness with combined wave and current; K_N is the Nikuradse roughness; ω_a is the absolute wave frequency; and ω_r is the relative wave frequency.

Then

$$\tau_{cw,max} = \tfrac{1}{2}\rho f_{cw} u_{bm}^2 m^*$$

19

Sediment Transport by Tidal Residual Flow in Bays

Tetsuo Yanagi

Department of Ocean Engineering, Faculty of Engineering, Ehime University, Matsuyama, Japan

This paper describes a series of detailed observations of the distribution of sediments contaminated by various pollutants, the distribution being the result of long-term tidal residual flow, rather than tidal currents. Observations were made in the Kasado Bay area, and results for vorticity and divergence compared with tidal rise and fall are presented. An hydraulic model experiment of the Seto Inland Sea is also described, showing the effects of closure of two entrances on circulation.

Introduction

In coastal waters where industrial effluents enter, the contamination of sediments are observed extending from the fixed sources of pollutants towards the flow direction of the tidal residual flow. The tidal residual flow is defined as the flow which is caused through the non-linearity of tidal current in relation to the boundary geometry and the bottom topography and whose period of fluctuation is longer than that of tidal current (Yanagi, 1976). Zimmerman (1982) reviewed the tidal residual flow in the world from the viewpoint of dynamics, diffusion and geomorphological significance. Yanagi and Yoshikawa (1983) summarized the generation mechanisms of tidal residual flow into seven types. The tidal residual flow is generally very weak compared with the tidal current but it plays a more important role than the tidal current in the long-term dispersion phenomena of the material in coastal waters (e.g. Yanagi, 1974).

In this paper some examples which show the relationship between the sediment transport and the tidal residual flow in bays are discussed. Then the three dimensional structure of tidal residual flow system and its relation to the distribution of bottom sediments are discussed. Finally, the possibility of control of the tidal residual flow pattern in bays is shown.

Tidal residual flow and heavy metal contamination in surface sediment

The Seto Inland Sea is situated in the western part of Japan and its length is about 500 km, its average width about 50 km and its average depth about 30 m (Fig. 1). The most predominant flow there is M_2 tidal current which is governed by M_2 tidal waves entering from the Pacific Ocean (Yanagi et al., 1982). The Seto Inland Sea is a very developed area in Japan and there are many oil, iron and chemical plants around this sea. Two examples of sediment contamination by heavy metals there are given. One is about the eastern part of Hiuchi-Nada (Sea of Hiuchi in Fig. 1 ②). There are large pulp plants in the southern coast of Hiuchi-Nada which is shown by the shadow area in Fig. 2(d). Figure 2 shows the bottom topography (a), the major and minor axes of M_2 tidal current ellipse (b), observed residual flow in the upper layer (c) and the heavy metal (Cu) concentration in fine particle fraction in the surface sediment whose diameter is smaller than 76 μm (d). The M_2 tidal current flows nearly along the coast. The residual flow is fairly weak and makes a counterclockwise circulation. Heavy metal (Cu) supplied from the pulp plants mainly spreads to the north-east

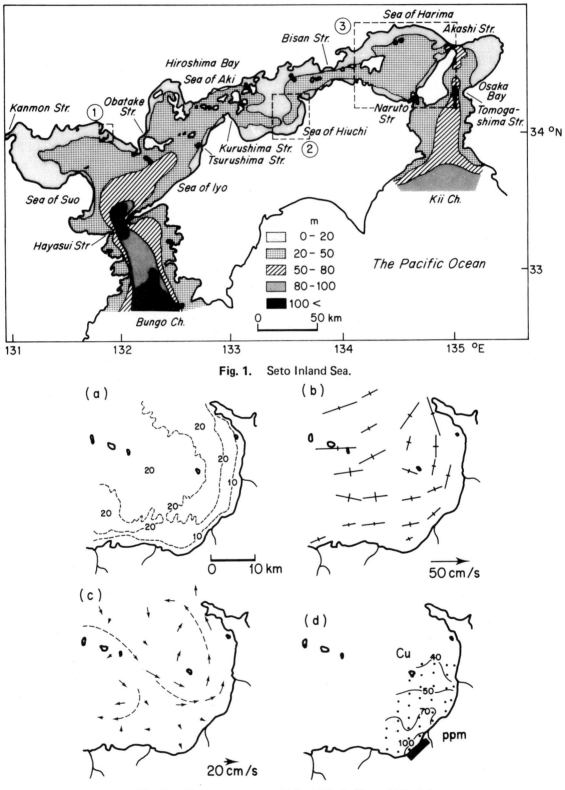

Fig. 1. Seto Inland Sea.

Fig. 2. The eastern part of Hiuchi-Nada (Sea of Hiuchi).

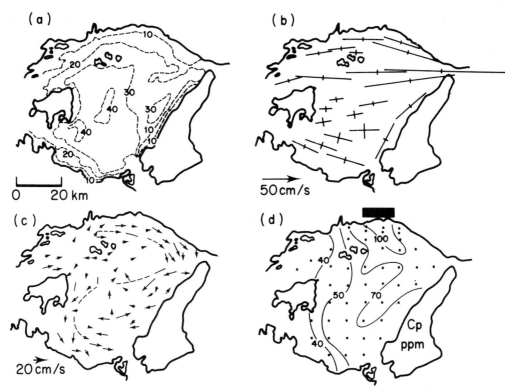

Fig. 3.   Harima-Nada (Sea of Harima).

direction after the residual flow. The residual flow is defined as the steady part in the harmonic analysis of observed tidal current and is thought to be caused by such things as the local wind stress on the sea surface, mean sea-level slopes, the horizontal density gradient and/or the nonlinearity of tidal current etc. The main component of the residual flow in Hiuchi-Nada is the tidal residual flow because there is no prevailing wind and a large river there (Yanagi and Higuchi, 1979). As the stratification does not develop there in summer, the tidal residual flow in the upper layer is considered to be nearly the same as that in the lower layer. From Fig. 2 we can see that heavy metal (Cu) supplied from land does not spread in semi-circular form due to the diffusion by fairly strong M_2 tidal current but spreads long and slenderly after the direction of residual flow.

Another example is about Harima-Nada (Sea of Harima Fig. 1 3). Figure 3 shows the bottom topography (a), the major and minor axes of M_2 tidal current ellipse (b), the observed residual flow in the upper layer (c) and the heavy metal (Cr) concentration in fine particle fraction in the surface sediment whose diameter is smaller than $46\,\mu m$ (d) (Figure 3 (d) is made from the data by Hitomi et al., 1977). There are oil, iron and chemical plants in the northern coast of Harima-Nada which are shown by the shadow area in Fig. 3(d). The M_2 tidal current flows in an almost

east—west direction. The residual flow has a clockwise· circulation in the northern part and a counterclockwise one in the southern part of Harima-Nada. The main components of residual flow in Harima-Nada are considered to be the tidal residual flow and the density current due to the river discharge from the northern coast (Yanagi and Higuchi, 1979). Heavy metal (Cr) supplied from land is seen to spread after the flow direction of residual flow.

The detailed behavior of heavy metal in the sea water is very complicated and it may be a very important problem to be studied in the field of chemical oceanography. It should be pointed out that the heavy metal concentration in the surface sediment is strongly affected by the residual flow patterns in the bay.

Quantitative relation between the degree of contamination and the area of contamination

The regularity of distribution of the heavy metal contamination is examined. The degree of contamination has a quantitative relation to the area of contamination (Fig. 4). The following empirical formula is obtained. obtained.

$$C_s - C_{\infty s} = K_s S_s^{-m_s} \qquad (1)$$

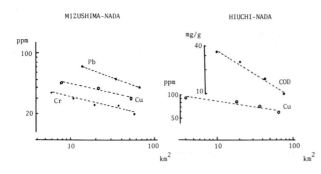

Fig. 4. Degree of contamination $(C_s - C_{\infty s})$ plotted against area of contamination (S_s).

Here C_s is the pollutant concentration in the surface sediment, $C_{\infty s}$ its background value and S_s the polluted area with the concentration higher than $C_s - C_{\infty s}$. The parameters K_s and m_s appear to become constant for given location and pollutant (Fig. 4 and Table 1). Similar relations are obtained about the effluent dispersion in water (Table 1).

$$C_w - C_{\infty w} = K_w S_w^{-m_w} \qquad (2)$$

Such a simple relation suggests that the dispersion of pollutants is controlled by a fairly simple diffusive process. As for Table 1, m_w has a large value and m_s has a small one. The difference of m_w and m_s may be explained as follows (M. Ioi, personal communication). The heavy metal or other suspended matter in the sea water are supposed to be adsorbed to the surface sediment according to Freundlich's adsorption formula, that is,

$$C_s - C_{\infty s} = \alpha(C_w - C_{\infty w})^{1/n} \qquad (3)$$

TABLE 1 Values of m in empirical formula of $C - C_{\infty s} = KS^{-m}$

Location	Hiuchi-Nada	Mizushima-Nada	Osaka Bay	Harima Nada
Scale	100 km^2	100 km^2	500 km^2	1000 km^2
Cl in water	1.14–1.24[a]			
COD in water	1.30–1.62[a]			
COD in sediment	0.61[b]		0.45[c]	
Heavy metal in sediment	0.15–0.29[b]	0.21–0.35[b]	0.15–0.30[c]	0.35–1.00[a]

[a]Hiraizumi *et al.* (1978)
[b]Yanagi (1973).
[c]Joh *et al.* (1974).

Here α and n are the experimental constants and $1/n$ is said to be between 0.2 and 1.0. Substituting equation (2) into equation (3) we have

$$C_s - C_{\infty s} = \alpha \cdot K_w^{1/n} \cdot S_w^{-m_w/n} \qquad (4)$$

After equations (1) and (4) m_s is expected to become $1/n$ times m_w. The difference of m_s value between COD and heavy metal is not explained now. It may depend on the nonconservative character of COD material.

Three-dimensional structure of the tidal residual flow

I will make clear the three-dimensional structure of the tidal residual circulation in this section. The most predominant tidal residual circulation is seen in Kasado Bay (Fig. 5b). Kasado Bay is about 20 km round, its

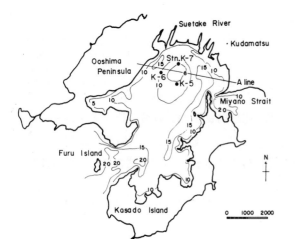

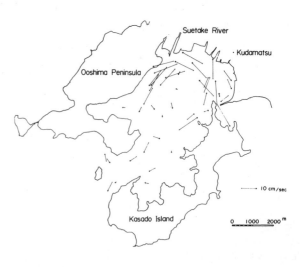

Fig. 5. (a) Map showing observation stations and line of cross-section in Kasado Bay. Numerals indicate depth in metres. (b) Observed residual flows in upper layer.

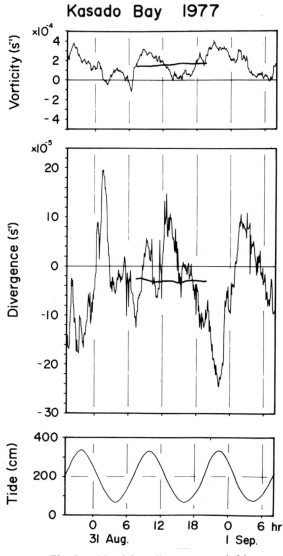

Fig. 6. Vorticity, divergence and tide.

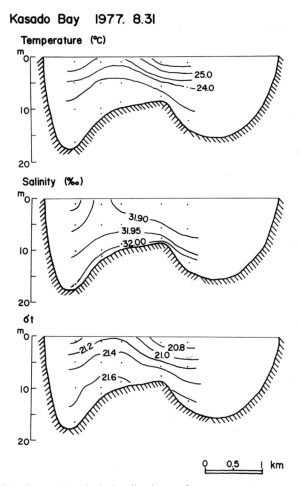

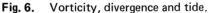

Fig. 7. Vertical distributions of water temperature, salinity and σ_t along the A line in Fig. 5(a).

average depth is about 13 m and there is a topographic high called 'Nakamochi' at the center of the bay (Fig. 5(a)). The flow in the bay is so distinctive that the tidal residual flow is stronger than the tidal periodic current (Yanagi, 1977). The tidal residual flow is caused by the horizontal geometry of the bay and makes a predominant counterclockwise circulation in the bay (Fig. 5(b)). After the current measurements there we found that (1) the tidal residual flow at spring tide is stronger than that at neap tide by about twice, (2) the tidal residual flows have almost the same flow direction in the upper and lower layers and its speed in the upper layer is twice as large as that in the lower layer, and (4) the horizontal counterclockwise circulation is accompanied with a vertical residual circulation which converges in the lower layer and diverges in the

upper layer (Yanagi, 1979). Figure 6 shows the calculated vorticity and divergence of the flow field which is obtained by three current meters set 1.5 m above the bottom at Stns. K-5, K-6 and K-7 in Fig. 5(a). In this figure the thin lines show the raw data taken at 5 min intervals and the thick lines show the residual component i.e. the running means averaged over 24 hours 50 minutes. During flood the sign of divergence is negative and that of vorticity is large and positive. On the other hand, the divergence is positive and the vorticity is small and positive during ebb. The averaged value of divergence is constant at $-3 \times 10^{-5} \mathrm{s}^{-1}$ and that of vorticity is also constant at $2 \times 10^{-4} \mathrm{s}^{-1}$. This result shows that there is a counterclockwise residual circulation which converges in the lower layer in Kasado Bay. The speed of upward residual velocity estimated by the continuity equation

$$w = \int_{0}^{z_1} \left(\frac{\partial u}{\partial x} + \frac{\partial v}{\partial y} \right) dz$$

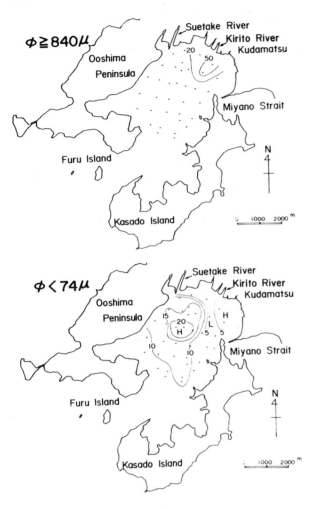

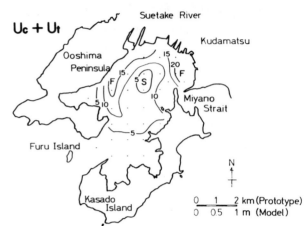

Fig. 9. Horizontal distribution of flow speed. Numerals indicate speed in cm s^{-1}.

Fig. 8. Horizontal distribution of weight percentage of coarser sediment (upper) and that of finer sediment (lower).

(z_1 denotes the height of current meter) is 4.5×10^{-3} cm s^{-1}. Vertical distribution of water temperature, salinity and σ_t along the A-line in Fig. 5(a) is shown in Fig. 7. It is seen that the bottom water of low water temperature and high salinity upwells due to the vertical residual flow in the central part of the bay. The weight frequency of mud particles coarser than 840 µm in diameter is shown in the upper part of Fig. 8 and that of mud particles finer than 74 µm in the lower part of Fig. 8. It can be seen that the coarser mud particles are mainly supplied from the Kirito River. The finer mud particles mainly accumulate in the central part of the bay. As the sedimentation process depends on the absolute value of velocity, we have to examine the distribution of flow speed in the bay. The distribution of flow speed of the tidal residual flow plus the major axis of M$_2$ tidal current from the results of a hydraulic model experiment is shown in Fig. 9. The absolute value of flow speed is small in the central part of the bay. The suspended mud particles flow out of the Kirito River and the coarser particles are deposited near the mouth of the river. On the other hand, the finer particles are transported within the whole bay by the horizontal counterclockwise tidal residual circulation and are converged in the central part of the lower layer by the vertical residual circulation. Finally such fine particles accumulate in the central part of the bay where the flow is weak. The coarser sediments are often found on submarine topographic highs, where the winnowing of fine sediments might be expected. Despite this, finer sediment accumulation occurs in the central part of Kasado Bay over a topographic high. We may consider that this fact strongly supports the existence of vertical residual circulation.

Control of the tidal residual flow

The tidal residual flow is mainly generated due to the horizontal geometry in the Seto Inland Sea. This fact suggests the possibility of control of the tidal residual flow by the change of horizontal geometry. Two examples of the drastic change of the tidal residual flow pattern induced by the variation of horizontal geometry are shown.

Let us consider the tidal residual flow in Kasado Bay. There is a large hydraulic model of the Seto Inland Sea whose horizontal scale ratio is 1/2000 and the vertical is 1/160 in the Government Industrial Research Institute, Chugoku, Japan. Some hydraulic model experiments were carried out to investigate the change of tidal residual flow pattern due to the variation of horizontal geometry of Kasado Bay. At

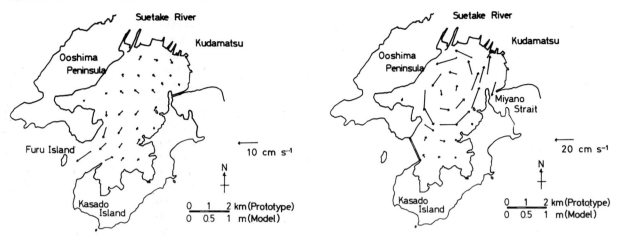

Fig. 10. (a) Tidal residual flow in the case where Miyano Strait is closed; (b) tidal residual flow in the case where Furushima Strait is closed.

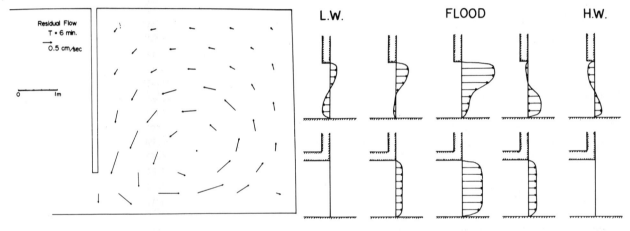

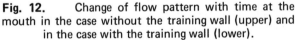

Fig. 12. Change of flow pattern with time at the mouth in the case without the training wall (upper) and in the case with the training wall (lower).

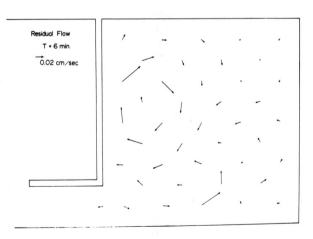

Fig. 11. (a) Tidal residual flow in the case without the training wall; (b) tidal residual flow in the case with the training wall at the mouth of the bay.

first the tidal residual circulation in Kasado Bay was found to be well reproduced in the hydraulic model. Then two experiments were carried out closing Miyano Strait or Furushima Strait. The tidal flow in the case where Miyano Strait is closed is shown in Fig. 10(a). There is not a distinctive residual flow in the bay in this case. The tidal residual flow in the case where Furushima Strait is closed is shown in Fig. 10(b). There is a remarkable counterclockwise residual circulation whose strength is about twice that in the prototype which is shown in Fig. 5(b) in the bay. These results of hydraulic model experiments suggest that the horizontal geometry of Miyano Strait rather than Furushima Strait is a governing factor in generating the tidal residual flow in Kasado Bay (Yanagi and Yasuda, 1977).

Similar hydraulic model experiments were carried

out using the fundamental basin model. The depth of model bay is 10 cm and the tide is generated by a tide generator of the plunger type which is situated in the left side of the basin. The period of tide is 6 minutes and the tidal range is 1.0 cm at the plunger. The tidal residual flow which is obtained by averaging the measured velocities over one tidal cycle is shown in Fig. 11(a). The tidal residual flows are easily seen to make a remarkable counterclockwise circulation in the bay. The flow pattern in the bay drastically changes with a training wall at the mouth. The tidal residual flow is very weak and does not form a remarkable circulation except for smaller eddies near the mouth as shown in Fig. 11(b). The reason why the strong tidal residual circulation does not occur in the experiment where the bay has a training wall at the mouth may be as follows. The difference of time series of flows of normal component to the mouth in two cases are shown in Fig. 12. As the time changes of water level in the bay are the same in both cases, the water mass fluxes at the mouth are therefore to be equal at any moment. The velocity shear is very strong in the upper figure and that is not so strong in the lower one. The vorticity transfer from the tidal periodic component to the tidal residual component is the divergence of the vorticity flux of tidal periodic current and strongly depends on the shear of tidal periodic current at the mouth. The difference of flow patterns at the mouth results in the difference of the tidal residual flow in the bay (Yanagi, 1976, 1978).

Conclusion

In this paper some examples are given which show the relationship between the sediment transport and the tidal residual flow, three dimensional structure of the tidal residual circulation and the possibility of control of the tidal residual flow pattern in the bay.

It is to be hoped the results of this study will help in the prediction and assessing of the degree of contamination of bottom sediment in coastal seas.

References

Hitomi, M. *et al.*, 1977. Bottom sediment in Harima-Nada. Data Report.

Hiraizumi, Y., Manaba, T. and Nishimura, H., 1978. Some regular patterns in the distribution of sediment contamination in the coastal waters along Japan. *J. Oceanogr. Soc. Jpn,* **34**, 222–232.

Joh, H., Yamochi, S. and Abe, T., 1974. Condition of heavy metal pollution in Osaka Bay. *Bull. Osaka Fish. Experiment Stn.,* **4**, 1–41.

Yanagi, T., 1973. Current and sediment in Mizushima-Nada and Hiuchi-Nada. *Bull. Coast. Oceanogr.,* **11**, 8–11.

Yanagi, T., 1974. Dispersion due to the residual flow in the hydraulic model. *Cont. Geophys. Inst. Kyoto Univ.,* **14**, 1–10.

Yanagi, T., 1976. Fundamental study on the tidal residual circulation. I. *J. Oceanogr. Soc. Jpn,* **32**, 199–208.

Yanagi, T., 1977. Tidal residual flow in Kasado Bay. *J. Oceanogr. Soc. Jpn,* **33**, 335–339.

Yanagi, T. 1978. Fundamental study on the tidal residual circulation *J. Oceanogr. Soc. Jpn,* **34**, 67–72.

Yanagi, T., 1979. Vertical residual flow in Kasado Bay. *J. Oceanogr. Soc. Jpn,* **35**, 168–172.

Yanagi, T. and Higuchi, H., 1979. Residual flow in the Seto Inland Sea. *Bull. Coastal Oceanogr.,* **16**, 123–127.

Yanagi, T., Takeoka, H. and Tsukamoto, H., 1982. Tidal energy balance in the Seto Inland Sea. *J. Oceanogr. Soc. Jpn.,* **38**, 293–299.

Yanagi, T. and Yasuda, H., 1977. Hydraulic model experiment of the tidal residual flow in Kasado Bay. *Rep. Gov. Indust. Res. Inst. Chugoku,* **2**, 31–40.

Yanagi, T. and Yoshikawa, K., 1983. Generation mechanisms of tidal residual flow. *J. Oceanogr. Soc. Jpn,* **39**, 156–166.

Zimmerman, J. T. F., 1982. Dynamics, diffusion and geomorphological significance of tidal residual eddies. *Nature, (London)* **290**, 549–555.

Section 6 — Storm Impact and Forecasting

Introduction

Sir James Lighthill (Chairman)
University College, London, UK

From the standpoint of the International Union of Theoretical and Applied Mechanics, of whose executive committee (the Bureau) I have been a member for fifteen years, I should like to express the satisfaction felt by the Union at the success of the Symposium on Seabed Mechanics held at the University of Newcastle-upon-Tyne. Because the Union covers equally the mechanics of solids and the mechanics of fluids (subject areas which, in some symposia on mechanics, are kept rather artificially separated from one another) the Union is particularly happy to have sponsored a meeting which so admirably blended those fields together. The Union's two symposium Panels, that concerned with Fluid Mechanics Symposia (of which I am the Chairman) and that concerned with Solid Mechanics Symposia (of which Professor Warner Koiter is the Chairman) had joined enthusiastically together to recommend the holding of the Seabed Mechanics Symposium; and that fine body of international experts in the study of Seabed Mechanics (whether from one or the other main standpoint or from both) who attended the meeting helped to ensure the excellence of the contributions right across the whole range of the Symposium's subject matter.

I was delighted, in particular, to chair the session on Nearshore Transport, Storm Impact and Forecasting: the second of those sessions where the primary concern was with the Fluid Mechanics effects upon sediment transport (although in these sessions, too, the influence of Solid Mechanics considerations was in much evidence). The previous session had covered many aspects of the influence of turbulence, of waves, of steady water motions, and of tidal residual currents, on the transport of sediment; whether as 'bed load' or as 'suspended load'.

It had clearly established the importance of boundary geometries as an influence in modifying the rate of transport. The next session went on to give three major surveys (involving the analysis of large amounts of field data, alongside data from physical modelling and theoretical or numerical modelling) of different aspects of sediment transport; and it was concluded by an account of many aspects of our knowledge of climatic variability and of its potential relevance to sediment transport problems.

Professor P. Holmes and Dr. P. C. Barber (Imperial College, London) opened the session with a most instructive account, entitled 'Nearshore hydrodynamics related to sediment transport', (not here reported) of an extremely large exercise involving a massive effort in the gathering of data in the field coupled with comparably massive modelling efforts all directed at the important objective of finding out how to maintain the integrity of King's Parade on the Wirral peninsula. They emphasized the need for proper instrumentation on an adequate scale of provision and level of sophistication, and the paper gave an illustration of how such adequate coverage can be achieved only through a very considerable measure of diversity; for example, through a combination of information from backscatter measurements, waverider buoys and, for calibration purposes, a tall instrumentation tower at a selected point 200 m offshore.

Much has been learnt, again, from a comparison of the data so gathered with the results of the extensive numerical modelling programme initiated by Holmes and Barber, utilizing a combination of coarse-grid and fine-grid modelling. This had quantified, for example, the very significant role played by the distribution of currents in modifying wave refraction, and had usefully identified the factors contributing to the observed tendency for waves to be reflected from the deep channel.

Finally, a major physical modelling exercise had been used in the latter stages of these studies to lead to good choices of both the location and the design of a breakwater system that would be effective without excessive cost. No attempt could, however, have been made to embark upon these local problems of design of structures to be placed relatively near King's Parade itself if the previous investigations covering an enormous horizontal area from the point of view of wave climate had not given the necessary supply of data on the distribution of waves incident on that area, their ensuing modification by currents and bottom topography, and the consequentially arising types of sediment movement.

The next presentation entitled 'Behaviour of fine sediments in estuaries and nearshore waters', given by Dr. W. R. Parker (not here reported), was concerned to stress the importance of studying sediments and their transport not only with a great deal of refinement across the horizontal dimensions of a large sea area but also with an at least equal degree of refinement and discrimination in the vertical direction. Dr. Parker's paper dealt with cohesive sediments; that is, broadly speaking, with the finer sediments having typical particle sizes of the order of 0.05 mm or less. The importance was stressed of cohesive forces at many levels in the vertical plane, ranging across all the main levels with which the Symposium was concerned (of a moving suspension, of a stationary suspension, of a consolidated bed and of a settled bed). Also indicated, was the wide range of different types of cohesive force that are brought into play in fine seabed sediments; where, for example, polymers of biological origin may be playing a major role (often just as important as that of the classical electric double layers) in generating forces of cohesion.

Dr. Parker criticized some laboratory experiments as 'over-refined' when they concentrate on very detailed measurements in systems that bring into play just one type of cohesive force. At the same time, he lamented the grossly inadequate volume of field data in this important subject area. Certainly, Dr. Parker has made some remarkable time-series observations of layer boundaries in the Severn estuary. Equally valuable was his discussion of the types of theoretical modelling which seem to give good results for the different types of layer; for example, his evidence for the appropriateness of a broadly 'pseudo-plastic' modelling of the mechanics of certain high-concentration layers.

This, again, was a lecture culminating in practical conclusions of substantial importance, including those

concerned with a comparison of different sounding methods for the determination of an effective bed location. The clear evidence presented for the superiority of a density criterion in this context was undoubtedly most valuable.

The Symposium next enjoyed an enthusiastic lecture and film presented by Dr. Silvester. He was concerned to criticize many classical methods aimed at the prediction of littoral drift (sediment drift along a coastline) on the grounds that they went too far in depicting such drift as a continuous process. He emphasized, rather, the special role of extremely infrequent, very powerful, storms in generating a large part of the total littoral drift that is observed; at least along coasts where tidal ranges are relatively low.

The particular mechanism which Dr. Parker proposes as underlying this effect of strong storms is an interesting one from the standpoint of fluid mechanics. It involves the quite well supported hypothesis that powerful storm conditions generate a strong backward density current which flows out from the beach's surf zone and, where a hydraulic jump decelerates it, generates a sand bar. For a certain time thereafter this bar partially protects the beach by absorbing wave energy. Later, the sediment is returned to the beach by obliquely incident waves at a location shifted along the coast from its original location to an average extent depending on the climatological distribution of the angle of incidence of those waves.

Although the discussion of Dr. Silvester's presentation emphasized the need to probe further the relative importance of the contributions to littoral drift from this mechanism and from other mechanisms that operate more continuously, nevertheless the existence of such a major contribution, in certain circumstances, from severe storms was fully accepted. This recognition gave a most valuable background against which it was possible to conclude that longer-term planning of engineering developments requiring an assessment of sediment-transport possibilities may depend upon an improved knowledge of climatic variability and, in particular, of the likelihood of any significant secular changes in the variations of climate.

Such a picture of the, admittedly limited, state of our knowledge of various aspects of climatic variation (including especially those aspects that may affect seabed transport) was given in the session's concluding contribution by Drs. Farmer and Kelly. This was, undoubtedly, a most interesting survey of an important subject; and one which succeeded admirably, like so many of the other good survey papers given at the Symposium, in making the audience fully aware of the subject's formidable difficulty and complexity!

20

Littoral Drift Caused by Storms

Richard Silvester

Department of Civil Engineering, University of Western Australia, Nedlands, Australia

The offshore bar constructed by storm waves is returned to the beach by subsequent swell waves in a matter of two or three weeks. During this process sediment suspension and littoral current are excessive, which results in a large pulse of littoral drift. Longshore transport at this time can be an order of magnitude greater than that throughout the remaining swell season. It is such pulsational movements that cause sand spits to form or enlarge across river mouths or embayments. Calculations of littoral drift from accretion assumed to be over a year can be grossly in error.

Wave climate

On oceanic margins waves can be classified as storm waves, which are still under the influence of wind within the fetch, or swell, and which are waves dispersing beyond the generation area (Silvester, 1974a).

Storm waves are multidirectional, owing to the generating process and the many fetches operating as a cyclone travel near to or cross the coast. They are high and tend to break because of wind pressure and steepening of short waves at the crests of longer components of the spectrum. Their duration throughout the year is very short, as storms on any coast last for only three or four weeks annually.

Swell, on the other hand, varies very little in its direction of approach to a specific section of coast. The storm centres generating waves are repetitive in location from decade to decade, so that swell is almost unidirectional and certainly within one seaward quadrant. Swell is also persistent in its duration, as it may take three weeks for storm waves to disperse to a distant shore when generated over one week. The longer-period components of the storm spectrum are predominant in swell, since the shorter waves are dispersed over wider fans from the fetch.

While height and period of storm waves vary along the length of the fetch, initially over time but then steady when fully arisen, it is the swell that experiences the greatest variation in these variables. The longest waves with little height arrive first, followed by the mid-periods with greatest energy, after which shorter and smaller waves follow. It thus becomes very difficult to assess an average height or period for computing littoral drift over any reasonable length of the year.

Littoral drift computations

It is the usual practice to assume some hypothetical swell wave arriving at some angle to the coast from which some longshore component of energy is related to the transport of sediment. Empirical coefficients are obtained from measurements of accretion against a structure or in a sand spit, purportedly to have occurred over 12 months. There are as many resulting formulae as there have been workers in this field, no doubt owing to the unique conditions operating on a particular coast during the period being observed (Silvester, 1974b).

Some analyses are more sophisticated than others, taking into account friction factors and permeability effects. However, all are based on a uniformly sloping profile offshore and up to the beach face. They also assume a constant or uniform distribution of sediment size in the active region of the waves. It will be shown later that these assumptions are far from reality for the short time periods when the bulk of the transport takes place.

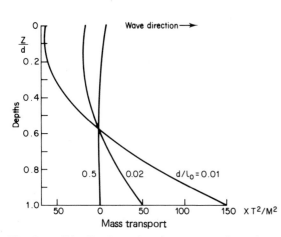

Fig. 1. Distribution of mass transport throughout the water column: X = transport (ft) per wave period T (s) for wave height H (ft) at Z from the mean water level (MWL) in depth, d, L_0 being deep water wave-length.

Shoreline processes

Swell waves propagating shorewards will be slowed down and, hence, increase in height. The orbital motions of the water particles vary from elliptical at the surface to rectilinear oscillations at the bed. They do not return to the same position after each wave cycle but suffer a net movement along the orthogonal or normal to the wave crest, as seen in Fig. 1. At the bed this 'mass transport', as it is called, is in the direction of propagation or shorewards and increases as shallow water is approached (Longuet-Higgins, 1953). This is the region where sediment is disturbed by the water oscillations and bed material is thus carried towards the breaker zone. Gravity acts in opposition to this shoreward motion, so that a final equilibrium profile is developed which is parabolic from the waterline (Sitartz, 1963). The greater the obliquity of the swell the steeper the slopes associated with this concave upward bed.

Once broken these waves proceed to and up the beach face in the form of a bore while carrying sediment in suspension. The uprush percolates the beach face and soaks down to the water table at about mean sea level (Fig. 2 top). This condition exists while there is reasonable time between each wave, which is the case for swell. Because of this percolation the downwash is much smaller than the uprush and, hence, the sand is left stranded on the beach face. Therefore, during periods of swell the beach will accrete and the material accumulating will be loose and, hence, aid percolation. Later, after continual swashing of the swell, the face becomes compacted and can even take vehicular traffic.

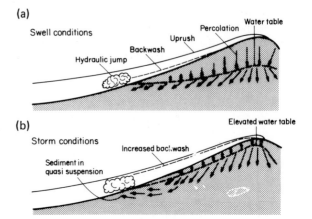

Fig. 2. Beach processes under (a) swell conditions and (b) storm conditions.

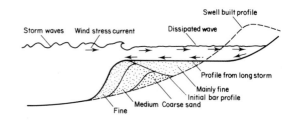

Fig. 3. Profile of final beach under storm conditions, showing distribution of sediment size.

Accretion will only take place while there is material offshore to be transported by mass transport and the bore within the surf zone. The closer the profile approaches the parabolic shape the less the build-up and, hence, the less the capability of the waves to transport sediment along the coast.

When storm waves reach this swell-built profile, they are steep and arrive almost every second. Much water is thrown onto the beach face, which quickly becomes saturated, which implies that the groundwater level is almost coincident with the beach face (see Fig. 2b). Less water can percolate and, hence, the downwash almost equals the uprush, causing erosion. This is aided greatly by the return of excessive groundwater to the sea, which, at the location of the hydraulic jump, is almost vertical, thus causing a 'quicksand' effect. It is little wonder that the beach subsides and is denuded quickly.

Sand-laden water proceeds seawards as a density current amplified by the strong shorewards drift at the surface due to the stress of the winds associated with the storm. However, as this current approaches deeper water, it is reduced in velocity, which causes its sedimentary load to be deposited. Material therefore accumulates in the form of a bar running parallel to the beach, which continues to build up until the depths over it are sufficiently small for incoming storm waves to be broken over it (see Fig. 3). At this stage erosion essentially ceases.

It can be concluded that the first storms of winter will have the greatest effect on the beach and the erosion at this time can be taken as a measure of the magnitude of recession for the year. Two situations can cause greater demand of beach material. If a second storm occurs with a higher water level, due to storm surge or spring high tide, the bar will have to be larger in order to break the waves. Also, if a second storm has longer duration, the partially dissipated waves reaching the receded water line, together with the circulating current in the swale between bar and beach, can take material to fill this depression (Silvester, 1979).

This natural process of placing sediment offshore in the form of a mound which limits the erosion that can take place in any storm sequence is beautiful to behold. If it did not occur, our shorelines could be spread evenly over the offshore area perhaps never to return to the beach or at least taking much longer to return. The mechanism is one that man can never hope to replicate, since such large volumes of sand are involved. Suggestions have been made of constructing submerged rubble mounds to accomplish this task but the oblique swell would scour the bed adjacent to them and result in large maintenance costs.

Return of bar

Actually part of this natural defence mechanism involves the replacement of the bar onto the beach when it is not required. The swell waves following the storm, either immediately or perhaps some weeks later, will break over this mound, causing excessive suspension of material. The bore, proceeding to the receded shoreline, carries this load and places it on the beach face. The berm is built up to the reach of the waves, which initially is not as great as later when the bulk of the mound has been removed. As the bar is worn down towards the original parabolic profile of the bed, the renourishment decreases in volume but the waves run further up the beach. Thus the final berm can be lower at the back of the beach than it is at the seaward edge. The present discussion is assuming little or no influence by tidal levels.

Assuming the swell waves are arriving obliquely to the coast, which is the case where beach erosion is a problem, they will break at a greater angle when encountering this sudden shoal than when arriving at the swell-built profile. This means that the littoral current is excessive at the same time as sediment suspension is optimum. The overall result is that a large

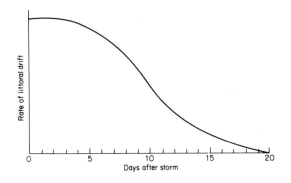

Fig. 4. Rate of sediment transport over time.

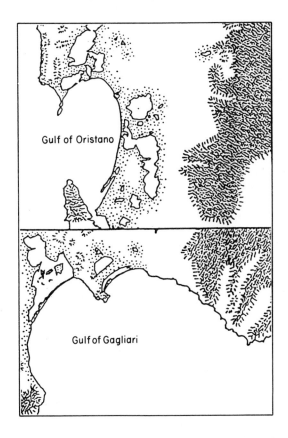

Fig. 5. Barrier beach across a coastal indentation.

pulse of littoral drift occurs during the period of bar return to the beach. This may extend over 2–3 weeks, depending upon the strength of the swell. Japanese engineers have concluded that 90% of the littoral drift can occur over 2 weeks and the remaining 10% over the rest of the year.

The above could be true for one storm during 12 months but if there are two, spaced, say 2–3 weeks

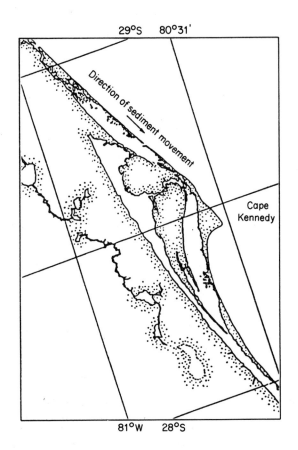

Fig. 6.   Coastal lagoons formed by spit formation in past geologic age.

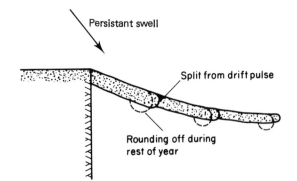

Fig. 7.   Formation of sand spit at change in shoreline alignment.

if a relation is required between wave energy and long-shore drift.

Geomorphological consequences

The writer has always wondered why sand spits are formed across deep embayments on the coast or lagoons are enclosed parallel to the shoreline as exemplified in Figs. 5 and 6. Why, for example, are such indentations not silted completely by slow filling from littoral drift? The reason appears to be provided by the above pulsational supply of sand to the waves, which can only pass it along the wave crests where it is deposited.

Take, for example, an abrupt change in the coastline towards which sediment is being transported by swell arriving obliquely, as in Fig. 7. A pulse of sand of large proportions is fed to the waves which can only form a spit parallel to the wave crests.

At the tip of the spit waves refract suddenly and have an upcoast energy component in opposition to the littoral current at this location (see Fig. 8). This causes deposition of sand at this point adding length to the spit, which grows parallel to the wave crests. This hypothesis of pulsation drift and the geomorphological result described above is quite novel and requires verification by monitoring accumulations of sand against structures or in spits while at the same time taking profiles of the upcoast offshore bar. Such measurements should be carried out daily over the 2—3 week period during the return of the bar material to the beach. This requires a large team of workers on stand-by during and immediately after a storm sequence.

During the remainder of the swell season, with very little drift taking place, the tip of the spit will be eroded and moved around to its lee side. After the next storm another pulse causes an elongation of the spit and the process is repeated. These barrier beaches, as they are called, are a predominant feature around the coastlines of the world and occur as small

apart, there will be two pulses of littoral drift. This will double the longshore transport for the year. If there are more than two cyclones near the coast, the volume of material moved will be equally amplified, which is not catered for by the calculations of drift previously alluded to. Even if the bar is partially replaced before a second storm, it is quickly reformed, but the act of renourishment in each case is accompanied by excessive littoral drift. The rate of transport is maximum immediately after the storm and reduces substantially as the parabolic profile is reached, as indicated in Fig. 4.

This implies that to obtain realistic measurements of transport, say by assessing accretion in sand spits or against obstructions, they should be carried out over two or three weeks after a storm with concomitant profiles of the bar. Generally such volumes are determined at an annual frequency on the basis that they have resulted from swell over such a period. It becomes obvious that they could be grossly in error. The wave incidence during this short period should be monitored

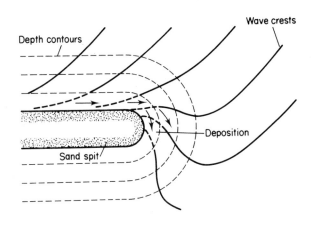

Fig. 8. Wave refraction and breaking at tip of spit, resulting in accretion parallel to incident wave crests.

features or are of mammoth proportions (Figs. 5, 6). On some coasts annual littoral drift has been assessed as 200 000 m³. If this actually occurs over a matter of a few weeks in the year, spits of prodigious proportions can result.

In previous geologic times the volume of sediment available from rivers has been much larger than that available today, owing to changes in climate and man's influence in building dams, so mitigating fluctuations in discharge. Hence, many of the barrier beaches previously constructed are now being eroded because of the lack of sediment supply. Some geomorphologists and environmentalists are blaming the general erosion on man and are having prohibitions put on defending the coast or retaining it on its present alignments (Kaufman and Pilkey, 1979). But the processes are natural and if facilities need to be protected some measures must be taken.

Silvester (1976) has suggested the construction of headlands, again a procedure used by Nature to stabilize the shoreline, which results in the formation of equilibrium-shaped bays which stop the longshore transport of sediment. If the predominant swell can be made to arrive normal to a coast, the material of the offshore bar is replaced back on the beach from whence it came, so providing a constant waterline.

Another process worthy of note is the sorting of sediment that takes place during the formation and return of Nature's protective bar. When the storm current is carrying sand offshore and is slowed down, the first particles to be dropped will be coarse, followed by medium-sized grains and finally the fine components (see Fig. 3). Hence, the bar will comprise coarse sand at the base and close to the shore; the middle mass will

consist of medium material; while the top and seaward face will be made up of the smallest particles. When swell commences to return the bar to the beach, it will act on the fine sand first, which will be placed at the back of the berm or against the dunes fronting the beach. The medium grains are then moved shorewards, while the coarsest contents are deposited in the seaward edge of the berm and in the final beach face. This distribution in the berm also helps the dispersement as discussed above in any following bar. Since littoral drift is maximum at the commencement of bar removal, it is seen that fine material is readily moved downcoast while coarse material remains essentially in the same position. This action accounts for variations in size distribution along a coast where fineness is accepted as a downcoast characteristic.

This variation in sand-grain size over the active part of the littoral drift cycle is not accounted for in littoral drift formulae. But the major omission in such computations is the change of width and shape of the surf zone when the bulk of material is being moved along shore. This affects both the suspension capacity of the waves and the littoral current generated.

Engineering implications

Where dredging is undertaken either continuously or sporadically, the fluctuation in littoral drift could be paramount in choosing the type and size of machine. The shoaling and perhaps spit formation can be sudden events which can make navigation hazardous. Knowledge of the time of the year when siltation is at a maximum could help in the planning of dredging programmes.

In some locations it is customary to dredge holes upcoast of an entrance in order to cater for later littoral drift. Predicting longshore transport accurately can aid such procedures. The occurrence of more than the usual storms on a coast could foreshadow swifter siltation of such man-made depressions.

The dredging of access channels to ports across the coastal boundary is continuous and costly. The section of such depressions close to the surf zone will receive a large influx of material just after each storm. This knowledge should help the surveying and dredging plans for the port.

Where littoral drift is left to bypass a river mouth naturally, it will do so by means of moving shoals or intermittent sand spits. Fluctuations in river discharge during storm sequences or just afterwards will deflect the position of these sedimentary features. Tidal oscillations will also disperse material either out of or into the mouth during these transient siltation periods.

Modifications to existing procedures have been suggested by Silvester and Ho (1972) and Silvester (1975, 1977), to inhibit the problems arising by pulsation littoral drift. For example, bypassing material across river or harbour mouths could be achieved by reflecting waves across the entrance and so form a short-crested system which induces suspension and, hence, transmission of sediment. Another approach on an eroding coast is to establish headlands between which zeta-shaped bays will form, for which the limiting indentation and curved shape are predictable. It would be possible to stop littoral drift altogether, the cost of which could be regained from the lease or sale of land gained from the sea.

Concluding remarks

Le Mehauté (1970) has promoted the idea that littoral drift is a nice uniform process, even less variable than transport of material down rivers. The concept of this transport as 'a river of sand' has detracted from the study of this mammoth transporting capacity of waves resulting from the severe meteorological fluctuations experienced for centuries. Concentration of mathematical theories to uniform conditions of persistent swell may satisfy the theoretician but cannot provide solutions of the real world, which are characterized by continual fluctuations of energy.

Many engineers cannot explain why shorelines migrate so readily and so consistently. The migration is brought about by variations in sediment supply along a coast. These can be due to fluctuating discharges of material from river mouths, the interruption of littoral drift by structures or dredged channels, or the concentration of wave energy on limited segments of coast. The volume of material removed from a beach, to be placed in circulation as an offshore bar, will vary with the storm wave energy, the water level existing at the time and the steepness of the offshore profile. This sediment will be swept downcoast by the subsequent swell, whose persistent direction of approach dictates the net sediment movement around the coastlines of the world (Silvester, 1962).

References

Kaufman, W. and Pilkey, O., 1979. *The Beaches are Moving*. Anchor Press, New York.

Le Mehauté, B., 1970. Comparison of fluvial and coastal similitude. *Proc. 12th Conf. Coastal Engineering*, Vol. 2, pp. 1077–1096.

Longuet-Higgins, M. S., 1953. Mass transport in water waves. *Phil. Trans. Roy. Soc., Ser. A*, **245**, 535–581.

Silvester, R., 1959. Engineering aspects of coastal sediment movement, *Proc. ASCE, J. Waterways Harbors Div.*, Paper No. 2168.

Silvester, R., 1962. Coastal sediment movement around the coastlines of the world. *Proc. Conf. Instn Civil Engrs*, pp. 289–315.

Silvester, R. 1974a. *Coastal Engineering*, Vol. I. Elsevier, Amsterdam.

Silvester, R., 1974b. *Coastal Engineering*, Vol. II. Elsevier, Amsterdam.

Silvester, R. 1975. Sediment transmission across entrances by natural means. *Proc. 16th Conf. IAHR*, Vol. 1, 145–156.

Silvester, R., 1976. Headland defense of coasts. *Proc. 15th Conf. Coastal Engineering*, Vol. 2, pp. 1394–1406.

Silvester, R., 1977. The role of wave reflection in coastal processes, *Proc. Coastal Sediments 1977, ASCE Charleston Conf.*, pp. 639–654.

Silvester, R., 1979. A new look at beach erosion control. Disaster Prevention Res. Inst., Kyoto, Ann. Rep. 22A, pp. 19–31.

Silvester, R. and Ho, S. K., 1972. Use of crenulate shaped bays to stabilize coasts. *Proc. 13th Conf. Coastal Engineering*, Vol. 2, pp. 1347–1365.

Sitartz, J. A., 1963. Contribution à l'étude de l'évolution des plages à partir de la consistance des profiles d'équilibre. *Trav. Centre Etud. Oceanogr.*, 10–20.

21

Climatic Variation in the North Sea Region

G. Farmer and P. M. Kelly

Climate Research Unit, School of Environmental Sciences, University of East Anglia, Norwich, UK

Climatic change has become of increasing concern in recent years. This is partially due to greater stress on the world's limited resources but it is also a reflection of increased awareness of environmental matters and, in particular, the realization that society may be about to bring upon itself a major climatic change. This paper presents a brief summary of the data available for studies of past and present climatic variation in the North Sea region. The possibility of long-term climate prediction to aid planning and design is also considered. It is concluded that, although current knowledge is not sufficient to support the production of definite long-term forecasts, study of the past record of climatic change is essential if reliable estimates of the potential range of future climatic conditions are to be obtained.

Introduction

Until relatively recently it was believed that climate records of sufficient length (30 years was often considered sufficient) would provide an adequate indication of the range of conditions in the recent past — and in the not too distant future. As longer instrumental climate records became available towards the middle of the present century, it became clear that this belief was ill-founded. Climate was observed to change on all time scales from the monthly or seasonal duration of short-lived climatic extremes up to the century-to-century changes seen in the longest instrumental records.

As society becomes more dependent on long-term planning and design, the importance of climatic change and its effects is increasing. Yet few long-term planning exercises incorporate the potential of climatic variation. In part, this is a hangover from the time when climatology was considered a simple exercise in book-keeping, with little thought required in the application of the accumulated data. It is still seen that way by many potential users. But it is also a reflection of the stage of development of the study of climatic change.

Knowledge is insufficient, at present, to answer many of the questions posed by those interested in the potential of climate-related impacts. Nevertheless, we do know that climate has changed in the historical period, and we can see changes in recent records. Climate is changing significantly at present, and this may be the first sign of a major rise in global temperatures caused by pollution of the atmosphere. We should, at the very least, be more aware of the scale of past variations in climate and consider seriously implications for the future.

In this paper we present an up-to-date summary, for the North Sea region, of the data available for the purpose of defining climatic change on the interannual time scale over the most recent three centuries and assess climatology's present ability to define the course of future climate in this area. Four major topics are discussed: the availability of data; methodology; the climatic record over the past 300 years; and the predictability of climate. We only discuss climatic change on time scales of less than 100 years or so and do not consider the swings between glacial and interglacial conditions.

In order to provide a focus for our discussion, we concentrate on the aspects of the climate system most relevant to sediment transport. It is our understanding that relatively little research has been undertaken concerning variations in sediment transport on the time scales under consideration here, and the intention of this review is to describe the main data sets and techniques available for this purpose. Sediment transport can be powered by currents, which may be wind-driven, and by storm events. Winds are, therefore, the dominant force. They themselves are dependent on the atmospheric pressure distribution, which is itself a function of global temperature patterns, among other factors. These, then, are the major points of reference of this survey.

Data availability

Most studies of climatic change and its effects are hampered by limited data availability. Reliable instrumental climate records are only available for the North Sea region for the past 2–3 centuries and these are mainly for countries around the North Sea. With care, indirect or proxy climate data (based on historical evidence or climate-sensitive biological indicators, for example) can be used to fill gaps in or to extend the instrumental record, but these data need to be used cautiously. Events may be poorly dated. Some climatic indicators respond immediately to climatic change; others react slowly, smoothing out shorter time-scale variations. Care must be taken in assessing the geographical representativeness of the data. Is the record purely a local index or can it be taken as a regional one? If little regard has been paid to the complexities of its derivation, it is unlikely that any conclusions of scientific value will result.

The reader is referred to Wigley *et al.* (1981) for a recent review of the uses and limitations of various forms of proxy data, with emphasis on the treatment of historical data sources. Concerned, as we are in this paper, with the recent past, we now restrict our attention to historical and instrumental data.

Countries surrounding the North Sea provide a very long record of weather and climate conditions. Daily weather diaries have been kept in various countries from as early as the fourteenth century, when William Merle was observing in Oxford and Lincolnshire. In the sixteenth century Tycho Brahe was recording in Denmark, and weather records are also available from Bavarian monasteries. Descriptions of the prevailing weather taken from diaries, monastic records, ships' logs, state papers, personal correspondence, and so on, can be collated to give a 'weather report', as shown in Fig. 1. This presents data for the winter of 1607/08, one of the severe winters of the so-called 'Little Ice Age' which affected Europe during the sixteenth, seventeenth and early eighteenth centuries. These qualitative descriptions can be interpreted using the experience of analysis of recent data to produce atmospheric circulation maps, as shown in Fig. 2. In this way a snapshot

of conditions during the most extreme seasons of the historical period can be built up.

The atmospheric data of most relevance to studies of long-term variations in factors such as sediment transport are those related to changes in storm frequency and wind direction and strength. Gottschalk (1971, 1975, 1977) and Lamb (1977, 1981a, b) present and discuss chronologies of periods of extreme storminess in the North Sea. Much of the historical material is anecdotal. For example, Lamb (1981b) describes conditions in the late thirteenth and early fourteenth centuries: 'The storm floods of the North Sea between AD1240 and 1362 caused sixty Danish parishes that produced over half the revenues of the diocese of Slesvig (now Schleswig) to be "swallowed by the salt sea" (Steenstrup, 1907), doubtless with much loss of life. And this presumably also occurred in northwest Germany and the Netherlands where the Jadebusen and the Zuyder Zee were formed at about that time.' Further details can be found in Gottschalk (1971). This type of information, while often fragmentary, is important because it enables the events of recent years to be placed in their long-term context, in terms of both magnitude and frequency.

More data concerning variations in the wind field over the North Sea and its immediate vicinity are available for the period since 1800 than for any other region in the world. Many of these time series have been analysed by Kelly (1975). The UEA Climatic Research Unit has recently started to collect and catalogue wind data for the United Kingdom and surrounding seas, and a listing of all readily available data will be published during 1984.

The longest records concern wind direction alone. Gray (1981) has produced a record of southwesterly wind frequency at London extending back to the late seventeenth century (Fig. 3). For the most part, it is based on two sources: the Booth catalogue of daily wind observations (1723–1880 and 1901–1910) and the Lamb catalogue of daily weather types (1861 to date). The latter is discussed in more detail below. The London record is of the monthly frequency of southwesterly winds, the dominant wind direction in this area. Similar records of wind frequency exist for Copenhagen (Fig. 4) and De Bilt. The Copenhagen record has been extended back to 1701 by C. Loader of the Climatic Research Unit, using naval observations from the early eighteenth century. Weiss and Lamb, in a study of changing wave height in the northern North Sea, used such records to produce a history of the winds over the North Sea for much of the past 1000 years (Weiss and Lamb, 1970; Lamb and Weiss, 1979). They also used recent German navy data to define variations in the North Sea winds in greater detail during the twentieth century. By using eighteenth century instrumental records J. A. Kington of the

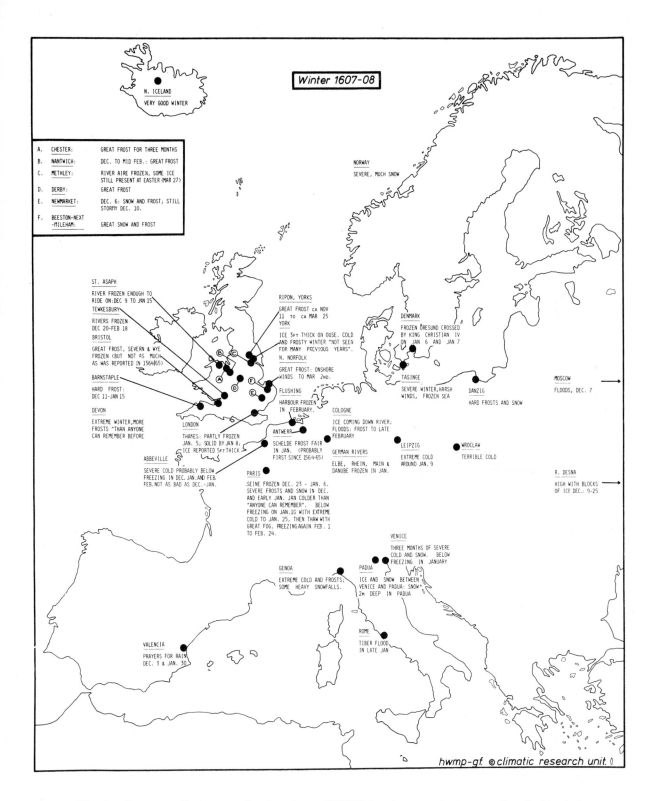

Fig. 1. Data compilation map for the winter 1607/08 for the countries surrounding the North Sea.

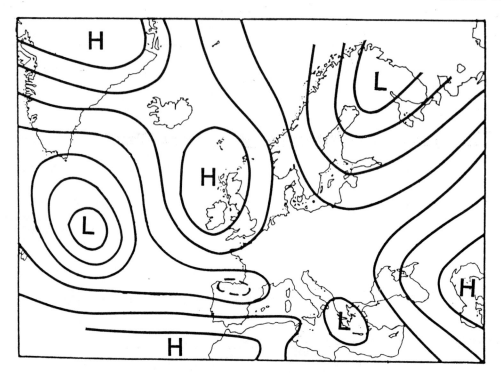

Fig. 2. Atmospheric circulation map for the winter 1607/08 interpreted using present-day analyses.

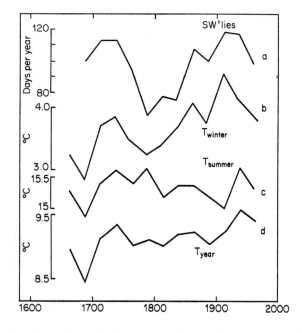

Fig. 3. Record of mean monthly south-westerly wind frequency in London, UK.

Climatic Research Unit has constructed daily synoptic pressure maps for the years 1781–1785 (Kington, 1980). These can be used to derive detailed information for that

period, and to provide comparisons with modern conditions.

Given the interest in the exploitation of North Sea resources, it is surprising that so little analysis of the potential of climatic change in this region has been undertaken. There is a fair amount of data available for studies of North Sea wind variations during the present century, although little has been done to collate the data from different sources, to check the data for errors, and to apply corrections when necessary and possible. Gridpoint surface pressure data, which can be used to calculate geostrophic winds for the region, are available as monthly averages back to 1873 and as daily data back to 1881 digitized on magnetic tape (obtainable from the UK Meteorological Office). Wind observations have been taken for many years by light-vessels along the margins of the North Sea, and data are also available from 'ships of opportunity'. In recent years offshore installations have provided a valuable additional source of direct observations of wind speed and direction, and of other environmental variables. Reliance on records of limited duration has already resulted in the re-estimation of return periods for extreme wave heights on the basis of data taken from such installations.

Smith (1982) has recently compiled wind observations taken at various UK sites, related them to pressure gradients and produced a synthetic index of

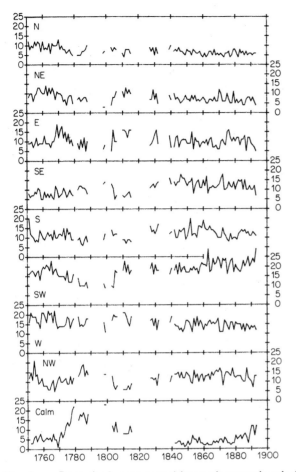

Fig. 4.   Record of mean monthly south-westerly wind frequency in Copenhagen, Denmark.

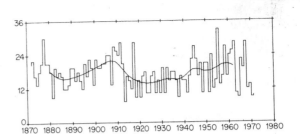

Fig. 5.   Record of storm surge liability in the North Sea.

type of interest during a particular period, a variety of indices can be devised for the study of long-term changes in the atmospheric circulation and their effects. For example, a count can be made of the number of days during which storm conditions were likely to affect the North Sea. A similar approach, using daily pressure charts, has resulted in a record of storm surge liability for the North Sea (Benwell, 1967; Hunt, 1970; see Fig. 5).

As far as future conditions are concerned, we are not aware of any publications on the subject of changes in the local atmospheric circulation, apart from Weiss and Lamb (1970) and Lamb and Weiss (1979).

Methodology

The selection of data is an important and often-neglected aspect of experimental design. Many investigators have relied on inappropriate measures of the state of the climate system which have little relevance to the physical processes being studied, often simply because these data were readily available. To cite a trivial example, annually averaged data are often used for the sake of convenience. Yet it is quite likely that in most cases of climate-related effects it is the seasonal timing of the changes that is important and this may not be reflected in an annual average. As already noted, there has been a tendency for investigators to use climatic data, and, in particular, proxy climatic data, with scant regard for their derivation, their interpretation and their reliability. Such data are often treated simply as a series of numbers to be marshalled in support of a particular hypothesis, the physical significance of the data being ignored.

The development of the science of climatic change has also been severely hampered by studies which lack scientific rigour, particularly in the area of empirical techniques. Analyses have been undertaken with no methodology, experiment or hypothesis in mind and with scant regard for statistical and/or physical significance. This has created a dense undergrowth of 'fact' which it is necessary to hack down before progress can be made. However, this is not the place to discuss statistical methodology in detail; the concerned

long-term wind variations for the region. One of the most detailed wind data sets for the North Sea itself has been produced by Dietrich *et al.* (1952) and updated by Schott (1970). Zonal and meridional winds, based on pressure gradients were presented for various sectors of the North Sea.

It may prove possible to extend these records further back in time using early instrumental pressure observations. Such data form the basis of the earliest portion of Lamb's catalogue of daily weather types. This is a valuable source of information concerning not only wind direction (or, more precisely, the steering of surface features of the atmospheric circulation), but also information about the type of atmospheric circulation feature affecting the region of the UK (for example, whether conditions are cyclonic or anti-cyclonic). The character of the atmospheric circulation over the region of the UK and surrounding seas is specified for every day of the period 1861 to date. By counting the number of occurrences of a weather

reader is referred to Pittock (1978) and Kelly *et al.*
(1982) for an account of the perils and pitfalls of this
field and for methodological guidelines. Mitchell *et al.*
(1966) and Barry and Perry (1973) contain concise
descriptions of the major statistical tools available for
the study of climatic change.

The climatic record

We focus, in this survey of recent climatic change, on
instrumental sources of data. A long record of temperature for central England, described in detail by
Manley (1974), shows the height of the so-called Little
Ice Age during the late seventeenth century and the
slow warming that occurred over the following 250
years (Fig. 3). Superimposed on this recovery from the
Little Ice Age appear short interludes of more severe
climate. The immediate cause of these trends was, it
is believed, a change in the character of the atmospheric circulation. The record of southwesterly wind
frequency at London exhibits long-term variations
similar to those seen in the temperature record. Warming
is associated with a greater frequency of southwesterly
winds advecting air from southern sources, and vice
versa. Southwesterly wind frequency at London is a
crude index of the character of the larger-scale westerly
flow across the North Atlantic and the variations seen
in the London record suggest a large-scale alteration in
the character of the atmospheric circulation. It is likely
that variations in the character of the winds and the
degree of storminess over the North Sea accompanies
these changes.

Since 1881 the hemispheric network of instrumental
observations has been sufficient to support the direct
estimation of larger-scale climatic trends (Jones *et al.*,
1982). The major climatic variation during recent
times has been the warming which has affected the
northern hemisphere throughout much of the past
100 years (Fig. 6). Cooling occurred during the 1940s,
1950s and early 1960s, but warming has become re-
established during recent years. Whether or not this
warming is a long-term phenomenon remains to be
seen.

The spatial patterns of climatic change can be complex. How representative of the North Sea region is this
particular climatic index? The variations seen in the
hemispheric average during recent years are not strongly
correlated with temperatures in the North Sea region
(Jones and Kelly, 1983), although they were during
the early years of the present century. The warming
of the 1970s and early 1980s has yet to be strongly
felt in the North Sea region. It is unwise to rely on
large-scale generalizations when considering local events.
Similarly, it is dangerous to place too much faith in
apparent correlations between distant, regional climatic
indicators.

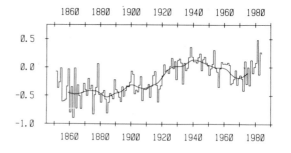

Fig. 6.   Climatic trends in the northern hemisphere
since 1850.

These changes in surface air temperature were
accompanied by fluctuations in other climatic parameters, by shifts in the atmospheric circulation (by
variations in wind direction, wind strength and storminess) and by changes in the ocean environment, many
of which could have affected features such as sediment
transport. What would be the level of importance of
such long-term variations in sediment transport produced by climatic change? There is an obvious need to
examine what long-term data exist concerning sediment
transport and/or to model the effects on transport of
the changes evident in the climatic record.

Climate prediction

We have shown that the climate of the North Sea region
has undergone significant changes in the historical
and instrumental periods. But what of the future? The
ultimate aim of studies of past climatic change is prediction. Reliable climatic prediction is, however, dependent on adequate understanding of the causes of
climatic change and this understanding is, at present,
severely limited. Most predictive techniques on the
shorter, seasonal and interannual time scales are essentially statistical in nature, although efforts have been
made to identify the physical processes underlying
successful rules. On longer time scales forecasts are now
largely theoretical in nature, based on the results of
numerical modelling and frequently only considering
the effects of one causal factor.

In the early 1970s a World Meteorological Working
Group, as reported in Lamb (1977), reviewed a set of
some two dozen predictions concerning the course of
climate over the coming 20–50 years. The forecasts
were close to unanimous — the cooling which had
affected the northern hemisphere over the previous
30 years was likely to continue. Although there might
be the occasional brief respite, there were considered
to be few indications that the cooling would end till
well into the twenty first century. In 1978 the US
National Defense University surveyed the opinions of

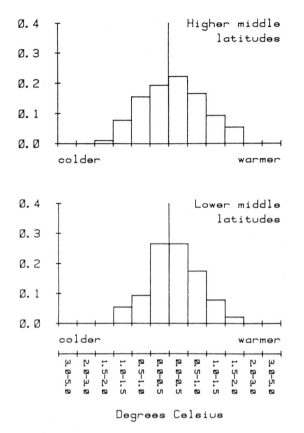

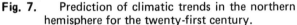

Fig. 7.   Prediction of climatic trends in the northern hemisphere for the twenty-first century.

the world's leading climatologists: What will our climate be like in the early twenty-first century? The consensus, although hardly unanimous (see Fig. 7), favoured warming. Why had the attitudes of climatologists changed over so short a time? To answer this question, we need to examine the bases of the two sets of forecasts.

The earlier predictions of cooling were largely based on empirical or statistical techniques. For example, long climatic records were examined for evidence of regular cycles and/or trends and these, when found, were extrapolated into the future. Little attention was paid to the physical causes of the variations, although most investigators favoured one or the other of two likely mechanisms (variations in either volcanic or solar activity). Whatever the data base or physical explanation advanced in support of these predictions, at their root was the observation that the warmth of the early twentieth century appeared without parallel during the present millennium and, possibly, during the present interglacial. If it was this unusual, it was unlikely to last for much longer, so cooling seemed likely.

Until the late 1970s the observational record (Fig. 6) seemed to favour the prediction. The warmth of the 1930s and early 1940s was followed by marked cooling

over much of the northern hemisphere right up to the late 1960s. In the mid-1970s Soviet climatologists attracted attention with claims that the hemisphere was warming again, but it was not until the early 1980s that this became generally accepted.

At the same time as climatologists were updating the data sets that eventually revealed the warming of the 1970s, scientists concerned with the physics and chemistry of the atmosphere were re-evaluating a claim originally made in the nineteenth century: that the increase in atmospheric carbon dioxide caused by the burning of fossil fuels (and, it has been realized relatively recently, deforestation) could lead to global warming. These warnings, although repeated during the 1950s, had been largely ignored. This was, in part, due to the attention of atmospheric scientists being focussed on short-term weather forecasting, particularly during the early twentieth century. Later the hemispheric cooling appeared to contradict the prediction of warming.

During the 1970s sophisticated computer models simulating the effects of increasing carbon dioxide on the Earth's climate indicated that the so-called carbon dioxide issue was a cause for concern from the theoretical point of view. During the early 1980s the warming seen in the final years of the record of hemispheric temperature (Fig. 6) suggested that the first signs of carbon dioxide effects may be appearing. Heightened awareness of the potential significance of society's pollution of the atmosphere explains the shift towards predictions of warming.

It is now accepted among the climatological community that rising atmospheric carbon dioxide levels may well produce a significant climatic change in the near future (Kellogg and Schware, 1981). Atmospheric carbon dioxide, by allowing through incoming solar energy but trapping outgoing terrestrial radiation, produces surface heating: the 'greenhouse' effect. The best estimate is that global mean temperatures will rise by $2-3\,^{\circ}C$ as the greenhouse effect is enhanced by a doubling of atmospheric carbon dioxide levels. In some areas the temperature rise will be greater than this globally averaged estimate; in others less. This climatic shift could occur by the year 2050. A climatic change of this magnitude has probably not occurred since the recovery from the last major glaciation and will be accompanied by major changes in other aspects of the climate system, such as the atmospheric circulation. Predictive efforts based on analysis of past records alone can no longer be considered to provide the key to the future.

It is still necessary to consider natural causes of climatic change. The rapid warming of the early twentieth century occurred at a time when explosive volcanic activity appears to have been notably reduced. Explosive eruptions inject dust and gas into the upper atmosphere and this forms a particulate veil which can last for months to years. This veil, by affecting the transmission of solar energy through the atmosphere, causes surface cooling. The major eruption of the Mexican volcano El Chichon during early 1982 is believed to have triggered climatic cooling likely to last 2–3 years (Robock, 1983).

Semi-empirical models of global temperature change of the past 100 years suggest that the record is one of slow carbon dioxide-induced warming with relatively short, but intense, periods of volcanically induced cooling superimposed on the warming trend (Hansen *et al.,* 1981; see Fig. 6). Future climatic change may well depend on the interplay between natural factors, such as changes in the degree of volcanic activity, and anthropogenic effects, such as may be produced by carbon dioxide increases.

It is unlikely that understanding of the causes of climatic change will ever be sufficient to permit detailed climate forecasts to be made on the longer time scales of planning and design. But this does not mean that we can ignore the issue. The potential of climatic change should be incorporated in long-term planning: for example, by consideration of the total range of past variations evident in the longest climate records rather than reliance on records of limited duration. Given the rate of current climatic change, it is likely that this range will encompass the variations experienced over the coming 20–30 years. After this time the impact of increasing levels of carbon dioxide may well surpass the range of our experience of 'natural' climatic change. The evaluation of what will happen to climate-related variables and activities if the predictions of carbon dioxide-induced warming are correct is essential.

Acknowledgements

The research reported on in this paper was supported by the Rockefeller Foundation, the Natural Environment Research Council and Shell (UK) Ltd. The Climatic Research Unit's collection of wind data is funded by the Central Electricity Generating Board.

References

Barry, R. G. and Perry, A., 1973. *Synoptic Climatology, Methods and Applications.* Methuen, London.

Benwell, G. R. R., 1967. Secular changes in the frequency of meteorological conditions favourable for sea surges along the east coast of Britain. *Technical Note No. 21,* Branch 11, Meteorological Office, Bracknell, UK.

Dietrich, G., Wyrtki, K., Carruthers, J. N., Lawford, A. L. and Parmenter, H. C., 1952. *Wind Conditions over the Seas around Britain during the Period 1900–1949.* German Hydrographic Institute, Hamburg.

Gottschalk, M. K. E., 1971. *Stormvloeden en Rivierverstromingen in Nederland,* I: *De Periode Voor 1400.* Van Gorcum, Assen.

Gottschalk, M. K. E., 1975. *Stormvloeden en Rivierverstromingen in Nederland,* II: *De Periode 1400– 1600.* Van Gorcum, Assen.

Gottschalk, M. K. E., 1977. *Stormvloeden en Rivierverstromingen in Nederland,* III: *De Periode 1600– 1700.* Van Gorcum, Assen.

Gray, B. M., 1981. *The Statistical Analysis of Rainfall and its Reconstruction from Tree Rings.* Ph.D. Thesis, University of East Anglia.

Hansen, J., Johnson, D., Lacis, A., Lebedeff, S., Lee, P., Rind, D. and Russell, G., 1981. Climate impact of increasing atmospheric carbon dioxide. *Science,* N.Y., 213, 957–966.

Hunt, R. D., 1970. Secular changes in the frequency of meteorological conditions favourable for sea surges along the east coast of Britain — bringing an earlier study up to date. *Technical Note No. 31,* Branch 11, Meteorological Office, Bracknell, UK.

Jones, P. D. and Kelly, P. M., 1983. The spatial and temporal characteristics of Northern Hemisphere surface air temperature variations. *J. Climatol.,* 3, 243–252.

Jones, P. D., Wigley, T. M. L. and Kelly, P. M., 1982. Variations in surface air temperatures. Pt 1, Northern Hemisphere 1881–1981, *Monthly Weather Rev.,* 110, 59–70.

Kellogg, W. W. and Schware, R., 1981. *Climate Change and Society: Consequences of Increasing Atmospheric Carbon Dioxide.* Westview Press, Boulder, Colorado.

Kelly, P. M., 1975. *Climatic Change in the North Sea Region.* Ph.D. Thesis, University of East Anglia.

Kelly, P. M., Jones, P. D., Sear, C. B., Cherry, B. S. G. and Tavakol, R. K., 1982. Variations in surface air temperatures. Pt 2, Arctic Regions, 1881–1980, *Monthly Weather Rev,* 110, 71–83.

Kington, J. A., 1980. Daily weather mapping from 1781. *Clim. Chnge,* 3, 7–36.

Lamb, H. H., 1977. *Climate: Present, Past and Future,* Vol. 2. Methuen, London.

Lamb, H. H., 1981a. An approach to the study of the development of climate and its impact in human affairs. In: T. M. L. Wigley, M. J. Ingram and G. Farmer, (eds.), *Climate and History.* Cambridge University Press, Cambridge, pp. 291–309.

Lamb, H. H. 1981b. Climatic fluctuations in historical times and their connexion with transgressions of the sea, storm floods and other coastal changes. In: A. Verhulst and M. K. E. Gottschalk, (eds.), *Transgressies en occupatiegeschiednis in de kustgebieden van Nederland en Belgie,* Rijksuniversiteit, Ghent, pp. 251–290.

Lamb, H. H. and Weiss, I., 1979. On recent changes of the wind and wave regime of the North Sea and the outlook. *Fachliche Mitteilungen,* 194.

Manley, G., 1974. Central England temperatures: monthly means 1659 to 1973. *Quart. J. Roy. Meteorol. Soc.,* 100, 389–405.

Mitchell, J. M., Dzerdzeevzkii, B., Flöhn, H., Hofmeyr, W. L., Lamb, H. H., Rao, K. N., and Wallen, C. C. 1966. Climatic change. Technical Note No. 79, World Meteorological Organization, Geneva.

National Defense University, 1978. *Climatic Change to the Year 2000.* **National** Defense University, Washington, DC.

Pittock, A. B., 1978. A critical look at long-term sun-weather relationships. *Rev. Geophys. Space Phys.* **16,** 400–420.

Robock, A., 1983. The dust cloud of the century. *Nature, Lond.,* **301,** 373–374.

Schott, F., 1970. *Montly Mean Winds over Sea Areas around Britain during 1950–1967.* ICES, Copenhagen.

Smith, S. G., 1982. An index of windiness for the United Kingdom. *Meteorol. Mag.,* **111,** 232–247.

Steenstrup, J., 1907. Danmarks tab til havet i den historiske tid. *Hist. Tidsskr.,* **8,** 153–166.

Weiss, I. and Lamb, H. H., 1970. On the problem of high waves in the North Sea and neighbouring waters and the possible future trend of the atmospheric circulation. *Fachliche Mitteilungen,* **160,** 6–13.

Wigley, T. M. L. Ingram, M. J. and Farmer, G. (eds.), 1981. *Climate and History, Studies in Past Climates and Their Impact on Man.* Cambridge University Press, Cambridge.

Part IV

Sediment – Structure Interaction

Section 7 — Influence of Wave/Current and Fluid/Sediment Types

22

Interaction Between Pipelines and the Seabed Under the Influence of Waves and Currents

E. W. Bijker and W. Leeuwestein
Delft University of Technology, The Netherlands

The prediction of the behaviour of submarine pipelines on an erodable seabed is made based upon a series of small scale flume tests with currents and waves. In these tests the influence of the basic parameters, such as pipe diameter, approach velocity and height of the pipe relative to the original seabed, have been studied. Results of measurements of velocities and bottom changes have proven to be enlightening, both for the understanding of the mutual interactions between the disturbed flow field around the pipe and the seabed morphology as well as to make a first prediction of scour depths near prototype pipelines by using the tests as a scale series of the prototype.

Introduction

During recent decades offshore pipelines have been widely used as a means of transport for oil and gas, and safety, for which the stability of the pipeline is essential, is receiving increasing attention. Three situations can be distinguished: (1) pipelines crossing areas where ships might anchor in cases of emergency; (2) pipelines crossing fishing areas; (3) pipelines in areas where no interference from human sources is to be expected.

In situation (1) either the pipeline must be buried so deep that anchors cannot reach it or there must be a cover layer which gives perfect protection against anchors. It is doubtful whether this latter goal can ever be achieved economically, in view of the large mass of the anchors of large vessels. In cases where the pipelines must be buried, two techniques can be used: (a) burying by jetting or fluidization; (b) dredging a trench and covering the pipeline after it has been placed in the trench.

When the pipeline is introduced into the seabed by a fluidization process, it is not necessary to supply sand around the pipeline after the operation: the pipeline merely sinks in the original seabed material. The packing of this material has been dramatically changed, however.

When sand has to be supplied after a jetting operation, or when the pipeline is laid in a predredged trench, the packing of the sand is also a point of concern. When an unacceptable chance of too-loose packing and subsequent liquefaction exists, the refilled trench has to be covered with coarse and heavy material. The design of this protective layer as a function of waves, currents and material characteristics is the subject of separate research.

This paper focuses on pipelines which can remain uncovered, or over which only a small sand cover is required. In this case it is worth while to study the possibility of laying the pipeline on the seabed and the behaviour of the pipeline under those circumstances.

Scour around pipelines

The development of local scour holes underneath pipelines on an erodable seabed is governed by a number of empirically defined independent parameters. General parameters, which describe the boundary conditions for the local scour process, are depth averaged flow velocity, v; pipe diameter, ϕ, height, d_0, of the pipe above the sea bottom; water depth, h; and grain diameter, D. Additional parameters under wave action are wave height, H, and wave period, T. From these primary parameters other parameters can be derived, such as bottom roughness, orbital velocities, u_b (undisturbed by the pipe) and u_p (underneath the pipe), and ultimate sediment transport

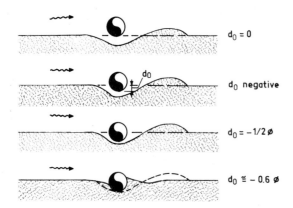

$d_0 = 0$

d_0 negative

$d_0 = -1/2 \, \phi$

$d_0 \leqq -0.6 \, \phi$

Fig. 1. Scour underneath a pipeline at different heights above the seabed due to unidirectional current.

capacities, S. In general, pipeline exposure will decrease significantly when the relative height, d_0/ϕ, of the pipe above the seabed becomes negative in the case of a partially buried pipeline (see Fig. 1). Scour underneath a pipeline will decrease and finally stop when d_0/ϕ reaches values of -0.5 to -0.7. The scour underneath the pipe reaches a maximum if a pipeline lies on the seabed and $d_0 = 0$ (Leeuwestein, 1983).

Scour around Pipelines due to Current Action

Pipelines which are laid on the seabed will, in general, be exposed to the lower part of the current profile. The pipeline thus protrudes a certain height into this profile and determines the flow pattern around it, resulting in the erosion and sedimentation of bottom material.

The fundamental cause of erosion around a pipeline is a local increase in transport capacity of the water passing the pipeline, while sedimentation takes place where this capacity decreases. These gradients in transport capacities exist only temporarily until the bottom has changed into its equilibrium configuration according to the instantaneous flow conditions. The transport capacity will be denoted by S, and subscripts o and p will be generally used to signify locations beyond the influence of the pipe and underneath the pipe, respectively. Although bed transport and suspended transport certainly should be treated separately, unless stated explicitly total transport will be used in this paper for simplicity. When an equilibrium situation has been established, both the depth of the scour hole, k_e, and the gap, e, between the pipe and the bottom of the scour hole remain constant. The transport capacities satisfy the equality

$$S_0 = S_p \tag{1}$$

When a scour hole is developing underneath a pipe, there is no balance of transport capacities:

$$S_p > S_o \tag{2}$$

A more detailed description of the erosion phase distinguishes two cases: (1) no transport upstream of the pipe ($S_{ou} = 0$), and transport only underneath and possibly just downstream of the pipe ($S_p, S_{od} \neq 0$); (2) transport at all locations ($S_p > S_o > 0$). In the second case there is a supply of sediment to the developing scour hole, with a possibility of exchange of sediment between the suspended and the bottom layer. Prediction of the rate of erosion and the ultimate scour depth on the basis of a sediment balance now seems to be possible, given a particular flow pattern around the pipe.

Up to now this fundamental approach to the computation of erosion underneath a pipe appeared to give rather unreliable results. The main reasons for this approach to be unsatisfactory are mainly:

(a) The transport formulae relating flow conditions to transport capacities usually have been developed and verified under uniform flow conditions. In contrast, the flow around a pipeline above the seabed or spanning a trench will be highly non-uniform. Relations between bottom shear stress and sediment transport, which form the basis of many sediment transport formulae, become invalid.

(b) Apart from the inapplicability to highly non-uniform flow conditions, the present transport formulae have only rather limited accuracy, even for uniform flow conditions.

(c) The flow around the pipe, for this particular purpose, cannot be characterized without the presence of turbulent vortices, especially downstream from the pipe, being taken into account. Incorporation of these turbulent influences in a flow transport computation will be necessary. Although numerical methods exist to compute this turbulent flow pattern around the pipe, an extension of such a method for computing (iteratively!) an equilibrium bottom profile seems very expensive. Three basic forms of erosion, being the direct consequences of this flow pattern, will be treated below.

The three types of erosion

There are three main locations of erosion or sedimentation on an erodable seabed around a current-exposed pipeline. Figure 2 illustrates the different phenomena on a seabed near an exposed pipeline. Luff erosion and sedimentation result from eddy formation upstream from the pipe due to the stagnation pressure and exchange of impulse from the main flow over the pipe to the eddy. Luff erosion is therefore likely to increase when the height of pipe exposure, d_e, increases and may thus initiate tunnel erosion, which will be treated below.

Lee erosion and sedimentation are caused by the turbulent wake downstream of the pipe, as illustrated

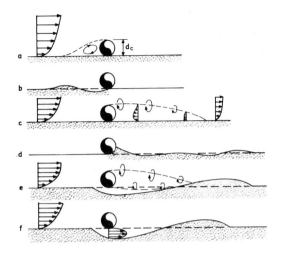

Fig. 2. Scour around current-exposed pipelines.

in Fig. 2(d, e), and by the re-emergence of the main flow several pipe diameters downstream of the pipe.

When a gap has been formed underneath the pipe and the flow under the pipe is sufficiently strong, vortex trails appear on both the top and bottom of the pipe, as illustrated in Fig. 2. These vortices periodically cause relatively high values of the bottom shear stress downstream from the pipe and thus significant erosion.

Tunnel erosion underneath the pipe is a direct consequence of the increasing velocities underneath the pipe relative to the velocities at a comparable height above the undisturbed seabed upstream (Fig. 2f). Within the relatively short distance of the length of the scour hole in the direct vicinity of the pipe, the boundary layer over the seabed cannot adapt to the increasing velocities underneath the pipe (van Ast and de Boer, 1973; Bijker, 1976). The result is an even higher bottom shear stress underneath the pipe than would be predicted from the increased current alone and a resultant greater transport capacity:

$$S_p > S_o$$

Erosion will develop until the flow around the pipe has changed sufficiently to restore the condition $S_p = S_o$. This is again the general criterion for a scour hole to be in its equilibrium phase when exposed to current action.

Especially when lee erosion is important, the erosion can be of very long duration, because the intensity of the vortices remains fairly constant over time. Only when the point of contact between these vortices and the bottom moves away from the pipe does dissipation weaken the vortices sufficiently to affect their erosion capacity. The major scour depth develops approximately as a logarithmic function of time.

Scour around Pipelines due to Wave Action

Pipelines on the seabed may also be exposed to the orbital velocities generated by surface waves. The time of exposure to these orbital velocities is relatively short compared with the duration of current action. The periodically changing direction of the orbital velocities requires an essentially different approach to the erosion process from that presented for pipelines exposed to currents.

When the flow around the pipe is oscillating, three main aspects of the erosion process depend on the periodicity: (1) friction along the pipe and seabed; (2) velocities above the seabed and underneath the pipe; (3) sediment transport and sediment transport distance. These three physical parameters successively depend on one another and are finally responsible for the establishment of an equilibrium scour hole with an equilibrium scour depth, as follows.

(a) Friction is introduced at the pipe and on the seabed, and by turbulent exchange of momentum its influence is transported through the boundary layers into the main flow. In an oscillating flow the boundary layers cannot fully develop and the penetration of friction influence into the flow is limited by the period of oscillation.

(b) Velocities above the seabed and through the scour hole underneath the pipe are therefore relatively less influenced by friction when the period of oscillation is smaller. Computational models for flow around a pipeline which is frictionless, as in the potential flow theory, have shown satisfying results for the computation of velocities underneath the pipe in an oscillating flow (Klein Breteler, 1982). The relative magnitude of these velocities is larger than those in a non-oscillating flow.

(c) Owing to the oscillating flow, the net sediment transports, S_o and S_p, will be zero, secondary influences such as non-sinusoidal waves and bed ripples being neglected. Model experiments with pipelines on a sand bottom, however, have shown that eddy formation near the pipe and neighbouring ripples can contribute to the local transport of sediment around the pipe in an oscillating flow (Klein Breteler, 1982). When bottom transport dominates, the transport, S_p, underneath the pipe is mainly determined by both velocities, u_p, underneath the pipe and the wave period, T, which together determine the local sand transport distance. These two parameters seem to be the most important and initially have been chosen as the subject of further study.

Scour mechanism

Experiments in a wave flume showed scour depths which were less than those measured under comparable current action. The scour depth appeared to increase with $\hat{u}_p$ —

dependent on wave height, H, and period, T. When bottom transport is considered, the following erosion mechanism will lead to an equilibrium scour depth. Beyond the pipeline's influence the orbital motion under a schematized sinusoidal wave can be characterized by a maximum orbital velocity $\hat{u}_b$ and an orbital excursion a, both just above the seabed. Underneath the pipe the orbital motion is amplified to a maximum velocity $\hat{u}_p$ and an orbital excursion a_p. Values for the amplification of velocities, $\hat{u}_p/\hat{u}_b$, range between 1 and 3, depending on pipe diameter, ϕ, wave parameters and the configuration of the pipe and scour hole (Klein Breteler, 1982; Leeuwestein, 1983). Owing to the orbital motion underneath the pipe, sediment is moving backward and forward along the bottom of the scour hole. The velocity amplitude decreases with increasing distance, x, from the centre of the pipe and beyond a certain distance the velocity amplitude, $\hat{u}(x)$, may no longer exceed a critical value. This critical velocity, u_{cr}, is a minimum velocity needed to cause sediment movement and depends on sediment parameters such as diameter, D, and relative density, Δ. If x_{cr} is the distance from the centre of the pipe determined by the condition

$$\hat{u}(x) = u_{cr} \qquad (3)$$

then, as long as the orbital motion underneath the pipe, characterized by a_p and $\hat{u}_p$, is so big that $a_p > x_{cr}$, erosion of bottom material from the scour hole can continue. Only when the scour depth has increased so far that $a_p \leqq x_{cr}$ and sediment is transported back towards the pipe instead of being thrown out of the scour hole, will the erosion stop. See Fig. 3.

In the case where, also beyond the pipe's influence, the bottom velocity, u_b, periodically exceeds u_{cr} (in

which case $x_{cr} \rightarrow \infty$), there will be a sediment transport (back) into the scour hole during a certain time within a wave period. Now the scour hole reaches its equilibrium when the sediment is swept out of the hole so far that a_p sufficiently exceeds x_u, the distance beyond which the orbital motion is undisturbed by the pipe (see Fig. 3).

Scour around Pipelines due to Both Current And Wave Action

The most general conditions for full-scale pipelines will include both current and wave action, each of which has its own angle of incidence relative to the pipeline direction. Experiments have been carried out at the Delft University of Technology which have shown a decreasing scour depth when waves were superimposed on a given constant current (van Ast and de Boer, 1973). Superposition of waves on an existing constant current with associated sediment transport increases the resultant time-averaged transport. A part of this sediment will settle in the scour hole and thus be responsible for the decrease of scour depth. According to these results, the scour depth under a combination of current and waves will lie between the upper limit valid for the current and a lower limit for the wave only (which is approximately 30 per cent of the upper limit). Experiments in progress must show whether the limits stated above still hold when resultant transport effects are negligible.

Prediction of scour depth underneath a pipeline

Since the prediction of scour development underneath a pipeline, which is based on hydraulics and computations of sediment transport capacities, requires at least some empirical data, a fully empirical approach remains, perhaps, attractive. Relations have been derived which describe the scour depth as a function of a number of independent parameters. Although these relations may be confirmed by a series of model scour measurements, often no physical understanding of the erosion process is provided. Application under circumstances which do not strictly reproduce the test condition should be done with great care.

Pipelines Exposed to Currents

For pipelines exposed to currents Kjeldsen (1974) gives an empirical function for the scour depth, k_e:

$$k_e = 0.972 \left(\frac{v^2}{2g}\right)^{1/5} \cdot \phi^{4/5} \qquad (4)$$

This function is valid for pipelines on a sand bed. Using results of scour model tests by van Ast and de Boer

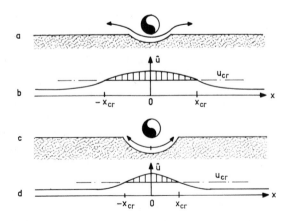

Fig. 3.   Scour underneath wave-exposed pipelines.

TABLE 1   Boundary conditions for scour tests with current-exposed pipelines

Name	Water depths (m)	Velocities ($m\,s^{-1}$)	Grain diameter (μm)	Pipe diameter (m)	Pipe above (+), on (0), below (−) bottom level
Kjeldsen (1974)	0.43, 1.43	0.20–0.52	74	0.06–0.05	0, --
van Ast and de Boer (1973)	0.21, 0.26	0.29–0.65	220	0.049, 0.088	+, 0, −
Jansen (1981)	0.36, 0.38	0.10–0.25	150	0.04, 0.05	+, 0, −
van Meerendonk and van Roermund (1981)	0.31	0.17–0.25	150	0.05	+, 0, −
Delft Hydraulics Laboratory (1982)	0.38, 0.40	0.28–0.58	90, 170	0.019, 0.075	+, 0, −

(1973), Kjeldsen (1974), Jansen (1981), van Meerendonk and van Roermund (1981) and Proot (1982) gives an empirical function similar to Kjeldsen's:

$$k_e = 0.929 \left(\frac{v^2}{2g}\right)^{0.26} \cdot \phi^{0.78} \cdot D^{-0.04} \qquad (5)$$

While these exponents for velocity head and pipe diameter do not differ signifantly, a relatively moderate influence of grain diameter, D, has been deduced. In equations (4) and (5) the pipeline is assumed to be fixed at the height of the original undisturbed (model) seabed. Both empirical functions suggest a scour depth increasing more sharply with pipe diameter, ϕ, than with average velocity, v. To give an impression of the primary validity range of functions (4) and (5), Table 1 lists the most important boundary conditions for the scour around the pipelines exposed to the currents.

Pipelines Exposed to Waves

In contrast to the scour depth under current action, the scour depth under wave action was scarcely influenced by the pipe diameter. Under the test conditions only local bottom transport has been observed around the pipe. Since the orbital excursion, a_p, underneath the pipe is an important parameter for the scour development, an empirical function for the scour depth will include the amplitude of the orbital velocity, $\hat{u}_b$, at the undisturbed seabed and the wave period, T. These wave parameters are basic parameters for determining the value of a_p. As the criterion for the start of sediment transport plays a role, the grain diameter, D, will also have to be included in the above-mentioned function, which can thus be formulated as

$$k_e = F(\phi, \hat{u}_b, T, D) \qquad (6)$$

Specification of this scour function is one of the objects of current research at the Delft University of Technology.

In the case of significant suspension of sediment, eddy formation contributes to the sediment movement around the pipeline, so that a single function such as equation (6) will probably not be enough to compute the scour depth. The vertical exchange of sediment may no longer be negligible.

Present research at the Delft University of Technology

Model tests which have formed the basis for predicting the scour around pipelines show a rather limited range of boundary conditions. Laboratory facilities and the instrumentation used can sometimes restrict the experimental variables, such as grain diameters, velocities and wave heights. At the Delft University of Technology a research programme is being carried out to answer questions about pipeline stability. Besides the morphological aspect, the hydrodynamic and soil mechanical aspects of this stability are studied. The model tests to investigate scour around pipelines are carried out with current, waves and combinations of both current and waves in flumes of 0.50 m and 1 m depth. Grain diameters used a range from approximately $100\,\mu m$ to $700\,\mu m$, while both height of the pipe above and depth of burial beneath the undisturbed bottom were varied. Measurements concerned velocities and bottom configurations around the pipe.

The experiments with current-exposed pipelines and flow velocities up to $0.40\,m\,s^{-1}$ have made possible prediction of the potential maximum scour depth due to current action. Verification tests will be carried out in one of the large-scale facilities of the Delft Hydraulics Laboratory, and a computational model is being used with the same objective. No final conclusions have been drawn from the experiments, although they indicate that calculations with equations (4) and (5) give underestimating values for the potential maximum scour depths during storm conditions. The flume tests have been carried out as a scale series, the principle of which will be treated below.

Any prototype pipeline with diameter ϕ exposed to a current velocity v can be scaled in a physical model with scale factors n_ϕ and n_v. Applying several values

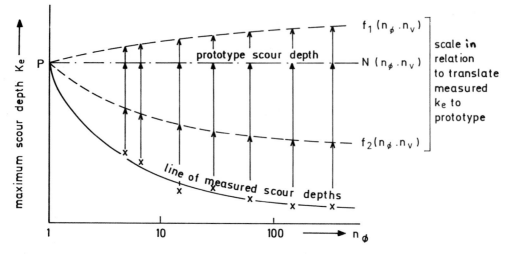

Fig. 4. Principle of a scale series.

for the scale factors for ϕ and v (the relevant independent parameters for the scour depth k_e), an extrapolation procedure leads to the prototype scour depth when both n_ϕ and n_v equal 1. Effectively, when both n_ϕ and n_v equal 1, also the scale factor n_{k_e} for the scour depth equals 1.

Functions such as equations (4) and (5) determine the scale relation $n_{k_e}(n_\phi, n_v)$, which should be used to translate any model test result to the prototype. By relating n_ϕ and n_v so that both equal 1 simultaneously, a basis can be obtained for direct extrapolation from the test results to the prototype (see Fig. 4). In the idealized case the correct scale relation, $n_{k_e} = N(n_\phi, n_v)$, is known and through this relation all measured scour depths (the full line in Fig. 4) converge to one and the same value for the prototype scour depth (given by the horizontal line in Fig. 4). Each assumed but incorrect relation $n_{k_e} = f(n_\phi, n_v)$, which besides must be uniquely defined for all ϕ and v, will lead to k_{ep} values given by non-horizontal lines (the dashed lines in Fig. 4). These lines, however, will intersect the vertical axis in the same point, P, as the horizontal (ideal) line, thus all pointing to the same prototype scour depth value. Unfortunately, no uniquely defined relations $f(n_\phi, n_v)$ exist, as the exponents in equations (4) and (5) show considerable variations with ϕ and v. The effect of this in Fig. 4 is that, using different lines with constant velocity, showing measured model scour depths, one gets different points P and so different prototype predictions.

This dependency on the choice of model velocity is unacceptable, so a somewhat different approach is used, taking into account all four lines of constant velocity (see Fig. 5); and besides, independent of any crude assumption for the scale relations $n_{k_e} = f(n_\phi, n_v)$, starting with four velocity scale factors, n_v, determined

by the four lines of Fig. 5, four pipe diameter scale factors, n_ϕ, are found through

$$n_\phi = n_v^p \tag{7}$$

Independent of the value of p, the condition $n_\phi(n_v = 1) = 1$ is satisfied, so that $n_{k_e} = 1$ for $n_\phi = n_v = 1$.

The value of p should now be chosen constant for a certain prototype prediction, which results through equation (7) in four values for n_ϕ and related values for ϕ. In fact, the value of p is in practice determined by the largest pipe diameter, ϕ_{max}, in the test with the highest velocity (the line with n_{v4} in Fig. 5). For that situation both n_ϕ and n_v are known, so by use of equation (7) the value p can be calculated.

Now four pairs of (n_v, n_ϕ) or (n_v, ϕ) are available, each with a value for k_e determined by the value for ϕ on the horizontal axis and for n_v along one of the lines of Fig. 5. Finally, the plot of ϕ (or n_ϕ) against k_e enables extrapolation to ϕ_p (or $n_\phi = 1$) now based on all four lines with constant velocity of Fig. 5.

Of particular relevance to the prototype pipeline are the time dependence of the scour process, pipe sagging and the three-dimensional effects of scour at the pipeline supports. Changes in span length will, together with the pipeline stiffness, finally determine the self-lowering behaviour of the prototype pipeline. These additional factors, too, are part of the present study of pipeline stability at the Delft University.

The same approach as has been discussed above for current-exposed pipelines is applied for wave-exposed pipelines, with n_T, the scale factor for the wave period, as an additional parameter. Ultimately, the final reports of the present research will be available as a basis for Dutch governmental pipeline concession policy towards the end of 1984.

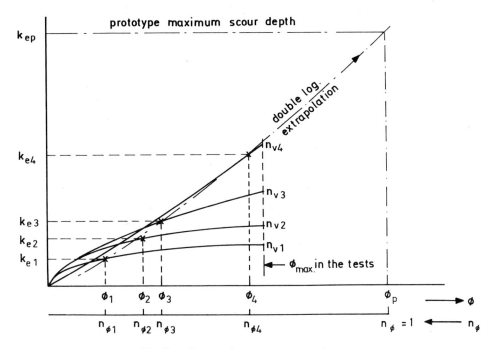

Fig. 5.   Extrapolation through n_v and n_ϕ (principle).

Conclusions and recommendations

The development of a scour hole underneath a pipeline is a very complicated process. Both erosion and sedimentation occur simultaneously as a result of the highly non-uniform and turbulent flow around the pipeline. A computational scour model based on theory should include turbulence and sediment concentration computations and will therefore be very complicated and time consuming. Obviously, model tests will be needed not only to improve the empirical scour functions, but also to gather the necessary physical data for it to be possible to apply model results to prototype pipelines. In order to account for scale effects, prototype or very large-scale model tests seem inevitable.

For a more fundamental approach, reliable relations must be found between non-uniform flow parameters and the sediment transport capacity. This approach requires detailed investigations of the turbulent boundary layer and the resulting shear stress as an intermediate step.

Symbols

a_p	$[L]$	orbital excursion underneath the pipe
d_0	$[L]$	height of pipe above seabed
d_e	$[L]$	height of pipe exposure above seabed
D	$[L]$	grain diameter
e	$[L]$	gap between pipe and bottom of scour hole
g	$[LT^{-2}]$	acceleration of gravity
h	$[L]$	water depth
k_e	$[L]$	depth of scour hole
S_o	$[L^3 T^{-1}]$	undisturbed sediment transport capacity
S_p	$[L^3 T^{-1}]$	sediment transport capacity underneath pipe
T	$[T]$	wave period
u_b	$[LT^{-1}]$	horizontal orbital velocity near bottom
u_{cr}	$[LT^{-1}]$	horizontal orbital velocity at initiation of sediment transport
u_p	$[LT^{-1}]$	horizontal orbital velocity underneath pipe
v	$[LT^{-1}]$	depth-averaged flow velocity
x	$[L]$	horizontal coordinate perpendicular to pipe
Δ	$[-]$	relative density of sediment
ϕ	$[L]$	pipe diameter

References

Bijker, E. W., 1976. Wave–seabed–structure interaction. *Proc. BOSS Conf.* pp. 830–844.

Delft Hydraulics Laboratory, 1982. Behaviour of the L10A/F pipelines. Publ. No. M1818.

Jansen, E. F. P., 1981. *Scour Underneath a Pipe.* [in Dutch]. Delft University of Technology, Coastal Engineering Group.

Kjeldsen, S. P., (1974). *Experiments with Local Scour around Submarine Pipelines in a Uniform Current.* Trondheim.

Klein Breteler, M., 1982. *Scour Underneath a Pipeline Due to Waves.* Delft University of Technology, Coastal Engineering Group [in Dutch].

Leeuwestein, W., 1983. *MaTS-pipelines, Scour, Vol. I.* Delft University of Technology, Coastal Engineering Group [in Dutch].

van Ast, W. and de Boer, P. L., 1973. *Scour Underneath a Pipeline due to Current and/or Waves.* Delft University of Technology, Coastal Engineering Group [in Dutch].

van Meerendonk, E. and van Roermund, A. J. G. M., 1981. *Scour Underneath Pipelines Due to Current.* Delft University of Technology, Coastal Engineering Group [in Dutch].

Proot, M. A., 1983. *Scour Underneath a Pipeline Due to Uniform Current.* Delft University of Technology, Coastal Engineering Group [in Dutch].

23

Local Erosion at Vertical Piles by Waves and Currents

A. Clark and P. Novak
University of Newcastle upon Tyne, UK

Although the study of local scour at cylindrical piers has been the subject of many studies, few resulted in clear design recommendations; furthermore, the results of these studies exhibit often wide differences pending on the type and range of parameters considered. There are very few studies available dealing with erosion under wave action. The paper analyses briefly the mechanics of local scour in the case of unidirectional flow both for velocities below and above critical velocities for sediment transport; it deals then with the mechanics of scour due to action of oscillatory flow and oscillatory flow superimposed on a unidirectional current. This is followed by a brief dimensional analysis of the problem. The major part of the paper is devoted to the description and analysis of experiments carried out in two flumes with different pile diameters, varying velocities and flow depths, different sediment sizes and wave amplitudes and lengths. The experimental results are presented in design graphs enabling a quick estimate of the scour depth as a function of flow depth H, pile diameter D and ratio of critical to flow velocity v_c/v for unidirectional flow. The added influence of waves can be estimated (within the experimental range) from graphs incorporating the influence of wave length and amplitude. The study results show that the maximum scour is in the region of 1.7D for $v_c/v \leqq 1$ with smaller values for $v_c/v > 1$. In the range $1 < (v_c/v) < 2$ the depth exerts little influence for values $D/H > 0.6$ and $D/H < 0.15$ with some influence in the intermediate range. If waves are superimposed on currents; the rate of scour formation increases substantially with the maximum scour increasing with wave amplitudes and wave lengths, but not exceeding the 1.7D value found for unidirectional flow alone.

Introduction

The study of local scouring of an erodible bed around flow obstructions has been undertaken by many researchers over several decades and some contributions have been made quite recently (Baker, 1981; Nakawa *et al.*, 1982; Suzuki, 1982). Substantial progress has been made, especially in the area of unidirectional flow, but many aspects of the problem remain unclear. To elucidate the process of local scour under unidirectional and oscillatory flow and further in order to achieve results which could serve as a basis for design, a series of experiments was conducted by the authors in the University of Newcastle upon Tyne. This work was carried out with the help of a Science and Engineering Research Council grant from the Marine Technology Directorate as part of a co-ordinated research effort in that area.

Before experimental work was undertaken on local scouring around obstructions subjected to simultaneously acting current/wave flow fields, the separate cases of unidirectional and oscillatory flow alone had to be investigated.

Literature survey

Previous publications exhibit a large degree of incompatibility between various experimental results, and between experiment and theory. The effect of water depth on scour is an example of one such controversial parameter.

For the case of continuous sediment transport in the approach flow most researchers claim that below a limiting depth of flow scour depth at circular vertical piles decreases with decreasing water depth; above this limiting depth scour depth would seem to become independent of flow depth. This limiting depth is usually quoted between one and three times the diameter of the obstruction. On the other hand, Larsen and others, influenced by the regimen theory, claim that for a constant approach velocity the scour depth increases with increasing water depth.

For clear water scour — i.e. scour without sediment transport in the approach flow — an increase in flow depth will increase the critical velocity, v_c; thus, the depth of scour decreases as the depth of flow increases. Even for this case, however, there is a limiting water depth, below which the depth of scour will decrease with decreasing water depth.

A detailed literature survey was undertaken, including work in the area of oscillatory flow scour (where very little published work is available), in order to determine general trends. A summary of this survey, documenting the widely varying results, forms an appendix of a previous publication of the authors (Clark *et al.*, 1982).

Theoretical background

Local scour is the localized change of level of an erodible bed near to an obstruction placed in what was originally an equilibrium situation.

In the case of *unidirectional flow* past a vertical obstruction in the form of a cylindrical pier, the scour will begin at two points located at approximately 45° to the direction of flow on the upstream side of the obstruction. Assuming adequate flow depth and velocity, further scour development results in two conical holes which meet at a common point upstream of the cylinder centre and have side slopes equal to the angle of repose of the saturated sediment. This period of erosion is very rapid and the upper edges of the two scour holes soon merge.

The next development, given sufficient strength in the flow field, is the formation of a horseshoe vortex in front of the obstruction. This vortex is the main cause of the inverted conical scour hole which develops in front of the pier. This scour hole will continue to grow until the vortex expands to a size at which it can no longer lift the bed material out of the depression. If the flow velocity is such that the entire bed is in motion ($v > v_c$), this type of scour will still occur as the vortex removes sediment from the scour hole it has initially formed. Irrespective of whether the bed is mobile or not, the 'final' scour depth will develop only after a certain time. The rate of scour depends on both geometrical and hydraulic parameters.

In the case of *oscillatory flow,* the scour is caused again by the increased bottom shear stress but the contribution of the horseshoe vortex is minimal, as the distance of fluid particle travel is usually not large enough for a sufficient boundary layer height — an essential prerequisite for a strong horseshoe vortex — to develop. In this case the scour is mainly caused by the acceleration of the flow past the obstruction and the flow pattern in the wake of the obstacle (Niedoroda and Dalton, 1982). It is obvious that in the case of oscillatory flow alone this process is symmetrical and occurs on both the 'upstream' and 'downstream' side of the pile.

In the case of *oscillatory flow superimposed on a unidirectional current,* the flow pattern is even more complicated. In the case of wave motion and the current acting in the same direction, the resulting horseshoe vortex is enhanced, the opposite being the case if the wave motion opposes the current. Depending on the relative strengths of the orbital velocity due to the oscillatory flow and the unidirectional flow velocity and on the period of wave motion, different scour patterns and magnitudes may arise and the combined wave and current climate usually — but not always — is likely to cause a bigger scour than would result from waves or currents alone, particularly if acting in the same direction.

Large obstructions (i.e. obstructions with a large width to flow depth ratio) will tend to inhibit the development of the horshoe vortex system and a totally different pattern of scour will result. Indeed aggregation may be the more serious problem. For this reason most but not all of our experiments have concentrated on circular piers which are 'slender', meaning that the ratio between the diameter of the obstructions and mean flow depth (D/H) is less than 0.5. This category includes such typical obstructions as piles, risers and platform legs.

Dimensional analysis

Even in the 'simple' case of local scour around slender obstructions in unidirectional flows, the number of parameters involved is fairly large and their relationships can be complex. An expansion of this work into the area of wave and currents combined introduces even more variables. In our work the following parameters were considered to be significant: D, H, v, a, L, y, g, d_s, d_{90}, α, μ, ρ, ρ_s, θ, ψ. Using the Buckingham π theorem with 15 variables and 3 dimensions this results in 12 π terms. H, v and μ were taken as the repeating variables; formulating and solving the resulting equations gives the following dimensionless groups:

$$y/D, \ H/D, \ L/D, \ a/D, \ v^2/gD, \ d_s/D, \ d_{90}/D,$$

$$\alpha, \ \theta, \ \psi, \ \frac{\rho D v}{\mu}, \ \frac{\rho_s D v}{\mu}$$

These can be rearranged and combined to give

$$y/D = f\left[D/H, \frac{v}{(gH)^{1/2}}, \frac{\rho vH}{\mu}, \frac{d_{90}}{d_s}, \alpha, \theta, \psi, \frac{a}{D}, \frac{L}{D},\right.$$

$$\left. \times \frac{u}{\{gd_s(S_s - 1)^{1/2}\}}\right] \qquad (1)$$

(where $S_s = \rho_s/\rho$); equation (1) formed the basis of the experimental work.

Apparatus

Experiments were conducted in the hydraulics laboratories of the Department of Civil Engineering. A 460 mm wide by 400 mm deep channel, 12 m in length, was used as the primary installation. This flume was fitted with a roughened bed for 7 m of its length and also with a sand reach of 2.5 m, allowing for a full development of the velocity profile.

The initial wave experiments were carried out in this flume and for this purpose a 'rolling seal' paddle wave generator was installed at the inlet end of the flume. This paddle was driven by a variable-speed electric motor with the drive connected to a variable cam-shaft.

At the outlet end of the flume a horizontal V-type beach was installed with packed expanded metal. The piers were located in the sand bed by means of a locking system. Flow conditions could therefore be stabilized before the introduction of the obstruction to the equilibrium field.

Measurement of scour both during and after the experiments was taken with a bed profile meter, which 'follows' the bed and causes little disturbance to the flow field around the obstruction. Wave properties were measured by using a twin-wire capacitance probe.

Further experimental work was undertaken in a substantially deeper (2000 mm) and wider (1000 mm) wave flume, specially constructed for this and other projects. Three rolling seal wave generators were installed to give a random wave capability. An expanded matting material was used as the beach.

Experimental range

Three pier diameters were used for the initial experiment: 25, 48 and 75 mm. Further experiments were restricted to 48 and 25 mm piers only, owing to some side-effects caused by the 75 mm diameter obstruction in the 460 mm wide flume.

Velocities in both flumes ranged from 0.5 to 2.5 times the flow velocity required for initial bed movement of the sediment bed. A set of results was also obtained for a superimposed wave pattern. Waves ranged from 5 to 20 mm in total height, with associated wave lengths of 1000–3000 mm. A uniform flow

depth of 200 mm was used in all the wave experiments in the small flume.

Three different sediments were used: sand of 0.8 and 1.5 mm diameter and Bakelite with a mean diameter of 0.7 mm. Experiments were allowed to run for a maximum period of 6 h, in which time equilibrium scour was usually judged to have occurred. If the scouring was still active, as it would be with a large flow depth and/or a low flow velocity, then curve fitting was used to determine an extrapolated value for the maximum expected scour, the recorded development of scour with time being used. A more detailed description of the apparatus used and of the experimental range for unidirectional flow alone was given elsewhere (Clark *et al.*, 1982).

Analysis of results

Unidirectional Flow

The analysis of the results was primarily carried out on a GEC 4070 minicomputer situated in the Faculty of Engineering and jointly funded by the University and the Science and Engineering Research Council. This machine, because of its graphics capability, provides a rapid method of analysing results and their correlations.

Assuming in equation (1) the ratio d_{90}/d_s and α, θ and ψ to be constant, an initial set of graphs was generated using the following sets of dimensionless numbers for each sediment:

'Pier Reynolds No.'	$= \rho vD/\mu$
Flow Reynolds No.	$= \rho vH/\mu$
'Pier Froude No.'	$= v/(gD)^{1/2}$
Flow Froude No.	$= v/(gH)^{1/2}$
Scour Depth/Flow Depth Ratio	$= y/H$
Scour Depth/Pier Diameter Ratio	$= y/D$

Although various trends could be seen, no simple interdependency was immediately apparent. By using the three-dimensional graphics capability available on the Faculty computer installation, an extra parameter could also be displayed. Various surfaces were constructed and from these the most satisfactory seemed to include a Froude number.

After refinement and careful study of our experimental results, a final surface and contour plan could be drawn for unidirectional flow results showing on the y axis the relationship D/H and on the x axis the ratio between critical velocity (for sediment transport) and flow velocity. (The critical velocity, v_c, for incipient motion at each depth and for each sediment used was computed and partly verified by separate experiment.) The ratio between scour depth y and pile diameter, D, formed 'contours' on the graph (Fig. 1).

The graph in Fig. 1 resulted from an analysis of

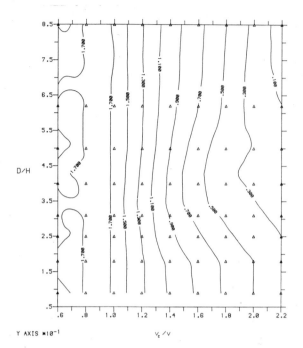

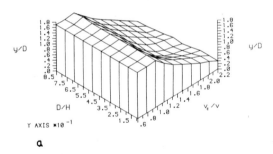

Fig. 1. Contours of the ratio between scour depth (*y*) and pile diameter (*D*) determined from experimental results.

a total of 117 experimental results, from which 29 were excluded on account of obvious experimental errors or because of different particle shape or because low velocities of flow did not allow even the attainment of a transition from smooth to hydraulically rough channel flow (some of the Bakelite experiments). These experiments, therefore, did not satisfy the necessary similarity criteria for local scour modelling; a more detailed analysis of these criteria is given elsewhere (Clark *et al.*, 1982). Figure 3 is an example of the plot of y/D against v_c/v for a given small range of D/H; the confidence limits placed on each experiment were 2–3 mm in the 'clear water' scour and 6–8 mm in the range $v_c/v < 1$. Figure 1 was constructed from a number of such 'sections' as shown in Fig. 3. The three-dimensional representation of Fig. 1 is given in Fig. 2(a–d); y/D is plotted vertically (*z* axis) with v_c/v and 10 D/H forming the horizontal plane (*x* × *y* axes). Figures 1 and 2 can be used directly in design for the prediction of the maximum scour at circular vertical piles in a unidirectional current for different ratios of v_c/v and D/H.

Waves and Currents

For the experimental conditions available no scour was obtained with waves only. Because of the characteristics of the wave generators, the best wave profiles were formed at a flow depth of 200 mm, which was

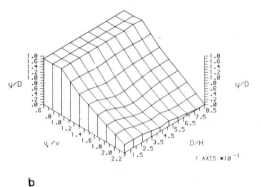

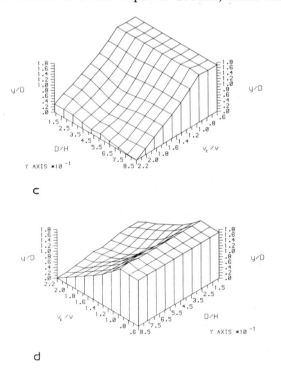

Fig. 2. (a–d). Three-dimensional representation of the ratio *y/D*.

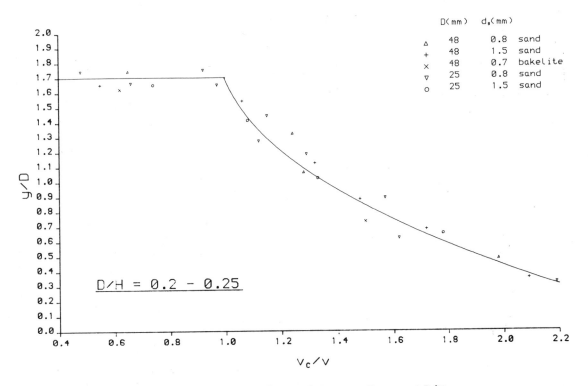

Fig. 3. Plot of y/D vs v_c/v for a small range of D/H.

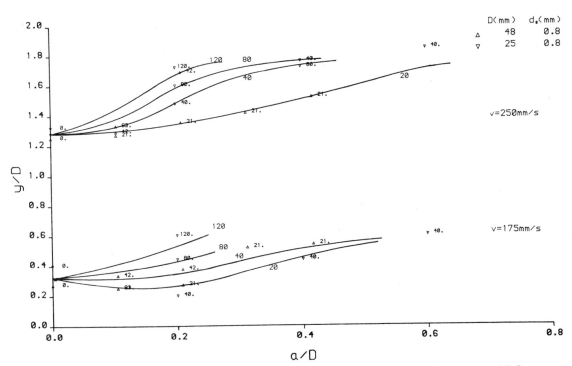

Fig. 4. (a, b). Experimental results showing plots of y/D vs. a/D for a sand grain size of 0.8 mm, and two piles of differing diameter.

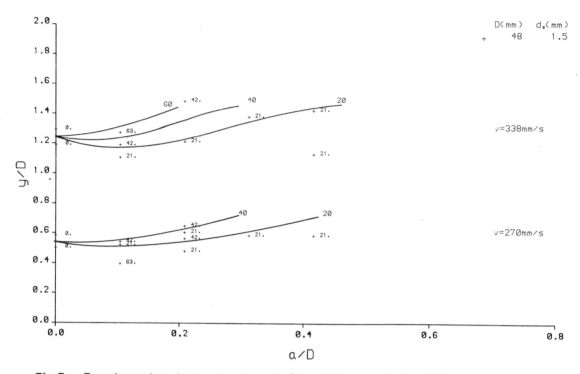

Fig. 5.   Experimental results showing plots of *y/D* vs *a/D* for a grain size of 1.5 mm and a single pile.

taken as the basis for studying the combined effect of waves and currents. Figures 4 and 5 show the result of these experiments in a non-dimensional form by plotting the ratio between scour depth, y, and pile diameter, D, against the ration between wave height a, and pile diameter, D, with ratio between wave length, L, and pile diameter, D, as the third parameter (figures against experimental points and contour lines). Figure 4 refers to experiments with sand 0.8 mm diameter and two piles (25 and 48 mm diameter) and Fig. 5 to results with sand 1.5 mm diameter and one pile diameter (48 mm); in both cases two different velocities of approach, v, were tested. Although necessarily there is some scatter in the results and although more experimental evidence would be desirable, the trend in the results is consistent and clear; taking Fig. 1 or 2 as the basis for scour without waves, the additional effect of superimposed waves on maximum scour can be judged from Figs. 4 and 5.

In principle it can be seen that for ratios of wave amplitude over pile diameter $a/D > 0.2$ the scour increases over the unidirectional flow scour; furthermore the scour increases to a smaller extent with an increase in wave length. There is not really sufficient experimental evidence to show that the rate of increase of scour with wave superimposition is independent of the unidirectional flow velocity, but in any case the effect of flow velocity on this rate of increase does not seem to be large and the maximum scour attained with wave

action did not exceed the maximum attainable scour, 1.7D, valid for unidirectional flow. It was observed, however, that the rate of scour development was substantially faster for currents and waves than for currents alone, even if the wave action was relatively very weak (small amplitudes).

It is relatively easy to compute for the given wave parameters the theoretical orbital velocity near the bed due to wave motion; adding this to the unidirectional current velocity near the bed, we can estimate for a logarithmic velocity flow profile a new mean velocity and compute from Fig. 1 the new increased scour depth, which can be compared with the experimental result. Alternatively, the depth for orbital velocity which, when added to the current velocity, would produce the given scour, could be estimated. Both approaches were tried, and although the first one looks promising more experimental results are required to prove its validity conclusively.

Conclusions

(1) From experiments with *unidirectional flow,* which confirmed the theoretical discussion, it may be concluded that:

(a) For values of $v_c/v \lesseqgtr 1$ the maximum scour is about 1.7 times the pile diameter and independent of flow depth;

(b) The influence of flow depth increases as the ratio v_c/v increases. In the range $1 < v_c/v < 2$, which covers all practically important cases of 'clear water' scour (i.e. no sediment transport in the approaching flow), the depth exerts no influence for values $D/H > 0.6$. For values $0.15 \lesssim D/H \lesssim 0.6$ a change in depth influences the value of the scour and for $D/H < 0.15$ again the scour remains practically constant (but, in general, at a different value than for $D/H > 0.6$).

(2) *If waves are superimposed on currents* and both act in the same direction, the final scour is the maximum possible under the prevailing conditions. From the experimental results it can be concluded that in these cases:

(a) The rate of scour formation is substantially faster than for the case of currents alone.

(b) An increase in the wave amplitude, a, over $a/D \cong 0.2$ and an increase of the wavelength, L, both increase the final scour over that corresponding to the current alone, with the effect of wave amplitude more pronounced than the effect of wavelength.

(c) The maximum scour does not exceed the value $1.7D$ applicable to the scour under the action of current alone for the ratios between critical and flow velocity $v_c/v \cong 1$.

List of symbols

a	wave amplitude
b	width of flume
D	pier diameter
d_s	mean diameter of sediment
d_{90}	90 percentile of grain size distribution
g	acceleration due to gravity
H	mean depth of flow
L	wavelength
S_s	specific gravity of sediment
v	mean cross-sectional velocity
v_c	threshold velocity for incipient bed motion
y	scour depth
α	angle of attack of current to waves
θ	angle of repose of saturated sediment
μ	dynamic viscosity coefficient
ρ	density
ρ_s	density of sediment
ψ	shape factor of sediment grains

Acknowledgements

The authors wish to acknowledge the support of the Marine Technology Directorate of the Science and Engineering Research Council in enabling this research to be undertaken. Interactive computing facilities available through the SERC were also utilized, and Mr M. J. Cope has helped greatly with the production of the figures on this facility.

References

Baker, C. J., 1981. New design equations for scour around bridge piers. Technical note. *J. Hydraul. Div., ASCE* (HY4) 507–11.

Clark, A., Novak, P. and Russell, K., 1982. Local scour at slender cylindrical piers: a review and experimental analysis. *Proc. Int. Symp. Engineering in Marine Environment*, Koninklijke Vlaamse Ingenieursveringing, Brugge, pp. 2.43–2.55.

Clark, A., Novak, P. and Russell, K., 1982. Modelling of local scour with particular reference to offshore structures. *Proc. Int. Conf. on Hydraulic Modelling of Civil Engineering Structures*, BHRA, Coventry, pp. 411–423.

Nakawa, H., Otsabo, K. and Vakagawa, M., 1982. Characteristics of local scour around bridge piers for nonuniform sediment. *Trans. Jpn. Soc. Civil Engrs*, **13**.

Niedoroda, A. W. and Dalton, C., 1982. A review of the fluid mechanics of ocean scour. *Ocean Engng.*, **9** (2), 159–190.

Suzuki, K., 1982. Study on the clear water scour around a cylindrical bridge pier. *Trans. Jpn. Soc. Civil Engrs*, **13**.

24

Grain Shape Effects on Sediment Transport with Reference to Armouring

B. B. Willetts

Department of Engineering, University of Aberdeen, UK

Evidence is drawn from the literature of unidirectional flow to show that grain shape affects transport behaviour down to grain size $100\,\mu$m and, in particular, that beds of grains of low sphericity adapt to flow in such a way that grain activity is diminished. Some evidence is offered that the adaptive process takes many minutes. Such 'armouring' is therefore relevant to tidal currents but not to currents generated by short waves.

The difficulties of shape description which have led to its neglect are asserted to require resolution. An agreed measure of shape is urgently needed so that shape data can begin to be accumulated. Rollability is suggested as an attractive basis for a parameter which might attain wide currency.

Introduction

Because the process mechanics of sediment transport contains persistent obscurities, it is common practice to resort to dimensional analysis in formulating arguments and presenting data. A typical statement concerning transport rate might read as follows:

$$\frac{q_s}{\rho v_*^3} = f\left\{\frac{\rho v_*^2}{(\gamma_s - \gamma)d}, \frac{\mu v_*}{\gamma_s d^2}, \frac{y}{d}, \frac{\gamma_s}{\gamma}, \sigma\right\} \qquad (1)$$

in which q_s = rate of sediment transport by weight per unit width; v_* = shear velocity; μ = fluid viscosity; ρ = fluid density; γ = fluid specific weight; γ_s = grain specific weight; d = equivalent diameter; y = stream depth; σ = a dimensionless shape parameter.

Faced with this daunting array of independent variables, investigators look eagerly for evidence that some of them have little influence and therefore can be discounted. The grain shape parameter, σ, is usually discarded early in this process. The justification for so discarding it seems to be twofold. First, there is the difficulty of assigning a value to a chosen shape parameter for any one irregular body. Second, the extension of the description to a population of grains has to be considered. Natural grain populations often contain diverse minerals and grain sizes. Both mineralogy and

grain size affect shape, so such a population is inevitably heterogeneous with regard to grain shape. This renders quite difficult the selection of a representative shape for the population as a whole. These problems appear to have prevented the assembly of a sufficient body of data about the effect of shape on transport processes to make possible an argued decision to discard it. Despite some evidence that in some circumstances shape has a pronounced influence on transport, it is nevertheless neglected because of difficulties of description.

One result of this neglect is that whereas study of many facets of seabed mechanics can draw heavily on data and ideas from earlier work on unidirectional stream flow, there is comparatively little such help available with regard to shape effects. It is the purpose of this paper to review the evidence that is available, with seabed circumstances in mind. It will examine three questions: (1) Is the influence of shape on transport behaviour significant? (2) If so, can it be expressed practically? (3) What is the relevance to seabed conditions?

Armouring

'Armouring' here is used to mean resorting of the active layer such that transport susceptibility diminishes.

The phenomenon is well known in river mechanics and has been the subject of recent studies which have attempted sub-classifications (Bray and Church 1980, 1982; Raudkivi and Ettema, 1982). The conception here is of a layer of grains immobile for long periods. Disruption, which occurs with a return period of the order of the interval between 'wet seasons', is followed by rapid reworking of the bed and re-formation of the armour. Most of the literature concerns beds in which the size range is very large (say 0.1–100 mm), having been established by a non-fluvial process (e.g. glaciation). Usually the armour is thought of as a single grain surface layer, but recently Borah *et al.* (1982) have written of such a layer underlying a population of more mobile grains which are active at more frequent intervals. (The tendency of heavy grains to collect at the foot of the slip face of bed-forms is said gradually to build an armour layer in the plane tangent to the bed feature troughs.)

Common seabed conditions involve less vigorous flows and therefore smaller particles. While the flow often has a periodicity which may involve reversal of direction (or large changes in direction), the periodic maxima have values which vary substantially only when circumstances are dominated by shallow-water wave orbits. There is therefore less scope for the creation, by an exceptional event, of a bed layer which is resistant to attack by subsequent flows over a long period. Peaks of tidal flow may produce a bed condition which is impervious to disturbance in the intervals between peaks, but of more general concern are the more subtle changes of sorting which produce diminished, rather than zero, activity. The term 'winnowing' has been used to describe such sorting changes.

Seabed circumstances introduce many complications to the sediment transport problem. Among them are: flow unsteadiness; seepage flows produced by the wave pressure field; colloidal effects (including flocculation) on fine particles; ageing changes in cohesive effects. Progress in discussion in the present state of knowledge demands that some of these be evaded. Accordingly, this paper will consider grains down to a lower limit of size $d_{50} = 100 \mu$m. Above this size, surface chemistry ceases to influence packing when disturbed grains re-settle (Mantz, 1977). The demarcation will also be found later to have some significance when shape is discussed. The corresponding threshold value of shear velocity (taken from the Shields diagram) is approximately 12 mm/s and the boundary Reynolds number $v_*k/v \approx 0.05$, in which k = boundary roughness size and v = kinematic viscosity.

Shape effects

With regard to fluvial armouring, grain shape has been observed to have a pronounced effect on the dislodgement susceptibility of larger sizes ($\sim$ 50–100 mm). Lane and Carlson (1954) found that grains of high sphericity sometimes had susceptibility equal to that of disc-shaped (natural) grains of 40 per cent their weight. That is to say, weight for weight the flatter grains were found much more stable than compact grains. The enhanced stability was associated with 'working' of the surface grains by the flow such that grain posture was adjusted by vibration and jostling to the most stable arrangement for the surface layer.

The principal feature of this stable arrangement is the imbrication of grains. They interleave with dip in the upstream direction, as shown in Fig. 1. Rusnak (1957) observed imbrication in grains down to sizes less than 100 μm.

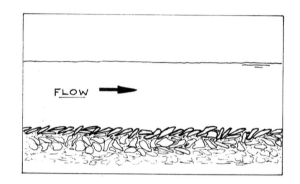

Fig. 1. Armouring by imbrication of plate-like grains in the surface layer.

Some results of Mantz (1980) in comparison with earlier ones of Taylor and Vanoni (1972) are illuminating. (The comparison was first made by Mantz.) He experimented with plastics discs (diameter 12.5 mm, thickness 0.25 mm, specific gravity 1.07) in two distinct circumstances. In one, the bed for subsequent transport experiments was formed by allowing discs to settle in still water; in the other, the discs were allowed to settle in a sub-threshold flow. In the latter case an imbricated bed was observed to form. The experiments of Vanoni and Brooks had been done with beds of natural grains; in some series as placed, in other series after having been 'armoured' by a sustained preliminary flow.

When plotted in terms of appropriate dimensionless groups, on the basis of nominal (volumetric) grain size, the results from beds of randomly placed discs and those from beds of unworked natural grains show remarkably good agreement, considering the great differences in bed material. However, the beds of imbricated discs and the 'armoured' beds of natural grains both give very different results from these, with much lower grain transport rates. This suggests that the influence of grain shape on transport susceptibility lies primarily in the ability of grains of low sphericity to

TABLE 1

	Original sand		Immediate (5 min) removals		15min–20 min removals	
	Size (d_{50})	Shape (sphericity)	Size (d_{50})	Shape (sphericity)	Size (d_{50})	Shape (sphericity)
Quartz sand	300 μm	0.73	290 μm	0.73	300 μm	0.75
Shell sand	270 μm	0.61	410 μm	0.57	330 μm	0.63

find attitudes of response in which they are difficult to dislodge.

Still in unidirectional flow, Meland and Norrman (1969) observed that, once dislodged, grains of low sphericity travelled faster than more compact grains. They drew attention to the sorting changes which this speed difference produces. The effect of shape on speed was greatest at low transport stages, but they obtained little data for grain sizes below 1000 μm.

The imbricated bed condition, which is an armoured condition, is effective for one flow direction only. Flow reversal, such as is common at the sea bed, exposes grains to flow in an orientation for which their resistance to disturbance is very much lower than the maximum found when flow direction corresponds to the imbrication pattern. Therefore flow reversal, with approximately equal current strength, may be expected to disturb (and perhaps transport) some grains and to build an imbrication pattern oriented to its new direction. Clearly the speed of this bed adjustment is relevant when flow reversals occur periodically. There is very little objective information about the sequence of events when sediment transport first begins. Table 1 summarizes the results of recent experiments in which quartz sand and a shell sand of much lower sphericity but of the same size distribution as obtained by sieving were compared in this respect (in unidirectional flow). In it are compared the size and shape of moving grains trapped during the first 5 min of exposure to a flow (grain transport stage approximately 1.2) and of those trapped between the fifteenth and twentieth minutes of the experiment. The shell sand can be seen to be making an adjustment of transport behaviour which is not complete by the end of 20 min. In contrast, the quartz sand appears to behave quite uniformly during the whole period. One might infer that beds mainly composed of particles of low sphericity are likely to armour during tidal cycles, but that the process is too slow to become effective in the time-scale of wave orbits. Indeed, for this reason, wave action is prone to act upon the bed in its most vulnerable condition.

The grain size below which shape ceases to be influential is difficult to determine. The well-known data collection of Rouse (1961) suggests that, for bodies fixed in a flow, shape effect disappears below Reynolds number $vd/v \approx 30$. More recent settling experiments (e.g. Komar and Reimers, 1978) suggest that drag is affected by shape down to Reynolds numbers less than 10 when grains are free to adopt a posture in response to the flow.

If the appropriate velocity for calculating Reynolds number is taken to be the velocity at one grain diameter up the velocity profile, then particle Reynolds number 10 corresponds to 100 μm grains at the threshold of motion. This gives a very rough indication of the lower limit of grain size below which threshold behaviour would be uninfluenced by grain shape. This coincides with the upper limit of colloidal influence, and is corroborated by Rusnak's observations of imbrication, reported above.

Shape description

The achievement of a grain shape description for a sand presents two difficulties, neither of which can be resolved completely. For any particular grain the shape irregularities defy accurate description. Furthermore, the grain population will contain individuals of diverse shape. On either count, complete description is quite impossible. The objective must therefore be a 'practical' description which conveys shape information about the whole population adequate for the purpose of discussing grain transport processes. The description must be simple, and it is highly desirable that it can be expressed by means of the numerical value of one parameter.

Two distinct approaches are possible. In one, grains are examined individually and the geometric parameters thereby derived for each grain are treated statistically to obtain representative population values for the same parameters. In the other, the performance of a sample from the population is measured in a test based on some shape-dependent behaviour. (Terminal settling velocity in water, submerged angle of repose and settled bulk density are possible bases for such a behaviour test.)

For routine engineering purposes the microscopic examination of individual grains in sufficient numbers to characterize a population is unreasonably time-consuming and therefore expensive. Methods for investigation of specimens in bulk are to be preferred, provided that the results obtained can be interpreted consistently. An ingenious shape appraisal technique based on 'rollability' has been introduced (Winkelmolen, 1969, 1971) and has been found to give results consistent with geometric shape parameters (Willetts and Rice, 1983). Were it to be widely adopted, it would make possible the inclusion of grain shape data in routine descriptions of sand populations.

Conclusions

Both fundamental considerations of fluid mechanics and the limited amount of field data available indicate that grain shape is important in transport mechanics. Its neglect hitherto has resulted rather from practical difficulties of description than from doubt that it is influential. It should be considered in any problem concerned with sediments of equivalent diameter greater than $100\,\mu m$.

Certain simple deductions can be made about shape effects at the seabed. They need to be tested and extended by means of field and laboratory observations. Satisfactory theoretical treatment of the subject may be decades away. However, it is very important that the collection of data be not delayed until theory can be written but be put in hand much sooner. Not only will a growing body of well-founded data assist in the development of theoretical treatment, but also it may make possible the solution of engineering problems by empirical methods meanwhile.

A useful data collection programme requires an agreed method of shape measurement, or no more than two cross-calibrated methods at most. It can be argued that lack of such an agreement has prevented the very large body of extant data on sediment transport in unidirectional flow from contributing significantly to knowledge about the influence of shape. Rollability seems a good candidate for acceptance as a 'standard shape test'. Whether or not it is so adopted, the matter of standardization should be given urgent attention.

The susceptibility to transport of any sediment deposit depends not only on grain and prevailing flow parameters, but also on its recent history. Transport rate programmes which ignore this will produce confusing data and sometimes spurious arguments. One of the mechanisms for modifying the susceptibility involves grain shape and orientation. Its influence cannot be relied upon to be trivial.

References

Borah, D. K., Alonso, M. and Prasad, S. N., 1982. Routing graded sediments in streams: formulations. *J. Hydraul. Div., ASCE,* **108** (HY12), 1487–1503.

Bray, D. I. and Church, M., 1980. Armoured versus paved gravel beds. *J. Hydraul. Div., ASCE,* **106** (HY11), 1937–1940.

Bray, D. I. and Church, M., 1982. Closure to discussion. *J. Hydraul. Div., ASCE,* **108** (HY5), 727–728.

Komar, P. D. and Reimers, C. E., 1978. Grain shape effects on settling rates. *J. Geol.,* **86**, 193–209.

Lane, E. W. and Carlson, E. J., 1954. Some observations on the effect of particle shape on the movement of coarse sediments. *Trans. Am. Geophys. Union,* **35** (3), 453–462.

Mants, P. A., 1977. Packing and angle of repose of naturally sedimented fine silica solids immersed in natural aqueous electrolytes. *Sedimentology,* **24**, 819–832.

Mantz, P. A., 1980. Low sediment transport rates over flat beds, *J. Hydraul. Div., ASCE,* **106** (HY7), 1173–1190.

Meland, N. and Norrman, J. O., 1969. Transport velocities of individual size fractions in heterogeneous bed load. *Geograf. Ann.,* **51** (Ser. A), 127–144.

Raudkivi, A. J. and Ettema, R., 1982. Stability of armoured layers in rivers. *J. Hydraul. Div., ASCE,* **108** (HY9), 1047–1057.

Rouse, H., 1961. *Fluid Mechanics for Hydraulic Engineers.* Constable, London, Ch. 11.

Rusnak, G. A., 1957. The orientation of sand grains under conditions of unidirectional fluid flow. *J. Geol.,* **65** (4), 384–409.

Taylor, B. D. and Vanoni, V. A., 1972. Temperature effects in low-transport flat-bed flows, *J. Hydraul. Div., ASCE,* **98** (HY8) 1427–1445.

Willetts, B. B. and Rice, M. A., 1983. Practical representation of characteristic grain shape of sands: a comparison of methods. *Sedimentology,* **30**, 557–565.

Winkelmolen, A. M., 1971. Rollability, a functional shape property of sand grains. *J. Sediment. Petrol.,* **41**, 703–714.

Winkelmolen, A. M., 1969. The rollability apparatus. *Sedimentology,* **13**, 291–305.

25

Potential Influence of Gas-induced Erosion on Seabed Installations

Martin Hovland and Ove T. Gudmestad

Statoil, Stavanger, Norway

Seabed erosion features potentially caused by migration of gas and fluid are reviewed. A possible explanation of the formation of these features is discussed and the influence of active migration processes on seabed installations is assessed.

Introduction

Over the past ten years geophysical and geochemical evidence has given strong indications that escape of gas and fluid through the seabed is common in certain provinces, many of them rich in hydrocarbons. Such indications are found in the form of gas-charged sediments, morphological seabed erosion features and sediment clouds in the water column (King and MacLean, 1970; McQuillin and Fannin, 1979; Hovland, 1981). In parts of the North Sea where the upper sediments are about 10 000 years old, a potential seepage process might span an even longer period. Indications are that it is still active in certain regions (Hovland, 1983).

This paper mainly discusses potential seepage and migration through the upper soils in certain parts of the North Sea area. A possible explanation for such phenomena is given and the potential influences upon the foundations of structures to be installed in such regions are discussed.

Seabed erosion features

At least two types of seabed erosion feature may possibly be caused by gas/fluid migration during recent geological history. These are seabed pockmarks and 'mottled seafloor'. Their distribution and occurrence indicate a widespread active process through the seafloor. Figure 1 shows the parts of the North Sea where pockmarks

occur. Pockmarks are crater-like depressions generally found in regions where the soil consists of soft, silty clays. The craters are rimless, which indicates that the eroded material is widely and evenly redistributed when it is brought into suspension. Pockmarks in the North Sea vary in diameter from 5 to 100 m, with an average of about 30 m. Their depths range from 1 to 15 m, with a typical depth of about 3 m. They are best detected by use of side-scan sonar (Fig. 2). Pockmarks are also often composite, with more than one centre. Their walls may bear evidence of slumping and coarse material (stones) has been found at their centre. Indications of shallow, mobile gas associated with pockmarks are shown in Fig. 3. Small faults in the upper sedimentary units may also be seen. Buried pockmarks are also shown in Fig. 3. These were once at the surface, but were buried during periods of higher sedimentation. In the Norwegian Trench buried pockmarks appear at two horizons below the present seafloor (Hovland, 1982).

The 'mottled seafloor' occurs in areas where the upper sedimentary layers consist of fine sands. The disturbed (mottled) sand patches are best located by use of side-scan sonar and shallow seismic profiles. The sonar picture appears as a light surface with dark (disturbed) patches and dark stripes (Fig. 4) which are associated with slight depressions. The disturbed zones vary from some metres to a kilometre in length.

Lineations, or strings of small pockmarks, are often

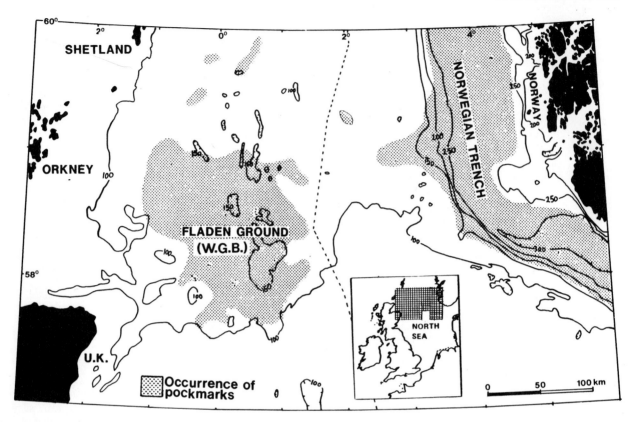

Fig. 1.   Distribution of pockmarks in the northern North Sea (WGB = Witch Ground Beds).

observed on the sonar pictures. The most prominent strings follow a NNW.–SSE. trend which coincides with the predominant bottom current direction in the Norwegian Trench. However, some pockmarks and clusters of pockmarks have several strings apparently emanating from their centre in several directions which may relate to stress patterns suggesting a doming of the seabed as an initial stage in pockmark development.

There may be an association of pockmark occurrences with seismic epicentres but the positions of epicentres through historical time is thought to be too poorly defined to provide a good possibility of accurate correlation. However, the density of pockmark distribution is currently being investigated in relation to faults in the substrata.

Calcium carbonate reefs and crusts have been found by using a remotely operated vehicle in pockmarks formed where the Holocene sandcover is thin over layered silts. It appears that gas-induced erosion there continues through the sandcover (1–2 m thick) and about 0.5 m into the silts; the calcium carbonate cements have been recorded at about 0.25 m thick. It is considered that it is probably the paucity of data that may suggest this feature is rare rather than its actual frequency of occurrence, despite its absence in several holes drilled

into pockmarks by the British Geological Survey.

Migration and seepage mechanism

The above evidence has led to the postulation that the formation of pockmarks arises from local erosion induced by the flotation and removal of clay and silt particles by emanating gas bubbles (King and MacLean, 1970). Further evidence that seepage and migration occur through the upper soils in certain areas is provided by observations of gas bubble plumes and sediment clouds in the water column (McQuillin and Fannin, 1979). However, very little is known about the dynamics of gas and fluid migration through marine soils. A tentative migration model has been developed on the basis of a careful analysis of high-resolution, shallow seismic records obtained from the Norwegian Trench and the Witch Ground Beds in the northern North Sea (Fig. 1). Both regions are heavily pockmarked, with up to 35 pockmarks per km^2 (McQuillin *et al.*, 1979; Hovland, 1982).

In the Norwegian Trench the two upper geological layers consist of:

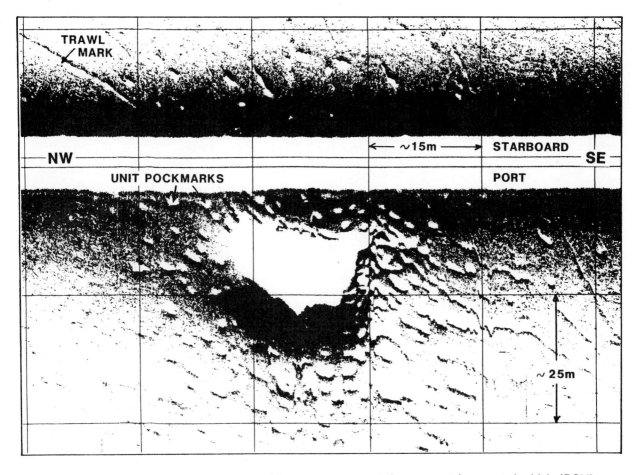

Fig. 2. A pockmark as it appears on a side-scan sonar mounted on a remotely-operated vehicle (ROV).

(a) Unit 1, which is a banded layer up to 50 m thick of soft, silty, normally consolidated clay. This clay has a low shear strength, typically increasing with depth as $Su_D \sim 1.8 \times Z \, kN/m^2$, where Z is depth in metres below seafloor. Unit 1 has a high water content. This unit overlies

(b) Unit 2, which is a homogeneous (not banded) clay containing small pebbles and chalk fragments. The increase in shear strength with depth is generally only slightly higher than that of Unit 1, but the water content is significantly lower.

The boundary between the two layers shows clearly as a strong reflector on shallow seismic records. It is characterized as a glaciated surface with numerous furrows and hummocks.

The origin and constitution of the gas and fluids possibly seeping through the seabed is unknown. The migrating medium could be biogenic (shallow) gas formed in shallow layers or thermogenic (deep-seated) gas and fluids from hydrocarbon reservoirs. In all respects,

however, it most probably originates from below the glaciated boundary between Units 1 and 2. Only a geochemical analysis of the gas will provide the answer to its origin.

Over a larger area (several square kilometres) the pockmark-producing process seems to be relatively uniform. On a local scale of some hundreds of square metres, however, the seabed is either pockmarked or undisturbed. A focusing or gathering effect must be responsible for this type of local erosion. Various features seen on shallow seismic records suggest that the migration through Unit 2 is different from that through Unit 1 (Fig. 5). Within the deeper layer (Unit 2) the postulated migration seems to be homogeneous and even. Migration of gas, which is either in solution or in saturation (bubble form), probably takes place through a mesh of minor fissures in the clay, which is of relatively low water content.

At the boundary of Unit 1 the banded, cohesive sediment of higher water content is encountered. This unit probably represents a high resistance to further

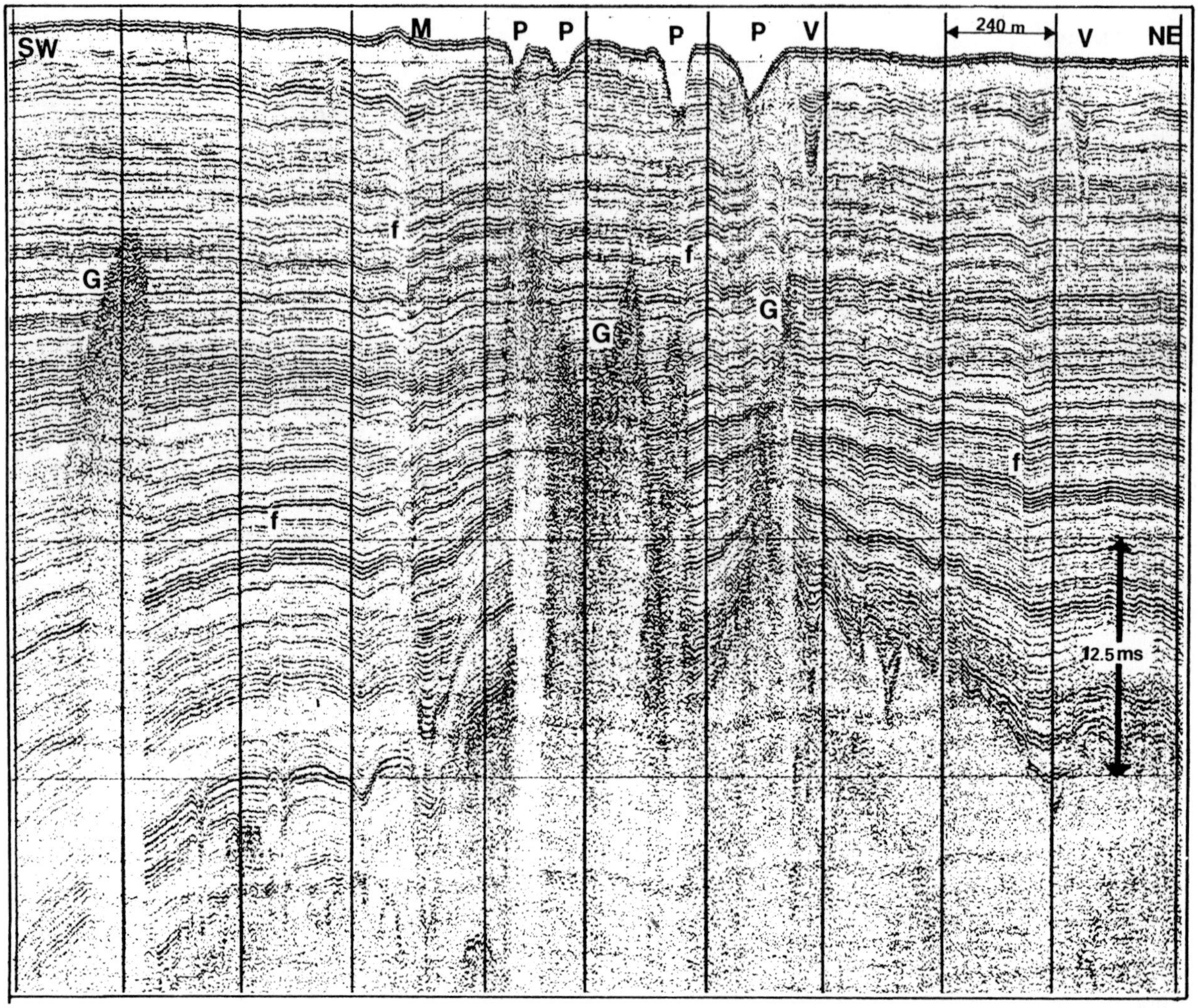

Fig. 3. Shallow seismic record across pockmarks (P), showing the presence of mobile gas (G) and numerous faults (f) or flexures. Buried pockmarks (V) and a mound (M) are also seen.

migration. The focusing effect probably acts within this upper, partly impermeable unit. Since the material in Unit 1 is layered, it is considered likely that most of the migration occurs up-dip (semihorizontally) along more porous bands. The vertical migration process through Unit 1 may act in one of two ways: (1) low-resistance vertical migration through weakness zones (faults and fissures), or (2) a locally increased pore pressure build-up which causes a doming of the seabed.

Indications of weakness zones in Unit 1 are seen on shallow seismic records, as shown in Fig. 3. Regarding the latter possibility, domes are reported in certain areas (Hovland, 1981). Such domes will act as migration traps which will rupture when the necessary pressure has built up. The rupture may not be dramatic, since

the soil shear strength in the Norwegian Trench increases linearly with depth. Only the upper 2–5 m of Unit 1 is often found to be slightly overconsolidated.

Although the initial formation phase of a pockmark may be sudden, when a dome actually ruptures, the subsequent outflow of gas, sediments and fluid is assumed to be of a gentle nature. In cases of vertical migration through natural weakness zones (faults and fissures), the entire mode of formation and development is assumed to be gentle.

In both cases free outflow to the water column is the result, possibly initiating local erosion in the form of a depression or crater as sediment is brought into suspension. Over time (length unknown) it may be deepened and modified, but is limited in depth by slumping.

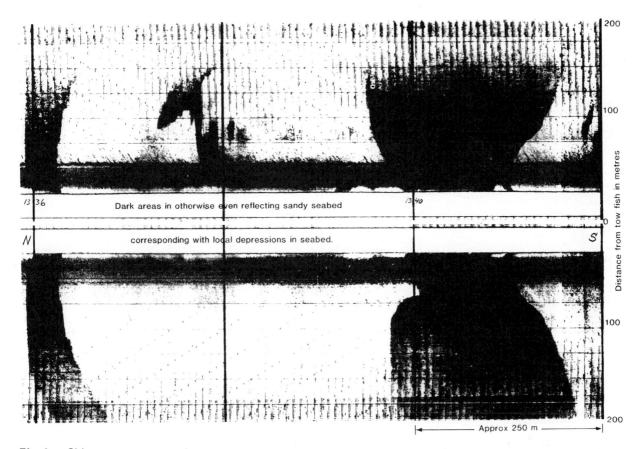

Fig. 4.   Side-scan sonogram of a 'mottled seafloor' where the dark patches and stripes indicate disturbances in the fine-sand cover.

The layers in Unit 1 may be regarded as successions of seafloors, whereby each reflector necessarily has been exposed to the water column at some stage of sedimentation.

In example A of Fig. 5 the layers are seen to become successively evened out above Unit 2. However, in mid-Unit 1 two domes appear, and tend to develop in the upper layers of Unit 1. At the present seabed two pockmarks have formed above these domes. Somehow the two features seem to be associated with each other. Domes underlying pockmarks are commonly observed in the Norwegian Trench (Hovland, 1982). Since the probability of this coincidence occurring repeatedly is very small, a suggested link between the two features is established in the figure to the right. The arrows indicate gas and fluid migration paths. While the migration in Unit 2 probably is uniform, it becomes focused during passage through the layers of Unit 1. The focusing domains are indicated in the right hand figure by broken lines.

In example B of Fig. 5 two buried (relict) pockmarks are evident in mid-Unit 1. These pockmarks developed at the time when this layer was exposed at the seafloor. Their influence over the subsequent migration pattern is indicated in the right-hand figure.

It may be noted that there are no obvious migration paths in these two cases. It is difficult to establish clear migration paths, probably because the process is three-dimensional while the seismic description is two-dimensional. It may be possible to determine the migration routes properly only by use of narrow-spaced profiles in two orthogonal directions. The main objective of such route determination would be to predict the possibility of future pockmark development in a selected construction area.

In regions where the upper layer of the seafloor consists of porous material (fine sand) there is no focusing effect. Sediment disturbances due to seepages are then governed by natural weaknesses in the underlying unit. These may be fracture zones or furrows which cause linear zones of seafloor erosion and sediment disturbance ('mottled seafloor'). They may also be structural (domes or hummocks) causing a patchy pattern of local erosion and disturbance.

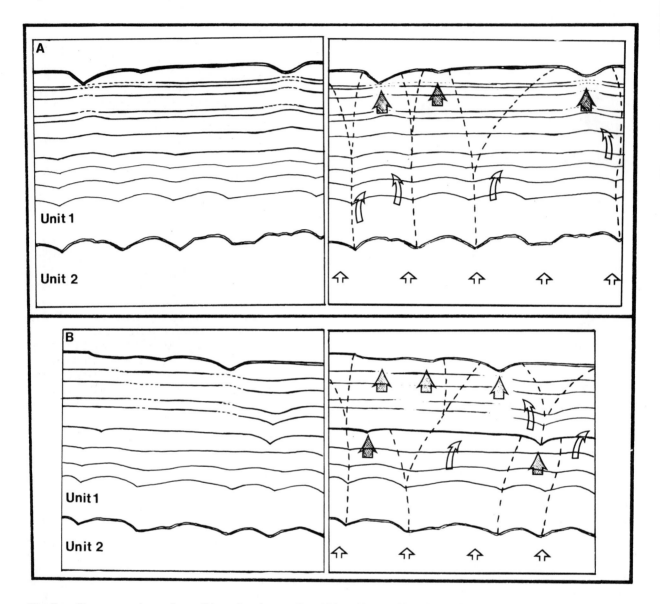

Fig. 5. Reconstruction of possible migration paths: (A) uniform migration through unit 2, focused migration through unit 1; (B) uniform migration through unit 2, two-step focusing through Unit 1.

Seabed construction

Prior to any seabed construction work in regions where there are indications of seepages (past or present) the mode and mechanism of the seepage processes should be assessed and possible consequences to the seabed installations should be evaluated.

Until now most construction work in the northern part of the North Sea has been carried out in areas of apparently low seepage activity. However, pipeline construction across the densely pockmarked Norwegian Trench was due to start in 1984. Plans are also under way for development of hydrocarbon findings in the Norwegian Trench with water depths of 300–340 m (Fay, 1983). Platforms which are being considered for this area are gravity-based platforms, conventionally piled platforms, anchored platforms and tension leg platforms (Huslid *et al.*, 1982).

The effects of seepage and migration processes on seabed installations can basically be divided into two

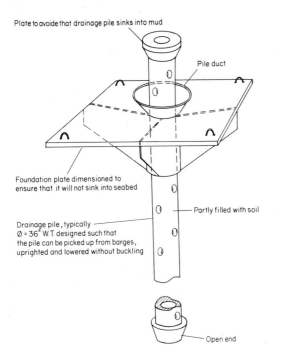

Plate to avoide that drainage pile sinks into mud

Pile duct

Foundation plate dimensioned to
ensure that it will not sink into seabed

Partly filled with soil

Drainage pile, typically
Ø = 36" W.T. designed such that
the pile can be picked up from barges,
uprighted and lowered without buckling

Open end

Fig. 6.   Drainage pile, schematic diagram.

categories: (1) the indirect effect of gas on the soil properties, e.g. the *in situ* soil shear strength versus shear strength as measured in the laboratory; (2) the direct effect on the behaviour of the foundation of platforms (gravity or piled) or on other seabed installations.

Indirect effects can be assessed only by careful *in situ* measurements of soil properties such as the amount of gas in the soil and the soil shear strength, followed by laboratory testing of retrieved samples under conditions as close to *in situ* as possible. *In situ* shear strength is normally measured by cone penetrometers or vanes. The cone factor could, however, be different for gaseous soils compared with soils free of gas. Equipment to carry out *in situ* tests of amount of gas, such as the ambient pressure sampler and the piezocone, are at present being tested. The ambient pressure sampler has so far been developed only for academic purposes, but could soon become operational on a commercial basis for sampling in deep waters. Indications of gas in the soil are also evident if the soil swells when it is taken to the surface. For deep waters even a small amount of gas will lead to large voids being formed in the sample when taken onboard the survey vessel. If there was free gas *in situ*, the soil might further appear under-consolidated, since gas will prevent consolidation. Such effects are, however, not seen in the Norwegian Trench.

Small amounts of gas in the soil may not have any major impact on the strength of the soil. This would

be the case in deep water, in particular, where the gas would be in solution with the pore water.

The potential direct effects on the foundation of seabed installations will be discussed below, following a brief discussion on possible methods to evaluate whether seepage processes could be active in certain areas.

A pockmark could be 'freshly' formed or active when its walls are steep and when there are numerous small 'unit' pockmarks surrounding the main depression. Such indications would appear on side-scan sonar images from detailed surveying carried out by use of remotely operated vehicles equipped with side-scan sonars. On stereo or still pictures there should further be indications of holes and disturbed sediments on the seabed. In order to assess the danger of formation of new pockmarks or seepages, a careful evaluation of the shallow seismic sections must be carried out. A prognosis of possible migration routes could be presented on the basis of closely spaced seismic profiles. There may be a high possibility that new pockmarks are forming in areas where indications of fresh pockmarks are found and where indications of gas-charged shallow sediments are seen on the seismic records. In regions of assumed seepage activity piles or cyclinders ('drainage' piles: Fig. 6) could be driven through the upper sedimentary layer in order to penetrate the semi-impermeable layers and thus introduce migration routes for the gas Migration and outflow from a particular area could thus, it is hoped, be controlled by artificial means. If such precautions are taken at the centre of an apparently 'fresh' pockmark, it may be termed 'active and controlled'. Even if the pile itself were filled with soil, it would introduce a weakness zone along which vertical seepage could occur.

For *pipeline* construction, the following problems related to seabed erosion features may arise:

(a) The pipeline may span dormant pockmarks. Depending on the span length, the pipe may be over-stressed. Vortex vibration could also occur due to current action. In the case of soft soil, however, the pipe will sink into the shoulder of the pockmark, thereby reducing the length of the free span.

(b) The pipeline may span or lie close to active pockmarks.

(c) The pipeline may cross areas where new pockmarks could develop.

Prior to pipe installation a detailed survey and interpretation procedure should be conducted to assess the danger of the above scenarios. Any pockmarks straddling the chosen route must be assessed for activity. Supports needed for structural purposes may be set up without any precautions in dormant pockmarks. Careful surveys of the status of the pipeline are necessary. If free spans were found to develop in areas of active seepage, piles could be driven as discussed above to act as controlled migration paths. A combination of pipeline supports and 'drainage' piles should thus be sufficient to control the stresses in the pipe. However, unless there is documented evidence of an active process with significant danger for erosion at a platform site or along a limited section of a pipeline route, it is not proposed that Statoil will put drainage piles along major sections.

A *gravity-base structure* could be sited on smaller pockmarks in the case where grouting is applied under the structure to provide full contact with the seabed. One would, however, wish to locate the gravity-base platform outside any pockmark and any site where there is a high probability of active pockmark formation. An active process could lead to removal of material along the edges of the platform, causing tilting and a possibility of sliding. Free gas could also lead to a reduced elasticity modulus of the soil, thus reinforcing cyclic loading effects such as pore-water build-up and associated settlements.

The seabed must therefore be carefully surveyed with shallow seismic and side-scan sonar prior to site selection. In the Norwegian Trench, with about 35 pockmarks per km^2, sufficient flexibility should still exist with respect to location of the platform. The gravity-base structure should further be constructed such that it prevents the accumulation of gas or low-gravity fluids within its base or within its skirts. The normal water drain system would probably be sufficient to act as a system to prevent build-up of pressure under the platform.

The gravity-base structure will introduce a load on the seabed which will influence the geotechnical characteristics of the upper sediment layer directly beneath and around the platform base. Owing to the pressure on the seabed the soil will be compacted, water will drain out and it is expected that the soil under the platform will be less susceptible to migration of gas (Foster and Ahlvin, 1954). Migration may, therefore, be led outside the 'influence area' of the platform. On the other hand, the production wells will act as migration paths. If there exists a possibility of active processes, the site could be surrounded by artificial migration paths, such as the shallow piles discussed previously, for reducing eventual outflow under the platform. The objective would be to introduce artificial weakness zones for subsequent gas or fluid seepages, to lead the activity away from the platform area. 'Drainage' piles could be installed some years prior to platform installation in order to vent the actual site.

For *piled structures* the holding capacity of the piles will depend upon the friction capacity of the pile and the end bearing capacity. However, in the case of tension piles the latter component is non-existent. The piles themselves will introduce migration routes which could reduce their friction holding power. A careful analysis of such effects is required prior to pile installation. Erosion effects at the top of the pile, i.e. at the seafloor, could also follow. For tension piles, in particular, even small amounts of seepage could be of concern. The use of tension piles in areas with soft soils and potentially active seepage processes should be carefully considered prior to choice of the tension leg platform concept in such areas.

Also in the case of piled structures, the method of artificial drainage could be employed in order to increase the holding capacity. Such drainage paths could be installed some years prior to platform installation, to reduce the risk of the piles acting as drainage paths. However, to date no piled structures have been built within pockmarked regions of the Norwegian sector and none have reported gas-induced erosion in the Forties field of the British sector, where pockmarks are common.

Model tests and full-scale field tests of the 'drainage' pile concept should be considered in any areas with very shallow gas. Tests would show whether the concept is useful and would also provide information on migration velocities through the soft soils.

Conclusions

Strong evidence supports the hypothesis that natural seepage of gas and low-density fluid occurs through the seabed in many regions of the world, among which are parts of the Norwegian Trench. The seepage process is probably not dramatic, but a possible migration mechanism should be carefully assessed at the site prior to any seabed construction. Precautions recommended depend upon the nature of the soil conditions and the migrating medium together with the consequences of the seepage process to the structure.

Acknowledgements

The authors express their thanks to Statoil for permission to publish this paper. However, the authors wish to emphasize that the statements made therein represent their personal views and conclusions.

References

Fay, C. E., 1983. New technology is needed to develop Norwegian Trench find. *World Oil,* January, 131–134; February, 57–62.

Foster, C. R. and Ahlvin, R. G., 1954. Stresses and deflections induced by a uniform circular load. *Proc. Highw. Res. Bd,* **33,** 467–470.

Hoyland, M., 1981. *A Classification of Pockmark-Related Features in the Norwegian Trench.* Publ. 106, Continental Shelf Institute, Norway, 28 pp.

Hovland, M., 1982. Pockmarks and the Recent geology of the central section of the Norwegian Trench. *Mar. Geol.,* **47,** 283–301.

Hovland, M., 1983. Elongated depressions associated with pockmarks in the Western Slope of the Norwegian Trench. *Mar. Geol.,* **51,** 35–46.

Huslid, J., Gudmestad, O. T. and Alm Paulsen, 1982. Alternate deepwater concepts for Northern North Sea extreme conditions. *Proc. BOSS-82,* Hemisphere Publishing/McGraw-Hill, pp. 18–49.

Judd, A. G., 1982. *The formation of pockmarks and associated features.* Internal Rep. to Statoil, 137 pp.

King, L. H. and MacLean, B., 1970. Pockmarks on the Scotian Shelf. *Geol. Soc. Am. Bull.,* **81,** 3142–3148.

McQuillin, R. and Fannin, N. G. T., 1979. Explaining the North Sea's lunar floor. *New Scientist,* **1163,** 90–92.

McQuillin, R., Fannin, N. G. T. and Judd, A., 1979. IGS pockmark investigations 1974–1978. Unit Rep. 98, IGS, Continental Shelf Division, Marine Geophysics, 50 pp.

Appendix to Parts III and IV

Sediment Transport Studies Workshop

P. Ackers

Binnie and Partners, London, UK

The preceding two days of the Symposium had covered a wide field of sediment mechanics: theory, laboratory experiment, large-scale field research, computational modelling, coastal morphology, muddy sediments, non-cohesive sediments, local scour at pipes and at piled structures, and sedimentary features of the sea floor. The symposium had brought together the geotechnical and hydrodynamic aspects in a very profitable forum, although now, at the end of the week, a large proportion of the geotechnical experts had left and few were available for the final workshop. Nevertheless, there had been a considerable exchange of views on the important geotechnical–hydrodynamic interactions. Geotechnics here embraced particle arrangement and the influence of cohesiveness on movement under hydrodynamic forces; and perhaps more significant, hydrodynamic pressures and the response of pore-water under tides and waves influencing geotechnical stability, in some cases leading to liquefaction.

The discussion during the workshop centred on:

(1) The physics of movement of non-cohesive sediments under waves and currents.

(2) Transferring laboratory-scale findings to the field situation, one question being whether laboratory research, e.g. with artificial roughness, was a sound basis for developing functional relationships that were to be applied at very different scale under naturally variable conditions.

(3) Fine sediment suspensions and muds. How much do we know now regarding their rheology and the action of hydrodynamic stress? Despite considerable advances, much remains to be learned from field and laboratory. Is mud a pseudo-plastic or a Bingham plastic, for example?

(4) Numerical simulations of fluid flow in one, two and three dimensions and the influence on sediment; the status and way forward.

(5) The interface with geotechnical aspects which is now recognised to be of considerable significance to offshore and coastal zone structures.

(6) The part natural armouring plays in stabilising sediments and the prospect of using it to engineering advantage.

(7) Engineering problems: for example, can we now adequately design a pipeline crossing a sandy foreshore? Should it be trenched? Will it naturally bury? What forces may arise?

(8) The future of research and design in these subjects.

A very useful exchange of views took place on all these topics.

Closing Address

Delayed Consolidation of Marine Clay:
Soil Mechanics from On-Shore to Off-Shore Seabed

Koichi Akai
Department of Civil Engineering, Kyoto University, Kyoto, Japan

Introduction

The reclaimed lands along the sea coast of Japan are located on alluvial plains where soft marine clay prevails, and the reclaimed lands themselves are composed either of soft clay or loose sand. Severe geotechnical problems, therefore, have been encountered during reclamation as well as during the use of these reclaimed lands.

Figure 1 shows the increase in extent of reclaimed lands in the last 30 years. It can be observed that the rate of land development had rapidly increased with the growth of natural economy for 20 years from 1955 to 1973, while the rate has gradually been reduced since the energy crisis in 1973. The decrease in the rate is also attributed to the shortage of shallow sea sites appropriate for reclamation and also to the environmental restrictions growing year by year (SMFE, 1977).

This paper briefly introduces the feature of land reclamation in Osaka Bay, mid-west Japan (Fig. 2). One of the main soil mechanics problems, in the author's opinion, the delayed consolidation (secondary compression) of marine clay is summarized herin.

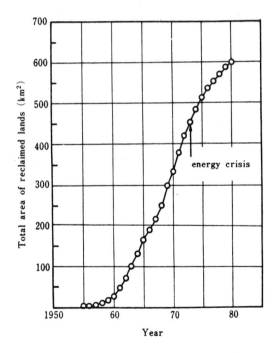

Fig. 1. Total area of land reclaimed in Japan since 1953.

Topographical and geological background around Osaka Bay

Figure 3 shows the development of reclaimed area around Osaka Bay. Especially after 1945, extensive reclamation projects have been undertaken all over the coast of Osaka Bay to satisfy the increasing demands for land as industrial, residential and port areas (SMFE, 1977).

A large epoch-making project in this area for the very near future is the construction of a new offshore airport in the bay, because the present Osaka International Airport (12 km north-west from the downtown) has

268   SEABED MECHANICS

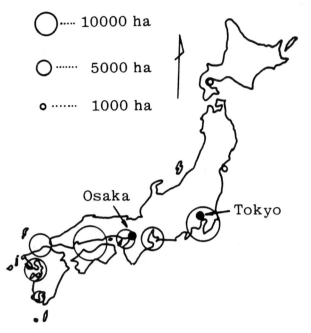

long suffered from a lack of landing capacity. The
location of the New International Airport is about
40 km southwest of the City of Osaka and 5 km from
the eastern coast of Osaka Bay (Fig. 4). An outline of
airport facilities is indicated in Fig. 5.

The geological survey revealed the following features
concerning the relations between the submarine topo-
raphy and the youngest deposits in Osaka Bay (Fujita
and Maeda, 1969).

(1) The unconsolidated deposits of Osaka Bay are
divisible into A and B layers (Fig. 6). The layer B is
inferred to be the latest Pleistocene sediments deposited
in the stage when the sea level retarded at 20 m below
the present level about 10 000 years ago. On the other
hand, the layer A is the Holocene sediments formed by
the quick transgression occured in the next stage.

(2) Submarine flat planes are classified as:
  (a) 10m, wave-cut terrace formed at the present
      sea level;
  (b) 10 m–20 m, depositional surface of A;
  (c) 40 m–50 m, approximate depositional surface
      of B.

Generalized profile of Osaka Bay shown in Fig. 6
indicates the relationship between the change of sea
level and the deposition of A and B.

Fig. 2. Locations and areal extent of major reclaimed
land in Japan since 1968.

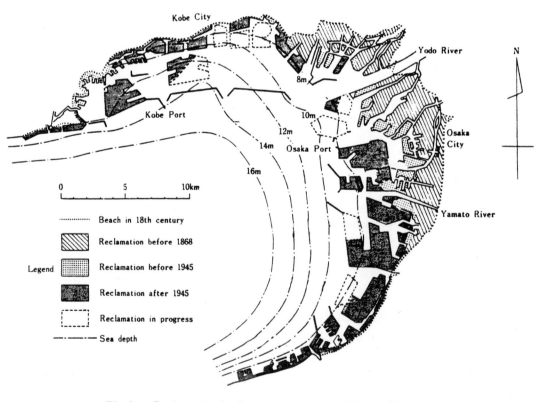

Fig. 3. Reclamation in the north-east coast of Osaka Bay.

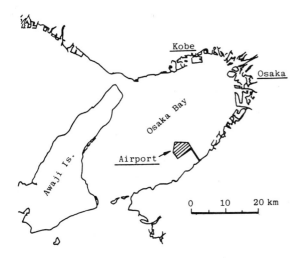

Fig. 4. Location of the New International Airport in Osaka Bay.

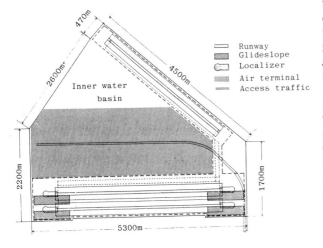

Runway
Glideslope
Localizer
Air terminal
Access traffic

Fig. 5. Generalized profile of Osaka Bay showing the relationship between the change of sea level and the deposition A and B.

Drained creep behavior of K_0-normally consolidated clay

As the first step to investigate the mechanism of secondary compression of marine clay, a series of drained creep tests have been performed by means of triaxial apparatus, taking kaolin as a soil sample.

The test procedure is as follows:

(1) The specimen (50 mm in diameter and 100 mm in height) is allowed to consolidate anisotropically under the effective stress ratio $\eta = 0.706$ which corresponds to the coefficient of earth pressure at rest $K_0^{NC} = 0.52$. The final consolidation pressure is $\sigma_v = 300\,kPa$.

(2) After the primary consolidation is over, the stress state is maintained as before; thus the vertical effective pressure $\sigma_v' = 300\,kPa$ and the horizontal effective pressure $\sigma_h' = 300 \times 0.52 = 156\,kPa$. Under this condition the sample is brought to the drained creep for about seven weeks.

(3) During the above mentioned, stages, the measured values are the axial strain ϵ_v, the volumetric strain v and the excess porewater pressure u. From the data of ϵ_v and v, one can calculate the average strain $v/3$, the deviatoric strain ϵ $(= \epsilon_v - v/3)$ and the horizontal strain ϵ_h.

Figure 7 indicates the test result. From the manner of pore pressure dissipation, it is known that the primary K_0-consolidation is arrived at $t_{100} = 90\,min$. During this period, K_0-condition is kept well, since there exists no lateral strain, $\epsilon_h = 0$. During $t > t_{100}$, on the other hand, the horizontal strain begins to increase in expansion, that is, the soil specimen deforms under non-K_0 conditions at the end of primary consolidation, while both the average strain and the deviatoric strain continue to increase. The ratio of the latter is $v/3 : \epsilon = 1 : 2$ during K_0-consolidation, which can be explained by the theoretical analysis of strain.

This ratio decreases gradually with time during the drained creep with constant effective stress ratio η. The deviatoric strain ϵ increases in proportion to logarithm of time for long period, whereas the mean strain $v/3$ ceases the incremental behaviour when $t = 10^4\,min$ is reached. It can be concluded, therefore, that the time-dependency of the deformation of kaolin is greater in ϵ than in $v/3$.

Change in state of soil during K_0-delayed consolidation

During primary consolidation, it is well-known that the mean effective stress p and the principal stress difference q increase gradually as the excess porewater pressure dissipates. At the moment of the end of primary consolidation, these stresses reach their final values.

On considering the change in state of soil during K_0-delayed consolidation, firstly suppose that the states of stress would not change at all. This corresponds to the stress condition in the drained creep test mentioned before. Then, the response of strains is considered to be as follows:—

(1) since $p = $ constant, the mean strain which can be assumed as inviscid $v/3 = 0$;

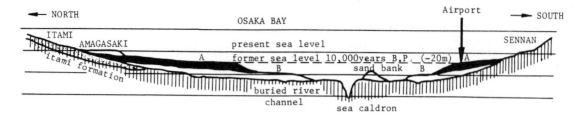

Fig. 6. Generalized profile of Osaka Bay showing the relationship between the change of sea level and the deposition *A* and *B*.

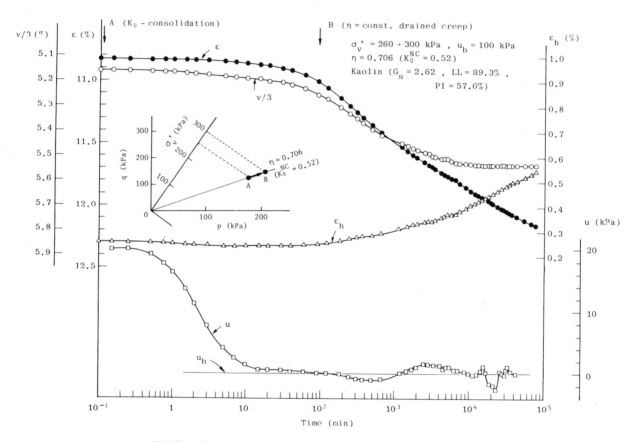

Fig. 7. K_0-consolidated drained creep test (p, q = constants).

(2) though q = constant, there occurs a deviatoric strain $\epsilon(t)$ with time lag, called the viscid creep;

(3) though q = constant again, there occurs a mean strain $v(t)/3$ with time lag, called the viscid dilatancy.

Owing to the restrictive conditions in terms of strain in K_0-condition, the ratio of the mean strain to the deviatoric strain should be $v(t)/3 : \epsilon(t) = 1:2$ at any time during delayed consolidation. If the former is smaller than the half of the latter just as in the test for kaolin mentioned above, an increase in p must be happened to compensate the shortage of mean strain. In addition,

a decrease in q takes place at the same time, owing to the restrictive condition in terms of stress; σ'_v = constant. Provided the former is larger than the half of the latter, on the other hand, a decrease in p accompanied by an increase in q should occur.

Thus, we have the possibility of three cases of change in K_0-value (increase, static, and decrease) during delayed consolidation of a normallly consolidated cohesive soil (Shertmann, 1983). This depends on the physical properties of soil under compression, namely, the creep and dilatancy characteristics, both of which

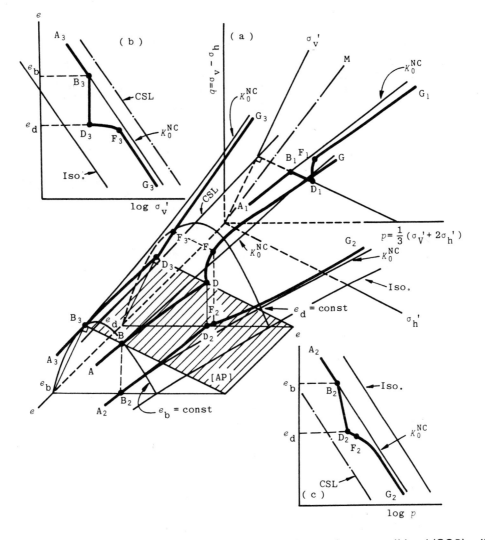

Fig. 8. State path in the $e-p-q$ space and their projections for quasi-overconsplidated (QOC) soil.

indicate high time-dependency.

In order to draw the spatial path for soil under delayed consolidation, one must introduce the σ_v'-axis beside the epq-axis and utilize the process shown in Fig. 8b where the path B_3D_3 during secondary compression on the $e-\sigma_v'$ plane is parallel to the e-axis. In Fig. 8a, the state path starts from point B on the state boundary surface (SBS) and moves into SBS on an inclined plane, called the aging plane [AP], to the present point D. It can be easily recognized that the soil under delayed consolidation has an elastic component with respect to deformation (Akai and Sano, 1981).

Figure 9 shows the axial strain-time curve as well as change in lateral pressure in the long-term K_0-consolidation test of kaolin for about one week. The test was performed in a K_0-triaxial cell, controlling

axial loading ($\sigma_v = 140 \rightarrow 300\,\text{kPa}$) and zero lateral strain of the test specimen during consolidation. It can be seen from this figure that the coefficient of earth pressure at rest, $K_0^{NC} = 0.52$, during primary consolidation, increases gradually in the period of secondary compression and reaches the final value of $K_0^{QOC} = 0.64$. The coefficient of secondary compression in terms of strain $\epsilon_\alpha = d\epsilon_v/d \log_{10}t$ is almost constant at the instant of beginning of secondary compression when the K_0-value starts to increase. After a long period, the consolidation-time curve turns concave upwards, and the secondary compression tends to end.

Deformation after loading on a soil under delayed consolidation is also of great interest, because of the amount of settlement caused in the young diluvial clay is of great importance in the analysis of the behavior

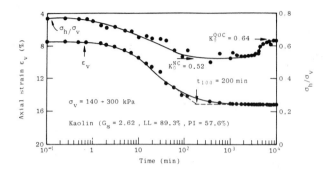

Fig. 9. Axial strain and lateral pressure behavior during K_0-consolidation of kaolin.

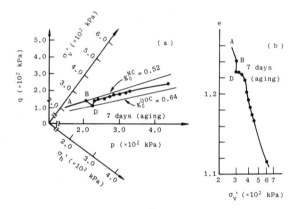

Fig. 10. Effective stress path and $e - \log \sigma_v'$ curve in the K_0-triaxial loading test for kaolin.

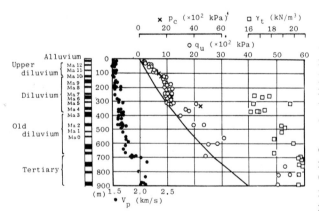

Fig. 11. Soil profiles in Osaka.

of reclaimed on-shore or offshore land. Figure 10 indicates the effective stress path and $e - \log \sigma_v'$ curve in the K_0-triaxial loading test for kaolin after a one-week period. At the first loading, one can recognize

a locking response where the stress path goes on, maintaining the constant effective mean stress, p. Owing to the complete confining condition of K_0-deformation, the deviatoric strain ϵ and the vertical strain ϵ_v (= the volumetric strain v) scarcely occur, resulting in almost no decrease in the void ratio, e. By successive increments of loading, the stress path gradually approaches the K_0^{NC}-line, reaching a fairly high K_0-value in this experiment. The general nature of state path DFG after K_0-loading is additionally shown in Fig. 8.

Case studies on delayed consolidation at on-shore seabed

The seabed of Osaka Bay, Japan consists of soil layers, (up from the sea bottom) an alluvial marine layer, a terrace deposit and a series of diluvial layers in which marine deposits are numbered as Ma12, Ma11, etc. (Ma0 from the top to the bottom layer) as shown in Fig. 11. In the City of Osaka, the so-called Tem-ma sand/gravel layer (Tg) which corresponds to an old alluvium or a terrace form has always been treated as the bearing layer of buildings and heavy structures. The bearing capacity of the clay layer just beneath this sand/gravel layer should therefore be investigated where the thickness of the upper stratum is not sufficient. These diluvial clay layers (Ma12 and Ma11) are believed to have experienced neither a great decrease in overburden load due to erosion nor a tectonic lateral pressure due to folding. Therefore, their overconsolidation characteristics are quite small compared with older diluviums (Ma10, etc.). Thus, we cannot neglect the possibility of deformation under a new loading on these layers. On the large-scale reclaimed lands along Osaka Bay, for instance, settlement continues even now, many years after the completion of reclamation. Although this kind of settlement is found to be occurring in the deep sediments, the mechanism is still not clear from the viewpoint of soil mechanics.

As a case study on delayed consolidation of marine clay, let us investigate the ground behavior of a recently reclaimed Osaka South Port. For the construction of Osaka South Port, an area totalling $9.3 \times 10^6 \, \text{m}^2$ was filled in, or about 5% of the total land area within the municipal boundary of Osaka. The average depth of sea was 7–10 m and the planned ground height of the reclaimed land is 5–7 m above sea level. The work began in 1958 by dredging soils of seabed and continued about 25 years, including soil stabilization mainly by means of sand-drains for the uppermost alluvial layer. The following research is oriented to the deformation behavior of the upper diluvium Ma12.

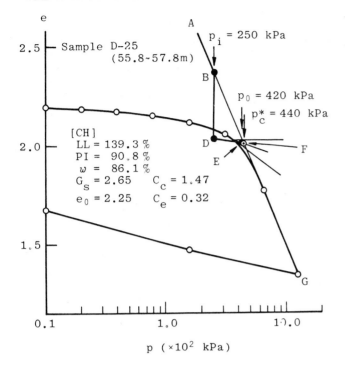

Fig. 12. Consolidation curve of the upper diluvial clay, Ma12.

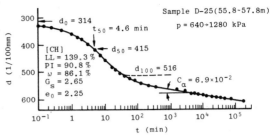

Fig. 13. Long-term consolidation test for Ma12.

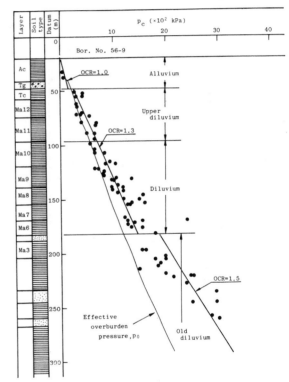

Fig. 14. Preconsolidation profile of seabed soils in Osaka Bay.

In respect of the compressibility of the upper diluvium Ma12, it has been made clear from the geological viewpoint that this layer had belonged to a kind of 'aged' normally-consolidated clay until the reclamation work began at this site. The term was used by Bjerrum (1967, 1972) who classified normally-consolidated clays into 'aged' and 'young' according to whether the clay has experienced secondary time effects during consolidation under the present effective overburden load.

Figure 12 indicates the consolidation curve in the oedometer test for a sample of Ma12 (depth $z = 55.8$–57.8 m). The initial effective overburden pressure before reclamation was $p_i = 250$ kPa and the corresponding state point is marked D in the figure. In this figure the process of sedimentation of this soil is assumed to be the virgin compression line AB and the period of aging is the vertical line BD for about the order of 10^4 years which corresponds to the end of diluvial age. The decrease in the void ratio e due to aging is, therefore, evaluated as large as $de = 0.36$, from the distance between two points B and D. According to Bjerrum, it is possible to draw time contours on this figure by which one can recognize the process of change in void ratio after sedimentation.

The present state of the clay sample tested is supposed to situate at point E in this figure, since the overburden pressure has been increased to $p_o = 420$ kPa due to filling up soils during the land reclamation. As is

seen from the figure, the quasi-preconsolidation pressure (point F) of the clay sample is $p_c^* = 440$ kPa. This is due to the precompression effect (p_c-effect) termed by Bjerrum which appears in the soil under delayed consolidation. The margin of preconsolidation ($p_c^* - p_o$) in this case is as small as 20 kPa; thus it can be recognized that the upper diluvium Ma12 has changed almost into the state of a normally-consolidated soil at present. Of course, any construction on this reclaimed land would cause an additional settlement along the virgin compression line FG, unless the load of structure is transmitted to the deeper layers than Ma12 by using some kind of deep foundation such as long piles.

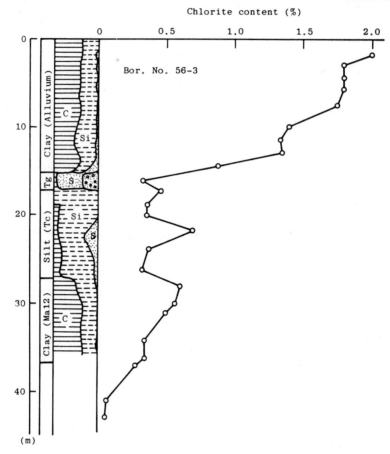

Fig. 15.   Variation in chlorite content in the pore water of alluvial marine clay with depth.

Another estimate of the amount of delayed consolidation has been performed using the long-term consolidation test using the same clay sample. Figure 13 shows the settlement–time curve for the sample of Ma12 above mentioned. The state of consolidation pressure from 640 kPa to 1280 kPa for this figure corresponds to the normally consolidated region, where the coefficient of secondary compression in terms of void ratio $C_\alpha = de/d \log_{10} t$ indicates the maximum value. Taking $C_\alpha = 3 \times 10^{-2}$/cycle as the estimated value for normally consolidated state, one obtains the decrease in void ratio after 10^4 years as $de = 0.42$, which is comparable with the value of 0.36 mentioned before using $e - \log p$ correlation (Fig. 12).

Soil mechanics problems in offshore seabed

The offshore seabed in Osaka Bay, 5 km far from the eastern sea coast, is being under investigation for the big project of construction of New International Air-

port in the Middle-West Japan. A huge amount of soil estimated to be about $5 \times 10^8 \, \text{m}^3$ is necessary to build up an artificial island onto the seabed with a water depth of about 20 m (Figs. 4 & 5).

Figure 14 illustrates the preconsolidation profile of seabed soil at the centre of the airport island. It is understood from the figure that the overconsolidation ratio (OCR $= p_c/p_0$) is estimated as follows: for the alluvial marine clay at the seabed surface and the terrace deposit; OCR = 1.0 (normally consolidated), for the upper diluvial strata (Ma12, Ma11) and the diluviums (Ma10–Ma6); OCR = 1.3 (slightly overconsolidated), and for the old diluviums (Ma3–); OCR = 1.5.

According to the design study, the increment of vertical stress is estimated as $\Delta p \simeq 380$ kPa which does not distribute with depth because of the large area of island. Therefore, the soil layers between the seabed surface and the bottom of the middle diluvium (Ma6), i.e., until about 180 m below sea level, would change their preconsolidation condition more or less into a normally consolidated state. The estimated total amount of settlement is about 6.4 m–8.5 m after 10

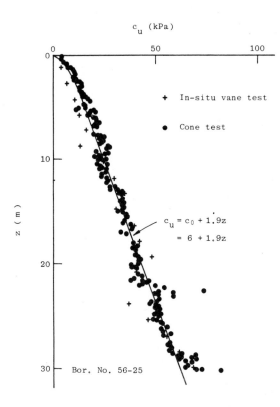

Fig. 16. Variation in *in situ* undrained strength of alluvial marine clay with depth.

alluvial clay which increases gradually with depth, as is typically observed in the seabed.

Conclusion

From the site investigations on overconsolidation characteristics of marine clay layers beneath the seabed of Osaka Bay, Japan, it has been found that the upper diluvium corresponds to a soil under delayed consolidation, where p_c/p_0 is a little greater than unity. It should be noted that an increase in the horizontal stress, σ'_h, during delayed consolidation would result in increase of K_0-value and OCR. The degree of increase depends on the physical properties of soil under compression; namely, the creep and dilatancy characteristics, both of which indicate high time-dependency.

Using the concept of a state boundary surface, it is shown that such soil has an elastic component with respect to deformation. In spite of the above characteristics, the time effect for soil under delayed consolidation generally differs from the loading-unloading effect for overconsolidated soil. State path for the former soil during both consolidation and loading is schematically drawn, based on the result of a series of K_0-secondary consolidation test and K_0-triaxial loading test for Kaolin.

years operation of the airport, though the consolidation of the alluvial marine layer at the seabed surface would be finished by the earliest operational time by means of soil stabilization.

As regards the characteristics of the alluvial clay at the seabed surface, one can recognize the tendency for a decrease of chlorite content in the porewater with depth as shown in Fig. 15. At the neighborhood of the seabed, the content is almost equal to that of the sea water (about 1.9%). The manner of decrease is similar to that of the liquid limit and the plasticity index of clay. Such phenomena are explained as resulting from the circumstance at sedimentation when the lower part of this clay layer was set in a brackish environment with fairly shallow water depth. Fig. 16 indicates the distribution of *in-situ* shear strength of this

References

Akai, K. and Sano, I. 1981. On the deformation behavior of soil under delayed consolidation. *Mem. Fac. Eng., Kyoto Univ.*, **43** 161–172.

Bjerrum, L. 1967. Engineering geology of Norwegian normally-consolidated marine clays as related to settlements of buildings (7th Rankine Lecture). *Geotechnique*, **17**(2), 83–117.

Bjerrum, L. 1977. Embankments on soft ground. *Proc. ASCE Conf. Performance of Earth and Earth-Supported Structures*, Vol. 2, pp. 1–54.

Fujita, K. and Maeda, Y. 1969. Latest Quaternary deposits in Osaka Bay. *Quaternary Res.* **8**(3), 89–100 (in Japanese).

Schmertmann, J. H. 1983. A simple question about consolidation, *J. ASCE*, **109**(1), 119–122.

SMFE, 1977. *Proc. 9th Inf. Conf.* Case History Volume, pp. 599–664.

List of Delegates

Dr. G. Almagor,
Marine Geological Survey of Israel,
30 Malkhe Yisrael Street,
Jerusalem, Israel

Mr. R. V. Ahilan,
University Engineering Department,
Cambridge University,
Trumpington Street,
Cambridge CB2 1PZ
UK

Mr. M. Adye,
Marine Technology Directorate,
Science and Engineering Research Council,
3–5 Charing Cross Road,
London WC2 CHW
UK

Mr. D. A. Ardus,
Marine Geology Unit,
British Geological Survey,
Murchison House,
West Mains Road,
Edinburgh EH9 ELA UK

Prof. Koichi Akai,
Department of Civil Engineering,
Kyoto University,
Kyoto 606,
Japan

Mr. Peter Ackers,
Binnie and Partners,
Artillery House,
Artillery Row,
London SW1
UK

Mr. S. Buchan,
Physical Oceanography Dept.,
University College of North Wales,
Menai Bridge,
Gwynedd, LL57 2HS
UK

Mr. J. R. Butler,
R5.4.2.,
British Telecom Research Labs.,
Martlesham Heath,
Ipswich, Suffolk
UK

Mr. A. R. Biddle,
McClelland Limited,
McClelland House,
Chantry House,
Chantry Place,
Harrow,
Middlesex, HA3 6NY
UK

Dr. J. S. Booth,
U.S. Department of the Interior,
Geological Survey,
Office of Marine Geology,
Woods Hole,
MA02543,
USA.

Mr. C. M. Chilton,
R339 Merz and McLellan,
Amberly,
Killingworth,
Newcastle upon Tyne NE12 0RS
UK

Dr. W. H. Craig,
Simon Engineering Laboratories,
University of Manchester,
Manchester M13 9PL
UK

Mr. R. Chilcott,
Ocean Engineering,
Department of Naval Architecture and Shipbuilding,
The University of Newcastle upon Tyne,
Newcastle NE1 7RU
UK

Dr. R. C. Chaney,
Department of Engineering,
Humbolt State University,
Arcata,
CA 95521,
USA

Mr A. Clark,
c/o Department of Civil Engineering,
The University of Newcastle upon Tyne,
Newcastle upon Tyne NE1 7RU
UK

Prof. Bruce Denness,
Ocean Engineering,
Department of Naval Architecture and Shipbuilding
The University of Newcastle upon Tyne,
Newcastle NE1 7RU
UK

Mr. T. J. Freeman,
Building Research Establishment,
Geotechnics Division,
Garston,
Watford, WD2 7JR
UK

Dr. G. Farmer,
Climate Research Unit,
School of Environmental Sciences,
University of East Anglia,
Norwich, NR4 7TJ
UK

Dr. M. Fukue,
Faculty of Marine Science and Technology,
Tokai University,
Orido,
Shizu
Shizuoka 424,
Japan

Dr. G. T. Houlsby,
Department of Engineering Science,
Oxford University,
Parks Road,
Oxford
UK

Prof. P. W. Holmes,
Department of Civil Engineering,
University of Liverpool,
4 Cambridge Street,
PO Box 147,
Liverpool, L59 3BX
UK

Dr. B. Hawkins,
Department of Geology,
University of Bristol,
University Road,
Bristol, BS8 1SS
UK

Mr. M. D. Howat,
Mass Transit Railway Corporation,
GPO Box 9916,
Hong Kong

Prof. R. Halliwell,
Department of Civil Engineering,
Heriot Watt University,
Riccarton,
Edinburgh, EH14 4AS
UK

Mr. M. Hovland,
R & D Dept.,
Statoil,
PO Box 300,
4001 Stavanger,
Norway

Prof. K. Ishihara,
Department of Civil Engineering,
University of Tokyo,
Bunkyo-ku,
Tokyo,
Japan

Mr. T. Kamei,
Department of Civil Engineering,
Tokyo Institute of Technology,
Ookayama,
Meguru-ku
Tokyo,
Japan

Dr. P. H. Kemp,
Civil Engineering Dept.,
University College London,
Gower Street,
London WC1
UK

Sir James Lighthill,
Provost,
University College,
Gower Street,
London WC1E 6BT
UK

Mr. P. C. Machen,
Ocean Engineering,
Department of Naval Architecture and Shipbuilding,
The University of Newcastle upon Tyne
Newcastle upon Tyne NE1 7RU
UK

Dr. A. W. Malone,
Geotechnical Control Office,
6th Floor,
Empire Centre,
Salisbury Road,
Tsimihatsui,
Kowloon,
Hong Kong

Professor D. McDowell,
Simon Engineering Laboratories,
University of Manchester,
Manchester, M13 9PL
UK

Professor Akio Nakase,
Department of Civil Engineering,
Tokyo Institute of Technology,
Ookayama,
Meguruku,
Tokyo,
Japan

Prof. P. Novak,
Department of Civil Engineering,
The University of Newcastle upon Tyne,
Newcastle upon Tyne, NE1 7RU
UK

Mr. C. P. Nunn,
Zetelogic Ltd.,
Unit 22, Berry Edge Venture Units,
Berry Edge Road,
Consett,
Co. Durham
UK

Prof. S. Okusa,
Department of Civil Engineering,
Faculty of Marine Science and Technology,
Tokai University,
Orido,
Shimizu,
Shizuoka 424,
Japan

Dr. R. Parker,
3 Curdleigh Lane,
Blagdon Hill,
Taunton, TA3 7SH
UK

Mr. M. Perlow Jr.,
Valley Foundation Consultants Inc.,
523 Second Avenue,
Suite 212,
Bethlehem,
PA 18018,
USA.

Dr. A. A. Rodger,
Department of Engineering,
Marischall College,
University of Aberdeen,
Aberdeen AB0 1AS
UK

Dr. A. G. Reid,
Department of Civil Engineering,
University of Strathclyde,
Glasgow,
UK

Dr. A. R. Reece,
Department of Agricultural Engineering,
The University of Newcastle upon Tyne,
Newcastle upon Tyne NE1 7RU
UK

Dr. A. F. Richards,
Fugro B.V.,
PO Box 63,
220 AB,
Leidschendam,
The Netherlands

Mr. M. J. Reardon,
Noble Denton and Associates Ltd.,
131 Aldersgate Street,
London EC1A 4EB
UK

Mr. I. Sano,
Department of Civil Engineering,
Kyoto University,
Sakyo-ku,
Kyoto 606,
Japan

Dr. P. J. Schultheiss,
Institute of Oceanographic Sciences,
Wormley,
Godalming, GU8 5UB
UK

Mr. S. E. J. Spierenburg,
TH-Delft,
Department of Civil Engineering,
Geotechnical Laboratory,
Stevinweg 1,
2628CN Delft,
The Netherlands

Mr. I. A. Savell,
Sir William Halcrow,
Burderop Park,
Swindon, SN4 0QD
UK

Dr. R. L. Schiffman,
Department of Civil Engineering,
University of Colorado,
Boulder,
CO 80309
USA

Dr. G. Sills,
Department of Engineering Sciences,
Soil Mechanics Group,
Oxford University,
Parks Road,
Oxford OX1 3PJ
UK

Dr. A. J. Silva,
Department of Ocean Engineering,
University of Rhode Island,
Narragansett Marine Laboratory,
Box No. 70,
Narragansett,
RI 02882,
USA

Dr. P. Strachan,
Ocean Engineering,
Department of Naval Architecture
    and Shipbuilding,
Newcastle upon Tyne University,
Newcastle upon Tyne, NE1 7RU
UK

Dr. J. F. A. Sleath,
University Engineering Department,
Trumpington Street,
Cambridge, CB2 1PZ
UK

Mr. R. R. Simons,
Civil Engineering Department,
University College London,
Gower Street,
London WC1E 6BT
UK

Prof. R. Silvester,
The University of Western Australia,
Department of Civil Engineering,
Nedlands,
W.A. 6009,
Australia

Mr. Gordon Senior,
Gordon Senior Associates,
31 Wolsey Road,
Esher, KT1 08NT
UK

Dr. R. K. Taylor,
Department of Engineering Geology,
The University of Durham,
South Road,
Durham
UK

Mr. R. Thomas,
Department of Engineering Sciences,
Soil Mechanics Group,
Oxford University,
Parks Road,
Oxford OX1 3PJ
UK

Dr. P. Thimakorn,
Asian Institute of Technology,
PO Box 2754,
Bangkok,
Thailand

Mr. S. U-He,
Tokuyama Technical College,
Takayjo, 3538,
Tokuyama-shi,
745 Japan

Mr. W. Leeuiwestein,
TH Delft,
Department of Civil Engineering,
Geotechnical Laboratory,
Stevinweg 1,
2628 CN Delft,
The Netherlands

Dr. B. B. Willetts,
Department of Engineering,
University of Aberdeen,
Marischall College,
Aberdeen AB9 1AS
UK

Dr. K. Yasuhara,
Department of Civil Engineering,
Nishinippon Institute of Technology,
Kanda 1633,
Fukuoka ken,
800-03,
Japan

Dr. T. Yanagi,
Department of Ocean Engineering,
Ehime University,
Bunkyo 3,
Matsuyama 790,
Japan

Mr. T. N. Burt,
Hydraulics Research Station,
Wallingford,
Oxford OX10 8BA
UK

Dr. A. J. Grass,
Department of Civil and Municipal
    Engineering,
University College London,
Gower Street,
London WC1E 6BT
UK